Dynamics and Randomness II

Nonlinear Phenomena and Complex Systems

VOLUME 10

The Centre for Nonlinear Physics and Complex Systems (CFNL), Santiago, Chile, and Kluwer Academic Publishers have established this series devoted to nonlinear phenomena and complex systems, which is one of the most fascinating fields of science today, to publish books that cover the essential concepts in this area, as well as the latest developments. As the number of scientists involved in the subject increases continually, so does the number of new questions and results. Nonlinear effects are essential to understand the behaviour of nature, and the methods and ideas introduced to treat them are increasingly used in new applications to a variety of problems ranging from physics to human sciences. Most of the books in this series will be about physical and mathematical aspects of nonlinear science, since these fields report the greatest activity.

Series Editors
Enrique Tirapegui (*Centre for Nonlinear Physics and Complex Systems, Santiago, Chile*)
Servet Martinez (*Centre of Mathematical Modeling, Universidad de Chile, Santiago, Chile*)

Advisory Editorial Board
Marc Etienne Brachet (*Ecole Normale Supérieure, Paris, France*)
Pierre Collet (*Ecole Polytechnique, Paris, France*)
Pierre Coullet (*Institut Nonlinéaire de Nice, France*)
Grégoire Nicolis (*Université Libre de Bruxelles, Belgium*)
Yves Pomeau (*Ecole Normale Supérieure, Paris, France*)
Daniel Walgraef (*Université Libre de Bruxelles, Belgium*)

The titles published in this series are listed at the end of this volume.

Dynamics and Randomness II

Edited by

Alejandro Maass

Servet Martínez

and

Jaime San Martín

Universidad de Chile,
Santiago, Chile

KLUWER ACADEMIC PUBLISHERS

DORDRECHT / BOSTON / LONDON

A C.I.P. Catalogue record for this book is available from the Library of Congress.

ISBN 978-90-481-6565-0   e-ISBN 978-1-4020-2469-6

Published by Kluwer Academic Publishers,
P.O. Box 17, 3300 AA Dordrecht, The Netherlands.

Sold and distributed in North, Central and South America
by Kluwer Academic Publishers,
101 Philip Drive, Norwell, MA 02061, U.S.A.

In all other countries, sold and distributed
by Kluwer Academic Publishers,
P.O. Box 322, 3300 AH Dordrecht, The Netherlands.

Printed on acid-free paper

Table of Contents

Foreword

This book contains the lectures given at the *II Conference on Dynamics and Randomness* held at the Centro de Modelamiento Matemático of the Universidad de Chile, from December 9th to 13th, 2002.

This meeting brought together mathematicians, theoretical physicists, theoretical computer scientists, and graduate students interested in fields related to probability theory, ergodic theory, symbolic and topological dynamics.

We would like to express our gratitude to all the participants of the conference and to the people who contributed to its organization. In particular, to Pierre Collet, Bernard Host and Karl Petersen for their scientific advise.

We want to thank warmly the authors of each chapter for their stimulating lectures and for their manuscripts devoted to a various of appealing subjects in probability and dynamics: to Jean Bertoin for his course on Some aspects of random fragmentation in continuous time; to Anton Bovier for his course on Metastability and ageing in stochastic dynamics; to Steve Lalley for his course on Algebraic systems of generating functions and return probabilities for random walks; to Elon Lindenstrauss for his course on Recurrent measures and measure rigidity; to Sylvie Méléard for her course on Stochastic particle approximations for two-dimensional Navier-Stokes equations; and to Anatoly Vershik for his course on Random and universal metric spaces.

We are also indebted to our sponsors and supporting institutions, whose interest and help was essential to organize this meeting: CNRS, ECOS-CONICYT, FONDAP Program in Applied Mathematics, French Cooperation, Millennium Scientific Initiative (P01-005) and Universidad de Chile.

We are grateful to Ms. Gladys Cavallone for their excellent work during the preparation of the meeting as well as for the considerable task of unifying the typography of the different chapters of this book.

We expect that, as in the previous School, the lectures, the informal discussions raised by them and the friendly ambiance in the School will help to many young latinamerican scientists to pursue their scientific careers.

Alejandro Maass   Servet Martínez   Jaime San Martín

SOME ASPECTS OF RANDOM FRAGMENTATIONS IN CONTINUOUS TIMES

JEAN BERTOIN

Laboratoire de Probabilités et Modèles Aléatoires
Université Paris 6
and Institut universitaire de France
175, rue du Chevaleret, F-75013 Paris, France
jbe@ccr.jussieu.fr

Abstract. Fragmentation processes serve as stochastic models for a mass that falls apart randomly as time passes. The purpose of these notes is to provide an elementary survey of some recent results in this area.

1. Introduction

In short, fragmentation processes are meant to serve as models for the evolution of an object that breaks down randomly into pieces as time passes. They have motivated an abundant literature in physics, computer science, ... We refer e.g. to the proceedings [14] for a number of examples arising in physics and chemistry.

One of the first probabilistic works in this area is due to Kolmogorov [20], who provided an explanation to a feature which is observed in mineralogy: the logarithms of the masses of mineral grains are often normally distributed. Specifically, Kolmogorov used the following Markov chain to model the formation of mineral grains. At the initial time we have a single block, say with mass $m_0 \in]0, \infty[$. At time 1, this block is broken randomly, which produces two smaller blocks, say with masses $m_{1,1} = V_0 m_0$ and $m_{1,2} = (1 - V_0)m_0$, where V_0 is a random variable with values in $[1/2, 1[$ which has a given distribution. Next, one iterates the procedure, i.e. at time $n + 1$, each of the 2^n blocks obtained at time n, say with masses $m_{1,n}, \ldots, m_{2^n,n}$, is again randomly broken into two pieces, independently of the past. More precisely, the split uses independent variables $V_{1,n}, \ldots, V_{2^n,n}$ which are all distributed as V_0. Roughly, Kolmogorov showed that there is

1

a simple rescaling of the empirical distribution of $\frac{1}{n}\ln m_{1,n}, \ldots, \frac{1}{n}\ln m_{2^n,n}$ which converges with probability one to the standard Gaussian distribution.

Kolmogorov's model can be viewed as a special instance of a branching random walk (see the proceedings [4] for references in this area). Indeed, if we write Z_n for the random point measure with atoms at $\ln m_{1,n}, \ldots, \ln m_{2^n,n}$, then $(Z_n, n \in \mathbb{N})$ is a branching process with values in the space of point measures on $\mathbb{R}$. More precisely, the particle located at $\ln m_{k,n}$ at time n gives rise at time $n+1$ to two particles located at $\ln m_{k,n} + \ln V_{k,n}$ and $\ln m_{k,n} + \ln(1 - V_{k,n})$, respectively, and different particles have independent evolutions. In this setting, Kolmogorov's result follows from the central limit theorem for branching random walks, and more generally, the theory of branching random walks provides many interesting properties of fragmentation in discrete times.

In this text, we shall focus on continuous times; in particular instants of splitting may be everywhere dense and the number of fragments infinite, which raises many interesting problems. Let us start by introducing the framework and basic definition. We will be working with the sub-space of decreasing numerical sequences

$$\mathcal{S}^{\downarrow} = \left\{ x = (x_1, \ldots) : x_1 \geq x_2 \geq \cdots \geq 0 \quad \text{and} \quad \sum_1^{\infty} x_i \leq 1 \right\}.$$

A configuration $x \in \mathcal{S}^{\downarrow}$ should be thought of as the ranked family of masses arising from the split of some object with unit mass. A portion of the initial mass may be lost during the split, which corresponds to the situation when $\sum_1^{\infty} x_i < 1$. We endow $\mathcal{S}^{\downarrow}$ with the uniform distance; it is readily checked that this turns $\mathcal{S}^{\downarrow}$ into a compact metric space.

A *fragmentation process* is a Markov process $X = (X(t), t \geq 0)$ with values in $\mathcal{S}^{\downarrow}$ that fulfills the so-called *fragmentation property*. Roughly speaking, this means that different fragments have independent evolutions, which can also be viewed as a version of the branching property. Specifically, let $\mathbb{P}_r$ stand for the law of X started from $(r, 0, \ldots)$, i.e. at the initial time, there is a single fragment with mass r. Then for every $s, t \geq 0$, conditionally on $X(t) = (x_1, \ldots)$, $X(t+s)$ has the same law as the variable obtained by ranking in the decreasing order the terms of the random sequences $X^{(1)}(s), X^{(2)}(s), \ldots$, where the latter are independent variables with values in $\mathcal{S}^{\downarrow}$, such that $X^{(i)}(s)$ has the same distribution as $X(s)$ under $\mathbb{P}_{x_i}$ for each $i = 1, \ldots$.

This text aims at presenting some key aspects of a class of fragmentations in continuous times which fulfill some natural scaling invariance property (which will be explained in the sequel). We shall first focus on the structure and basic properties in the so-called homogeneous case, when the scaling is the simplest. Then we shall consider the self-similar case that corresponds to more general types of scaling, using a transformation which reduces the self-similar case to the homogeneous one. Finally, we shall present some results on the asymptotic behavior of such fragmentations. The interested reader may find proofs and technical details in the references [5, 8, 9, 10, 11, 12, 13].

2. Homogeneous Fragmentations

A fragmentation process will be called *homogeneous* provided that for every $r > 0$, the distribution of $(rX(t), t \geq 0)$ under $\mathbb{P}_1$ is $\mathbb{P}_r$.

Throughout this section, we shall be working with a homogeneous fragmentation process X. Our first purpose is to specify the structure and to characterize the distribution of a homogeneous fragmentation. We shall then state basic properties of such processes in terms of their characteristics. To give a rigorous presentation, it is convenient to start by discussing the notion of subordinator, which might look at first sight as a digression, but is in fact closely related to our topic.

2.1. SUBORDINATORS

A *subordinator* is a real-valued process $\xi = (\xi_t, t \geq 0)$ with non-decreasing sample paths and independent and stationary increments. Thus a subordinator can be thought of as the continuous time analog of an increasing random walk. We refer to [6] for a detailed account on these processes, but here we shall only need the following well-known fact. The canonic decomposition of the non-decreasing path ξ as the sum of its continuous and pure-jumps components, is given by the so-called Lévy-Itô decomposition. If we suppose for simplicity that the process starts from $\xi_0 = 0$, then

$$\xi_t = \mathrm{d}t + \sum_{0 < s \leq t} \Delta \xi_s , \qquad t \geq 0 , \tag{1}$$

where $\mathrm{d} \geq 0$ is the so-called drift coefficient and the jump process $(\Delta_s := \xi_s - \xi_{s-}, s > 0)$ a *Poisson point process* with values in $]0, \infty]$ [1]. This means that for every Borel set $B \subset]0, \infty]$, the process which counts (as a function of time) the jumps of size in B is a Poisson process with intensity say $\Lambda(B)$, and to disjoint Borel sets correspond independent counting processes. The map $B \to \Lambda(B)$ is a Radon measure on $]0, \infty]$, called the characteristic

---

[1] When $\Lambda(\{\infty\}) > 0$, the process ξ jumps at ∞ and then stays there forever at some instant which has an exponential distribution with parameter $\Lambda(\{\infty\})$. One says that ∞ is a cemetery point, and that $\Lambda(\{\infty\})$ is the killing rate.

measure of the Poisson point process, and is also known as the Lévy measure of the subordinator.

The necessary and sufficient condition for a measure Λ on $]0, \infty]$ to be the Lévy measure of some subordinator is that

$$\int_{]0,\infty]} (1 \wedge y)\Lambda(dy) < \infty. \tag{2}$$

More precisely, (2) ensures the convergence of the random series in (1).

2.2. SPLITTING MEASURE AND EROSION

The structure of homogeneous fragmentations bears striking similarities with that of subordinators. Roughly, homogeneous fragmentations result from the combination of two different phenomena: a continuous erosion and sudden dislocations. The erosion is a deterministic mechanism, analogous to the drift for subordinators, whereas the dislocations occur randomly according to some Poisson point process, and can be viewed as the jump-component of the fragmentation.

More precisely, let us first focus on dislocations. We call *splitting measure* a measure ν on $\mathcal{S}^* = \mathcal{S}^{\downarrow}\backslash\{(1,0,\ldots)\}$ (i.e. $\mathcal{S}^*$ is the space of decreasing numerical sequences $x = (x_1,\ldots)$ with $\sum x_i \leq 1$ and $x_1 < 1$), which fulfills

$$\int_{\mathcal{S}^*} (1 - x_1)\nu(dx) < \infty. \tag{3}$$

Then, consider a Poisson point process $((\Delta(t), k(t)), t \geq 0)$ with values in $\mathcal{S}^* \times \mathbb{N}$, with characteristic measure $\nu \otimes \#$, where $\#$ denotes the counting measure on $\mathbb{N} = \{1, 2, \ldots\}$. One can construct a pure jump process $(Y(t), t \geq 0)$ which jumps only at times $t \geq 0$ at which a point $(\Delta(t), k(t))$ occurs. More precisely, the jump (i.e. the dislocation) induced by such a point can be described as follows. The sequence $Y(t)$ is obtained from $Y(t-)$ by replacing its $k(t)$-th term $Y_{k(t)}(t-)$ by the sequence $Y_{k(t)}(t-)\Delta(t)$, and ranking all the terms in the decreasing order.

For instance, if

$$Y(t-) = \left(\frac{2}{3}, \frac{1}{4}, \frac{1}{12}, 0, \ldots\right) \quad, \quad k(t) = 2 \quad \text{and} \quad \Delta(t) = \left(\frac{3}{4}, \frac{1}{4}, 0, \ldots\right)$$

then we look at the 2-nd largest term in the sequence $Y(t-)$, which is $\frac{1}{4}$, and split it according to $\Delta(t)$. This produces two fragments of size $\frac{3}{16}$ and $\frac{1}{16}$, and thus

$$Y(t) = \left(\frac{2}{3}, \frac{3}{16}, \frac{1}{12}, \frac{1}{16}, 0, \ldots\right).$$

Of course, it may happen that $Y_{k(t)}(t-) = 0$, and in that case we have $Y(t) = Y(t-)$.

Next, call erosion coefficient an arbitrary real number $c \geq 0$, and set $X(t) = e^{-ct}Y(t)$. The process $(X(t), t \geq 0)$ is again a homogeneous fragmentation, and conversely any homogeneous fragmentation process can be constructed like this. In conclusion, the distribution of a homogeneous fragmentation is entirely specified by its erosion rate and its splitting measure.

Observe that the condition (3) resembles (2) for the Lévy measure of a subordinator. Informally, if we tried to construct a fragmentation process using as above a Poisson point process with intensity $\nu \otimes \#$ for some measure ν such that the condition (3) failed, then there would be some many dislocations that the initial mass would be immediately reduced to dust.

In the case when the erosion rate is zero and the splitting measure has a finite total mass, the logarithm of a homogeneous fragmentation can be viewed as a special case of family of branching processes in continuous times considered by Uchiyama[24]. In this direction, the results that we shall present below can be viewed as extensions of classical theorems on branching random walks (see the recent proceedings [4] and the references therein) to the situation where branching may occur continuously.

2.3. THE INFINITESIMAL GENERATOR

It is easily checked that the semigroup $(P_t, t \geq 0)$ of a homogeneous fragmentation fulfills the Feller property, i.e. P_t maps $\mathcal{C}(\mathcal{S}^{\downarrow})$, the space of real-valued continuous function on $\mathcal{S}^{\downarrow}$, into itself (recall that $\mathcal{S}^{\downarrow}$ is compact), and $\lim_{t \to 0+} P_t f = f$ for every $f \in \mathcal{C}(\mathcal{S}^{\downarrow})$. Let us denote the domain and the infinitesimal generator of $(P_t, t \geq 0)$ by $\mathcal{D}$ and $\mathcal{G}$, respectively, i.e.

$$f \in \mathcal{D} \text{ and } \mathcal{G}f = g \iff f, g \in \mathcal{C}(\mathcal{S}^{\downarrow}) \text{ and } \lim_{t \to 0+} \frac{P_t f - f}{t} = g.$$

There is a simple and useful expression for the infinitesimal generator for a large class of functions; see [13]. More precisely, let $\varphi : [0, 1] \to]0, \infty[$ be a function of class $\mathcal{C}^1$ with $\varphi(0) = 1$. Define the map $f : \mathcal{S}^{\downarrow} \to \mathbb{R}$ by

$$f(x) = \prod_{i=1}^{\infty} \varphi(x_i), \qquad x = (x_1, \ldots) \in \mathcal{S}^{\downarrow}.$$

Then it can be shown that $f \in \mathcal{D}$ and

$$\mathcal{G}f(x) = \sum_{i=1}^{\infty} f(x) \left(cx_i \frac{\varphi'(x_i)}{\varphi(x_i)} + \int_{\mathcal{S}^{\downarrow}} \nu(dy) \left(\frac{f(x_i y)}{\varphi(x_i)} - 1 \right) \right),$$

where $c \geq 0$ is the erosion rate and ν the splitting measure.

2.4. THE TAGGED FRAGMENT

Although the law of a homogeneous fragmentation process is characterized by its erosion rate and splitting measure, in general we do not know how to describe explicitly e.g. the distribution of the process at a fixed time. However, some partial but quite useful information can be derived from a so-called size-biased sampling, and it turns out that the law of the latter is simple to formulate.

Roughly speaking, suppose that we start from an object with unit mass, and let us tag a point at random according to the mass distribution of the object, independently of the fragmentation. Then for any time $t \geq 0$, denote by $\chi(t)$ the mass of the fragment at time t which contains the tagged point (if there is no such fragment, then $\chi(t) = 0$).

Note that $\chi(t)$ is a size-biased pick from the sequence $X(t) = (X_1(t), \ldots)$, i.e. there is the identity in law

$$\chi(t) \stackrel{\mathcal{L}}{=} X_K(t) ,$$

where K is an integer valued variable whose conditional distribution given $X(t)$ is

$$\mathbb{P}(K = k \mid X(t)) = X_k(t), \quad k = 1, \ldots .$$

It can be shown that the process

$$\xi(t) = -\ln \chi(t), \qquad t \geq 0$$

is a subordinator (cf. section 2.1). The law of a subordinator is determined by its one-dimensional marginals, and hence by its Laplace transform. In the present case, we have

$$\mathbb{E}(\chi(t)^q) = \mathbb{E}(\exp(-q\xi(t))) = \exp(-t\Phi(q)), \qquad t, q > 0,$$

where the function Φ is given in terms of the erosion rate c and the splitting measure ν by the identity

$$\Phi(q) := c(q+1) + \int_{S^*} \left(1 - \sum_{i=1}^{\infty} x_i^{q+1}\right) \nu(dx), \qquad q > 0. \qquad (4)$$

3. Self-Similar Fragmentations

We now turn our attention to more general class of fragmentations, where roughly speaking, the fragmentation rate depends as a power function of the mass.

3.1. DEFINITION AND EXAMPLES

A fragmentation process $X = (X(t), t \geq 0)$ will be called *self-similar* if it fulfills the scaling property. This means that there is an index of self-similarity $\alpha \in \mathbb{R}$, such that for every $r \in [0, 1]$, the distribution of $(rX(r^\alpha t), t \geq 0)$ under $\mathbb{P}_1$ is $\mathbb{P}_r$. Of course, in the special case when $\alpha = 0$, we simply recover the definition of a homogeneous fragmentation.

As connections between the homogeneous and the self-similar cases will have an important role in our analysis, and in order to avoid a possible confusion, it is convenient from now on to denote self-similar fragmentations with index α by $X^{(\alpha)}$, and to keep the notation X for the homogeneous case.

A simple example of a self-similar fragmentation is provided by the following modified version of Kolmogorov's chain which was described in the Introduction. Consider a particle system in which particles are specified by their sizes, that evolves in continuous time according to following dynamics. A particle with size $r > 0$ waits an exponential time with parameter r^α and then splits into two particles with respective sizes rV and $r(1 - V)$, where V is a random variable with values in $[1/2, 1[$ which has a fixed distribution and is independent of the past of the system. If we write $X^{(\alpha)}(t)$ for the sequence of the sizes of particles at time t, ranked in the decreasing order, then $(X^{(\alpha)}(t), t \geq 0)$ is a self-similar fragmentation with index α.

This particle system was considered by Brennan and Durrett [15] as a model for polymer degradation. Before that, Filippov [16] studied a more general sub-family of self-similar fragmentations in which each split may produce an infinite number of particles. More recently, motivated by the analysis of the so-called standard additive coalescent, Aldous and Pitman [3] logged the continuum random tree along it skeleton and constructed an interesting self-similar fragmentation with index $\alpha = 1/2$; see also [7] for an alternative construction based on the Brownian excursion. Further examples using stable random trees have been studied by Miermont [22, 23].

3.2. CHANGING THE INDEX OF SELF-SIMILARITY

We shall present a simple transformation that changes a homogeneous fragmentation X into a self-similar one, $X^{(\alpha)}$. In this direction we shall implicitly assume that the fragmentation starts from a single fragment with unit mass.

It is convenient to use an interval-representation of fragmentation. This means that given a fragmentation $X = (X(t), t \geq 0)$, we may construct a nested [2] family of open sets in the unit interval, $(G(t), t \geq 0)$, such that

---

[2]This means that $G(t) \subseteq G(s)$ for $s \leq t$.

for every t, $X(t)$ is the ranked sequence of the lengths of the intervals components of $G(t)$. Next, for every $y \in]0, 1[$, let $I_y(t)$ denote the interval component of $G(t)$ that contains y if $y \in G(t)$, and $I_y(t) = \emptyset$ otherwise. We write $|I|$ for the length of an interval $I \subseteq]0, 1[$, and for every $y \in]0, 1[$ we consider the time-substitution

$$T_y^{(\alpha)}(t) := \inf \left\{ u \geq 0 : \int_0^u |I_y(v)|^{-\alpha} dv > t \right\}.$$

Because the open sets $G(t)$ are nested, we see that for every $y, z \in]0, 1[$, the intervals $I_y(T_y^{(\alpha)}(t))$ and $I_z(T_z^{(\alpha)}(t))$ are either identical or disjoint, so the family $\left\{ I_y(T_y^{(\alpha)}(t)), 0 < y < 1 \right\}$ can be viewed as the interval components of an open set $G^{(\alpha)}(t)$. It is straightforward that the family $(G^{(\alpha)}(t), t \geq 0)$ is nested. More precisely, if we write $X^{(\alpha)}(t)$ for the ordered sequence of the lengths of the interval components of $G^{(\alpha)}(t)$, then $(X^{(\alpha)}(t), t \geq 0)$ is a self-similar fragmentation with index α.

Any self-similar fragmentation $X^{(\alpha)}$ can be constructed from some homogeneous one X as above, and this construction can be inverted. In particular, the distribution of $X^{(\alpha)}$ is entirely determined by the index of self-similarity α, and the erosion coefficient $c \geq 0$ and the splitting measure ν of the homogeneous fragmentation X.

As in the homogeneous case, self-similar fragmentations are Feller processes on $\mathcal{S}^{\downarrow}$, and the time-substitution described above enables us to provide an explicit expression for the infinitesimal generator $\mathcal{G}^{(\alpha)}$ of $X^{(\alpha)}$ from the one which was found in the homogeneous case in section 2.4. Specifically, using the same notation as there, one gets

$$\mathcal{G}^{(\alpha)} f(x) = f(x) \sum_{i=1}^{\infty} x_i^\alpha \left(cx_i \frac{\varphi'(x_i)}{\varphi(x_i)} + \int_{\mathcal{S}^{\downarrow}} \nu(dy) \left(\frac{f(x_i y)}{\varphi(x_i)} - 1 \right) \right),$$

with the convention $0 \times \infty = 0$.

Recall from Section 2.4 that the so-called process of the tagged fragment $(\chi(t), t \geq 0)$ for a homogeneous fragmentation can be described as the exponential of a subordinator. The construction above of a self-similar fragmentation from a homogeneous one enables us to derive the law of the tagged fragment $\chi^{(\alpha)}(t)$ for a self-similar fragmentation.

Specifically, let $\xi = (\xi_t, t \geq 0)$ be a subordinator with Laplace exponent Φ given by (4). Introduce the time-change

$$\tau(t) = \inf \left\{ u : \int_0^u \exp(\alpha \xi_r) dr > t \right\}, \qquad t \geq 0,$$

and set $Z_t = \exp(-\xi_{\tau(t)})$ (with the convention that $Z_t = 0$ if $\tau(t) = \infty$). Then the processes $(Z_t, t \geq 0)$ and $(\chi^{(\alpha)}(t), t \geq 0)$ have the same law. In particular, this shows that the tagged fragment is a decreasing self-similar Markov process (see Lamperti [21]).

3.3. ON THE MASS OF SMALL FRAGMENTS FOR POSITIVE INDICES

In this section, we shall work with a self-similar fragmentation with positive index of self-similarity $\alpha > 0$. We also suppose that the total mass is conserved, or equivalently that the erosion rate c and the splitting measure ν fulfill

$$c = 0 \quad \text{and} \quad \nu \left(\sum_{i=1}^{\infty} x_i < 1 \right) = 0.$$

In [11], we obtained a simple condition on the splitting measure ν which ensures a interesting properties for the asymptotic behaviors of the mass of small fragments at a fixed time. In this direction, introduce the function

$$f(\varepsilon) := \int_{\mathcal{S}^{\downarrow}} \sum_{i=1}^{\infty} x_i \mathbf{1}_{\{x_i < \varepsilon\}} \nu(dx).$$

We further assume that f is regularly varying as $\varepsilon \to 0+$ with index $\beta \in \,]0,1[$, and that there exists an integer k such that $\nu(x_k > 0) = 0$. The latter condition means that with probability one, any dislocation breaks a mass into at most $k - 1$ fragments. Then

$$\lim_{\varepsilon \to 0+} f^{-1}(\varepsilon) \sum_{i=1}^{\infty} X_i(t) \mathbf{1}_{\{X_i(t) < \varepsilon\}} = \int_0^t \sum_{i=1}^{\infty} \left(X_i^{(\alpha)}(s) \right)^{\alpha + \beta} ds,$$

with probability one, and this quantity is positive and finite.

It may be interesting to note that the limit there is random, except when the index of self-similarity α and the index of regular variation β fulfill $\alpha + \beta = 1$, because then

$$\sum_{i=1}^{\infty} \left(X_i^{(\alpha)}(s) \right)^{\alpha + \beta} \equiv 1.$$

3.4. LOSS OF MASS FOR NEGATIVE INDICES

It is intuitively clear that the behavior of a self-similar fragmentation should heavily depend on the index α. In particular, for $\alpha < 0$, small masses are subject to intense fragmentation, and thus might be entirely reduced to dust. That this intuition is indeed correct was first established rigorously by Filippov [16] (see also the recent paper by Jeon [19] and the reference therein) in the case when the dislocation measure is finite.

So we suppose in this section that $\alpha < 0$, and we introduce

$$\zeta = \inf \left\{ t \geq 0 : X^{(\alpha)}(t) = (0, \ldots) \right\} .$$

Then it can be shown that $\zeta < \infty$ with probability one; however so far, little is known about the law of the instant ζ when a mass has been entirely reduced to dust. Haas [18] has shown that the tail distribution of ζ decays exponentially fast.

If we further suppose that $\alpha < -1$, then at each fixed time $t > 0$, with probability one there are only a finite number of fragments with positive size, i.e.

$$\text{Card} \left\{ j \in \mathbb{N} : X_j^{(\alpha)}(t) > 0 \right\} < \infty .$$

We stress that in general, no matter what the value of α is, there may exist random instants t at which

$$\text{Card} \left\{ j \in \mathbb{N} : X_j^{(\alpha)}(t) > 0 \right\} = \infty .$$

For instance in the case when the splitting measure fulfills

$$\nu \left(x_j > 0 \text{ for all } j \in \mathbb{N} \right) = \infty ,$$

then with probability one, there occur infinitely many sudden dislocations in the fragmentation process $X^{(\alpha)}$, each of which produces infinitely many masses. This does not induce any contradiction when $\alpha < -1$, because informally, as the index of self-similarity is negative, we know that small masses vanish quickly.

The phenomenon of loss of mass for negative indices can also be viewed via the so-called fragmentation equations. More precisely, the latter describe the evolution of the mass density in a medium which particles split as time passes. Typically, we are given a splitting measure ν which fulfills (3) and some index α, and we suppose that for every $y > 0$ and $x = (x_1, x_2, \ldots) \in \mathcal{S}^*$, a particle with mass y splits at rate $y^\alpha \nu(dx)$ to yield a sequence $x_1 y, x_2 y, \ldots$ of sub-masses. If we represent the density of particles with mass dy at time t by a measure $\mu_t(dy)$ on $]0, \infty[$, then the dynamics of the system are described by the equation

$$\partial_t \langle \mu_t, f \rangle = \int_{]0,\infty[} \mu_t(dy) y^\alpha \int_{\mathcal{S}^*} \nu(dx) \left(\sum_{i=1}^{\infty} f(x_i y) - f(y) \right) , \qquad (5)$$

where $f :]0, \infty[\to \mathbb{R}$ is a generic smooth function with compact support, and the notation $\langle \mu_t, f \rangle$ stands for the integral of f with respect to $\mu_t(dx)$.

See Aizenman and Bak [1], and also [17], [18] for recent contributions. This equation has a unique solution which is given by

$$\langle \mu_t, f \rangle = \mathbb{E}\left(\sum_{i=1}^{\infty} f(X_i^{(\alpha)}(t))\right),$$

where $X^{(\alpha)}$ is a self-similar fragmentation with index α, splitting measure ν and no erosion. Suppose now that sudden dislocations induce no loss of mass, i.e. $\nu\left(\sum_{i=1}^{\infty} x_i < 1\right) = 0$. Informally, if we were allowed to take $f(y) = y$ in (5), then we should get that the right-hand side is zero, i.e. that the total mass $\langle \mu_t, \mathrm{Id} \rangle$ is constant. This is not correct as the function Id does not have compact support, and in fact we know that for $\alpha < 0$, the map $t \to \langle \mu_t, \mathrm{Id} \rangle$ is a strictly decreasing function. We refer to Haas [18] and the references therein for much more in this area.

4. Asymptotic Behavior for Large Times

We start this section by studying the asymptotic behavior of a homogeneous fragmentation when time tends to ∞. We refer to [5] for results on the small time asymptotic behavior, which are of course of much different nature.

4.1. HOMOGENEOUS CASE

We first present an extension of Kolmogorov's result which was outlined in the Introduction. In this direction, we shall work here with a homogeneous fragmentation X such that the erosion rate is $c = 0$ and $\nu\left(\sum_{i=1}^{\infty} x_i < 1\right) = 0$. In other words, no mass is lost by the fragmentation process. Next, we introduce the first and second right-derivatives of Φ at 0,

$$m := \Phi'(0+) \quad , \quad \sigma^2 := -\Phi''(0+);$$

or equivalently in terms of the splitting measure

$$m = -\int_{\mathcal{S}^*}\left(\sum_{i=1}^{\infty} x_i \ln x_i\right) \nu(dx) \quad , \quad \sigma^2 = \int_{\mathcal{S}^*}\left(\sum_{i=1}^{\infty} x_i (\ln x_i)^2\right) \nu(dx).$$

Because m and σ^2 are the mean value and the variance of the subordinator ξ evaluated at time 1, the central limit theorem applies whenever the variance σ^2 of ξ_1 is finite, which we assume from now on. We then know that $\frac{\xi_t - mt}{\sigma\sqrt{t}}$ converges in law when $t \to \infty$ to the standard normal distribution.

By an argument of propagation of chaos, this enables us to derive the following result about the asymptotic behavior of the fragmentation. Let us represent the sequence $X(t)$ by the random probability measure

$$\rho_t(dy) := \sum_{i=1}^{\infty} X_i(t)\delta_{a_i}(dy), \qquad \text{where } a_i = \frac{\ln X_i(t) + tm}{\sigma\sqrt{t}},$$

where δ denotes the Dirac point mass. So ρ_t can be viewed as a rescaled empirical distribution of the logarithm of the fragments at time t. Then it can be shown that when $t \to \infty$, ρ_t converges to the standard normal distribution, i.e. for every $y \in \mathbb{R}$ we have with probability one that

$$\lim_{t\to\infty} \rho_t(] - \infty, y]) = \frac{1}{\sqrt{2\pi}} \int_{-\infty}^{y} \exp\left(-a^2/2\right) da.$$

Roughly, we have just seen above that the size of a typical fragment at time t is of order e^{-mt}, and the standard deviations derive from the central limit theorem. It is also natural to investigate the size of abnormally large and of abnormally small fragments, in the sense that for each $a < 0$, we should like to estimate the number of fragments of size of order e^{at} at time t. One can show that this number is approximately $e^{tC(a)}$ when t is large, in the sense that the following limit exists:

$$\lim_{\varepsilon\to 0+}\lim_{t\to\infty} \frac{1}{t}\ln\left(\text{Card}\left\{i \in \mathbb{N} : e^{(a-\varepsilon)t} \leq X_i(t) \leq e^{(a+\varepsilon)t}\right\}\right) = C(a).$$

One can express $C(a)$ in terms of a and the characteristics of the fragmentation as follows. First, denote by $\overline{p} > 0$ the location of the maximum of the function $p \to \Phi(p)/(p+1)$ on $[0, \infty[$, and by $\overline{m} = \Phi(\overline{p})/(\overline{p}+1)$ the value of this maximum. We have also $\overline{m} = \Phi'(\overline{p}) < m$. Second, introduce

$$\underline{p} := \inf\left\{p \in \mathbb{R} : \int_{S^*} \sum_{i=2}^{\infty} x_i^{p+1}\nu(dx) < \infty\right\}$$

(note that the sum above starts from $i = 2$), so that $-1 \leq \underline{p} \leq 0$ and Φ can be continued analytically on $]\underline{p}, 0]$. We set $\underline{m} = -\Phi'(\underline{p}+)$, this quantity is always greater than or equal to m, and may be infinite. Next, define the convex decreasing function Λ on $]0, \infty[$ by

$$\Lambda(p) = \begin{cases} -\Phi(p) & \text{if } \underline{p} \leq p < \overline{p}, \\ -(p+1)\overline{m} & \text{if } p \geq \overline{p}. \end{cases}$$

Finally, introduce the Fenchel-Legendre transform of Λ,

$$\Lambda^*(a) = \sup_{p\geq 0} (ap - \Lambda(p)).$$

Then we have for $a \in] - \underline{m}, 0[$

$$C(a) = -(\Lambda^*(a) + a).$$

It is easy to see that $C(-\overline{m}) = 0$, which means that there exists fragments of size $\approx \exp\{-t\overline{m}\}$; however the number of such fragments is sub-exponential, i.e. smaller than $e^{\eta t}$ for every $\eta > 0$. Moreover, $C(a) = -\infty$ for $a > -\overline{m}$, so that the largest fragment has a size of order $\exp\{-t\overline{m}\}$ when $t \to \infty$.

A much more precise asymptotic result in this direction was recently obtained in [12]: For every continuous function $f : \mathbb{R} \to \mathbb{R}$ with compact support, then for each $a \in] - \underline{m}, -\overline{m}[$, it holds with probability one that when $t \to \infty$

$$\lim_{t \to \infty} \sqrt{t} e^{-tC(a)} \sum_{i=1}^{\infty} f(at - \ln X_i(t)) = Y(a) \int_{-\infty}^{\infty} f(y) e^{(p+1)y} dy,$$

where $Y(a)$ is a certain a.s. positive random variable depending on a, and p the unique real number such that $\Phi'(p) = -a$.

4.2. ASYMPTOTIC BEHAVIOR FOR POSITIVE INDICES

We now conclude this survey by considering the asymptotic behavior of a self-similar fragmentation in the case of positive indices, that is $\alpha > 0$ (indeed, we already know from section 3.4 that for negative indices, the entire mass disappears in finite times a.s., so the asymptotic behavior is trivial). As fragmentation rate is higher for large fragments than for small ones, we should expect a homogenization type phenomenon when time tends to ∞. We shall assume that the homogeneous fragmentation X from which $X^{(\alpha)}$ is constructed does not loose mass, i.e. its erosion rate is $c = 0$ and its splitting measure ν fulfills $\nu \left(\sum_{i=1}^{\infty} x_i < 1 \right) = 0$. We further assume that the Laplace exponent Φ given by (4) has a finite right derivative at 0, which is denoted by $m = \Phi'(0+)$. Roughly, the effect of the time-change constructed in Section 3.2 is that the rate of decay of fragments is now polynomial with exponent $1/\alpha$ and not exponential as in the homogeneous case. More precisely, introduce the random probability measure

$$\rho_t^{(\alpha)}(dy) = \sum_{i=1}^{\infty} X_i^{(\alpha)}(t) \delta_{t^{1/\alpha} X_i^{(\alpha)}(t)}(dy).$$

One can show that when $t \to \infty$, $\rho_t^{(\alpha)}$ converges weakly towards some deterministic probability measure $\rho_\infty^{(\alpha)}$ on $]0, \infty[$. Specifically, the latter is determined by its moments which are expressed in terms of Φ by the following identity:

$$\int_{]0,\infty[} y^{\alpha k} \rho_\infty^{(\alpha)}(dy) = \frac{(k-1)!}{\alpha m\, \Phi(\alpha)\cdots\Phi((k-1)\alpha)} \quad \text{for } k = 1, \ldots,$$

with the convention that the right-hand side above equals $1/(\alpha m)$ for $k = 1$.

This limit theorem was obtained first by Filippov [16] in the special case when the splitting measure is finite, and then re-discovered by Brennan and Durrett [15] in the setting of the particle system which was described in Section 3.1. The general case is treated in [10]; it relies crucially on limit theorems for self-similar Markov processes.

References

1. Aizenman, M. and Bak, T. (1979) Convergence to Equilibrium in a System of Reacting Polymers, *Comm. Math. Phys.*, **Vol.65**, pp. 203–230.
2. Aldous, D.J.(1999) Deterministic and Stochastic Models for Coalescence (Aggregation, Coagulation): A Review of the Mean-Field Theory for Probabilists, *Bernoulli*, **Vol.5**, pp. 3–48.
3. Aldous, D.J. and Pitman, J. (1998) The Standard Additive Coalescent, *Ann. Probab.*, **Vol.26**, pp. 1703–1726.
4. Athreya, K.B. and Jagers, P. (Eds.) (1997) *Classical and Modern Branching Processes*, Springer-Verlag, New York.
5. Berestycki, J. (2002) Ranked Fragmentations, *ESAIM, Probabilités et Statitistique*, **Vol.6**, pp. 157–176. Available via
   `http://www.edpsciences.org/ps/OnlinePSbis.html`
6. Bertoin, J. (1999) Subordinators: Examples and Applications. In Bernard, P. (Ed.) *Lectures on Probability Theory and Statistics*, Ecole d'Été de Probabilités de Saint-Flour XXVII, Lecture Notes in Maths **Vol.1717**, Springer-Verlag, Berlin.
7. Bertoin, J. (2000) A Fragmentation Process Connected to Brownian Motion, *Probab. Theory Relat. Fields*, **Vol.117**, pp. 289–301.
8. Bertoin, J. (2001) Homogeneous Fragmentation Processes, *Probab. Theory Relat. Fields*, **Vol.121**, pp. 301-318.
9. Bertoin, J. (2002) Self-Similar Fragmentations, *Ann. Inst. Henri Poincaré*, **Vol.38**, pp. 319–340.
10. Bertoin, J. (2002) The Asymptotic Behavior of Fragmentation Processes, Preprint. `http://www.proba.jussieu.fr/publi.html`
11. Bertoin, J. (2002) On Small Masses in Self-Similar Fragmentations, Preprint. `http://www.proba.jussieu.fr/publi.html`
12. Bertoin, J. and Rouault, A. (2003) Additive Martingales and Probability Tilting for Fragmentations, Preprint. `http://www.proba.jussieu.fr/publi.html`
13. Bertoin, J. and Rouault, A. Analytic Methods for Fragmentation Processes, in preparation.
14. Beysens, D., Campi, X. and Pefferkorn, E. (Eds.) (1995) *Fragmentation Phenomena*. World Scientific, Singapore.
15. Brennan, M.D. and Durrett, R. (1987) Splitting Intervals. II. Limit Laws for Lengths, *Probab. Theory Related Fields*, **Vol.75**, pp. 109–127.
16. Filippov, A.F. (1961) On the Distribution of the Sizes of Particles which Undergo Splitting, *Th. Probab. Appl.*, **Vol.6**, pp. 275–293.
17. Fournier, N. and Giet, J.-S. (2002) On Small Particles in Coagulation-Fragmentation Equations, Preprint.

18. Haas, B., Loss of Mass in Deterministic and Random Fragmentations, *Stochastic Process. Appl.* To appear. http://www.proba.jussieu.fr/publi.html
19. Jeon, I. (2002) Stochastic Fragmentation and Some Sufficient Conditions for Shattering Transition, *J. Korean Math. Soc*, **Vol.39**, pp. 543–558.
20. Kolmogoroff, A.N. (1941) Über das logarithmisch normale Verteilungsgesetz der Dimensionen der Teilchen bei Zerstückelung, *C. R. (Doklady) Acad. Sci. URSS*, **Vol.31**, pp. 99–101.
21. Lamperti, J. (1972) Semi-Stable Markov Processes, *Z. Wahrscheinlichkeitstheorie verw. Gebiete*, **Vol.22**, pp. 205–225.
22. Miermont, G. (2003) Self-Similar Fragmentations Derived from the Stable Tree I: Splitting at Heights, Preprint. http://www.proba.jussieu.fr/publi.html
23. Miermont, G. Self-Similar Fragmentations Derived from the Stable Tree II: Splitting at Hubs, in preparation.
24. Uchiyama, K. (1982) Spatial Growth of a Branching Process of Particles Living in R^d. *Ann. Probab.*, **Vol.10**, pp. 896–918.

METASTABILITY AND AGEING IN STOCHASTIC DYNAMICS

ANTON BOVIER

Weierstrass-Institut für Angewandte Analysis und Stochastik
Mohrenstrasse 39, D-10117 Berlin, Germany
`bovier@wias-berlin.de`

Abstract. In these notes I review recent results on metastability and ageing in stochastic dynamics. The first part reviews a somewhat novel approach to the computation of key quantities such as mean exit times in metastable systems and small eigenvalues of the generator of metastable Markov chain developed over the last years with M. Eckhoff, V. Gayrard and M. Klein. This approach is based on extensive use of potential theoretic ideas and allows, at least in the case of reversible dynamics, to get very accurate results with comparatively little effort. This methods have also been used in recent joint work with G. Ben Arous and V. Gayrard on the dynamics of the random energy model. The second part of these lectures is devoted to a review of this work.

Contents

1. Introduction

In these lectures I will review a somewhat new approach to the old issue of metastability and the rather more recent issue of "ageing" in the framework

17

A. Maass et al. (eds.), Dynamics and Randomness II, 17–79.
© 2004 *Kluwer Academic Publishers.*

of stochastic dynamics, or more precisely Markov processes. This approach
addresses, in the context of metastability, the issue of precise asymptotics,
that, although frequently considered in the physical literature for decades,
had been lacking a fully rigorous mathematical treatment. Apart from potential intrinsic interest, this question is strongly motivated by the second
aspect, ageing, where, as will become clear, this will become indispensable
if anything is to be understood at all.

Metastability as well as ageing (at least in the context we will be interested in) can be seen as special aspects of the far wider issue of understanding the dynamic behaviour of a complicated system on particular time
scales in terms of a simplified dynamics with a reduced state space. Such a
coarse grained description will always have to be chosen in accordance with
the problem at hand. The phenomena we are trying to capture here concern
the existence of two (or more) *time scales* at which the systems considered
behave in distinct ways: on the short time scales, the system performs some
local motion that appears to be reasonably recurrent, suggesting that equilibrium is reached, while at the larger time scale, the systems undergoes
dramatic changes that seem to alter its character entirely. Our purpose
is thus to characterize the long time dynamics in terms of a process that
describes only the sequence of these global changes and ignores all of the
local short time motion. Our task is then to devise in a given setting such a
coarse grained description and to derive the properties of the corresponding
process and from the underlying microscopic dynamics.

The setting sketched above is common in a wide variety of natural phenomena. From the most classical setting of dynamical phase transitions
in solid state physics, conformational states of macromolecules, macro-
climatological states, or more speculatively, macro-economic states, there
is hardly a branch of science where one is not confronted with the same
type of general situation. While in the classical setting of metastability the
number of relevant "macro-states" is finite or even small, systems with a
large, resp. infinite, number of such states tend to "age", meaning in the
broadest sense that their long time behaviour is to be described in terms
of a more complicated process than just a Markov chain with finite state
space. Ageing is observed any many important materials like glasses, glassy
polymers, bio-molecules and plastic, to name just a few.

The study of metastability in the context of probabilistic models dates
back at least to the work of Eyring, Polanyi, Wigner, and Kramers [28, 29,
62, 45] in the context of chemical reactions. An early textbook reference is
[35]. Kramers was the first to use a model of a particle in a drift field subject
to forcing by Brownian motion, i.e. stochastic differential equations.

Stochastic differential equations form indeed one major context in which
metastability was studied extensively. In the physics literature, the main

tools were based on perturbation theory in the spirit of the quantum mechanical WKB method (a selection of references is [42, 43, 50, 51, 59]). Mathematical rigorous work was based mostly on the large deviation methods developed by Wentzell and Freidlin [31] who developed a systematic approach to the problem in the mid 1970's.

Over the last few year, in collaboration with M. Eckhoff, V. Gayrard, and M. Klein [8, 9, 10, 11] we have developed a somewhat novel approach to the metastability problem that is to a large extend based on potential theoretic ideas and makes extensive use of capacities and associated variational formulas. It has the virtue of being widely applicable while yielding rather precise results that improve on the exponential asymptotics obtained with large deviation methods. This approach has also allowed us, in collaboration with G. Ben Arous and V. Gayrard, to get some results on systems that show ageing, in particular in the Random Energy Model [1, 2]. In these lectures I want to review the key aspects of these developments.

2. A General Setting for Metastability and some Key Issues

The most general setting we will consider can be described as follows. We consider a Markov process X_t on a measure space $\Gamma \supset X_t$ with discrete or continuous time t. In continuous time the process is characterized by its *generator*, L, and in discrete time by the transition matrix, P. We will usually assume that the process is uniquely ergodic with invariant measure Q. We will denote the law of this process by $\mathbb{P}$. Moreover, we will denote by $\mathbb{P}_x$ the law of the process conditioned to verify $X(0) = x$. We will denote by τ_D, $D \subset \Gamma$, the first entrance time of $X(t)$ in D, i.e.

$$\tau_D \equiv \inf\{t > 0, X(t) \in D\} \tag{2.1}$$

We would like to say that the process is exhibiting metastability, if Γ can be partitioned into subsets S_i such that the process starting anywhere in S_i explores any S_i on a time-scale T_{fast} that is much shorter than the typical times it takes T_{slow} to leave S_i. Of course such a statement is imprecise and makes strictly speaking no sense. In particular the requirement that the process explores all of S_i must be qualified. There are a number of possible attempts to formalize this notion (see e.g. [37]) and the choice we will give below is certainly only one among a number of possibilities; however, it is useful inasmuch it implies a number of key properties of metastable systems while being sufficiently flexible to be applicable in a broad range of situations. We stress, however, that we do not necessarily consider this as the ultimate choice.

To make precise statements, we will also need a small parameter. Thus instead of a single Markov process, we will from now on always think of a family of processes X_t^ϵ (where in principle all key objects like $\Gamma, \mathbb{P}, \mathbb{Q}$ are allowed to depend on ϵ.

Provisional Definition 2.1. A family of Markov processes is called metastable, if there exists a collection of disjoint sets $B_i \subset \Gamma$ (possibly depending on ϵ), such that

$$\frac{\sup_{x \notin \cup_i B_i} \mathbb{E}_x \tau_{\cup_i B_i}}{\inf_{x \in \cup_i B_i} \mathbb{E}_x \tau_{\cup_{k \neq i} B_k}} \to 0, \quad \text{as} \quad \epsilon \downarrow 0 \tag{2.2}$$

Remark. Note that both numerator and denominator represent time scales that depend on the choice of the sets B_i. In fact, metastability may hold with respect to different decompositions on different time scales in the same system. This is often important in applications.

For many purposes it is useful to chose the B_i as small as possible. E.g. it will be very useful to have the property that

$$\frac{\sup_{x \in B_i} \mathbb{E}_x \tau_{B_k} - \inf_{x \in B_i} \mathbb{E}_x \tau_{B_k}}{\inf_{x \in B_i} \mathbb{E}_x \tau_{B_k}} \to 0, \quad \text{as} \quad \epsilon \downarrow 0 \tag{2.3}$$

Note that this requirement should be balanced against the main requirement (2.2).

Definition 2.1 defines metastability in terms of physical properties of a system we would like to consider as metastable. The problem is that it is not immediately verifiable, since it involves derived quantities, i.e. mean first hitting times, that are nor immediately computable. Our first problem will thus be to re-express the mean hitting times in terms of more manageable quantities.

The second problem will be to derive further properties of metastable systems. Since the definition implies frequent returns to the small starting set B_i before transit to another set B_j, this suggests an exponential law for the transit times. This also suggests that we may expect to describe the process of successive visits to distinct B_i asymptotically as a Markov process.

The most fundamental result we want to achieve, however, is a characterization of the spectrum of the generator, resp. the transition matrix of a metastable process. Our purpose here is to derive spectral information from the Definition above (an not the converse, which is much simpler).

3. Markov Processes and Potential Theory

The intimate relation between Markov processes and potential theory is well-known since the work of of Kakutani [39] and is the subject of numerous textbooks (see in particular the fundamental monograph by Doob [24], and for the discrete case [58]). This connection has found numerous and widespread applications both in probability theory and in analysis (see e.g. [25, 57] and references wherein). Our approach to metastability relies heavily on this connection. In the following pages we give a brief survey of some key facts (all of which are classical) that we will need later. While it would be possible to treat the continuous and the discrete case in unified way, this introduces some notational absurdities (in the discrete case) that we would rather avoid. Since all formulas in the case of diffusion processes can be found in [10], we will present here the formulas in the discrete case, i.e. when Γ is a discrete set.

Thus let Γ be a discrete set, $\mathbb{Q}$ be a positive measure on Γ, P a (irreducible) stochastic matrix on Γ. We will denote by L the generator of the process in the case of continuous time, and set $L = \mathbb{1} - P$ in the case of discrete time. We assume that L is symmetric on the space $L^2(\Gamma, \mathbb{Q})$.

Green's function. Let $\Omega \subset \Gamma$. Consider for $\lambda \in \mathbb{C}$ and g a real valued function on Ω the Dirichlet problem

$$
\begin{aligned}
(L - \lambda)f(x) &= g(x), & x \in \Omega \\
f(x) &= 0, & x \in \Omega^c
\end{aligned}
\tag{3.1}
$$

The associated Dirichlet Green's function $G_\Omega^\lambda(x, y)$ is the kernel of the inverse of the operator $(L - \lambda)^\Omega$, i.e. for any $g \in C_0(\Omega)$,

$$
f(x) = \sum_{y \in \Omega} G_\Omega^\lambda(x, y)g(y)
\tag{3.2}
$$

Note that the Green's function is symmetric with respect to the measure $\mathbb{Q}$, i.e.

$$
G_\Omega^\lambda(x, y) = \mathbb{Q}(y)G_\Omega^\lambda(y, x)\mathbb{Q}(x)^{-1}
\tag{3.3}
$$

Recall that the spectrum of L (more precisely the Dirichlet spectrum of the restriction of L to Ω, which we will sometimes denote by L^Ω), is the complement of the set of values λ for which G_Ω^λ defines a bounded operator.

Poisson Kernel. Consider for $\lambda \in \mathbb{C}$ the boundary value problem

$$
\begin{aligned}
(L - \lambda)f(x) &= 0, & x \in \Omega \\
f(x) &= \phi(x), & x \in \Omega^c
\end{aligned}
\tag{3.4}
$$

We denote by H_Ω^λ the associated solution operator.

Equilibrium Potential and Equilibrium Measure. Let $A, D \subset \Gamma$. Then the equilibrium potential (of the capacitor (A, D)), $h_{A,D}^\lambda$, is defined as the solution of the Dirichlet problem

$$\begin{aligned}
(L - \lambda)h^\lambda_{A,D}(x) &= 0, & x \in (A \cup D)^c \\
h^\lambda_{A,D}(x) &= 1, & x \in A \\
h^\lambda_{A,D}(x) &= 0, & x \in D
\end{aligned} \tag{3.5}$$

Note that (3.5) has a unique solution provided λ is not in the spectrum of $L^{(A \cup B)^c}$.

The equilibrium measure, $e^\lambda_{A,D}$, is defined as the unique measure on A, such that

$$h^\lambda_{A,D}(x) = \sum_{y \in A} G^\lambda_{D^c}(x, y)e^\lambda_{A,D}(y) \tag{3.6}$$

(3.6) may also be written as

$$e^\lambda_{A,D}(y) = -(L - \lambda)h^\lambda_{A,D}(y) \tag{3.7}$$

In fact, (3.7) defines a measure supported on both A and D that we will henceforth call the equilibrium measure. Note that this measure (in the case $\lambda = 0$) is zero on the interior of $A \cup D$, i.e. on the part of $A \cup D$ which is not connected by a non-zero element of the transition matrix P.

Note that (3.6) implies a representation formula of the Poisson-kernel H^λ_Ω, namely

$$(H^\lambda_\Omega \phi)(x) = \sum_{z \in \Omega^c} \phi(z)G^\lambda_{\Omega \cup z}(x, z)e^\lambda_{z, \Omega^c \setminus z}(z) \tag{3.8}$$

which is the discrete analog to the usual Poisson-Green's formula.

Capacity. Given a capacitor, (A, D), and $\lambda \in \mathbb{R}$, the λ-capacity of the capacitor is defined as

$$\mathrm{cap}^\lambda_A(D) \equiv \sum_{y \in A} Q(y)e^\lambda_{A,D}(y) \tag{3.9}$$

Using (3.7) one derives after some algebra that in discrete time

$$\begin{aligned}
\mathrm{cap}^\lambda_A(D) &= \tfrac{1}{2}\sum_{x,y} Q(x)\left[p(x,y)\left\|h^\lambda_{A,D}(x) - h^\lambda_{A,D}(y)\right\|^2 - \lambda\left(h^\lambda_{A,D}(x)\right)^2\right] \\
&\equiv \Phi^\lambda\left(h^\lambda_{A,D}\right)
\end{aligned} \tag{3.10}$$

where $p(x, y)$ are the transition probabilities (in discrete time) respectively transition rates (in discrete time). Φ^λ_Ω is called the Dirichlet form (or energy) for the operator $L - \lambda$ on Ω.

A fundamental consequence of (3.10) is the variational representation of the capacity when λ is real and non-positive, namely

$$\text{cap}_A^\lambda(D) = \inf_{h \in \mathcal{H}_{A,D}} \Phi^\lambda(h) \tag{3.11}$$

where $\mathcal{H}_{A,D}$ denotes the set of function

$$\mathcal{H}_{A,D} \equiv \{h : \Gamma \to [0,1] : h(x) = 0, x \in D, h(x) = 1, x \in A\} \tag{3.12}$$

Probabilistic Interpretation: Equilibrium Potential.
If $\lambda = 0$, the equilibrium potential has a natural probabilistic interpretation in terms of hitting probabilities of this process, namely,

$$h_{A,D}(x) \equiv h_{A,D}^0(x) = \mathbb{P}_x[\tau_A < \tau_D] \tag{3.13}$$

The equilibrium measure has a nice interpretation in the discrete time case as well:

$$e_{A,D}(y) = \mathbb{P}_y[\tau_D < \tau_A] \tag{3.14}$$

if $y \in A$. In particular, if $A = \{y\}$,

$$\mathbb{P}_y[\tau_D < \tau_A] = e_{y,D}(y) = \frac{\text{cap}_y(D)}{\mathbb{Q}(y)} \tag{3.15}$$

If $\lambda \neq 0$, the equilibrium potential still has a probabilistic interpretation in terms of the Laplace transform of the hitting time τ_A of the process starting in x and killed in D. Namely, we have for general λ, that

$$h_{A,D}^\lambda(x) = \mathbb{E}_x e^{\lambda \tau_A} \mathbb{1}_{\tau_A < \tau_D} \tag{3.16}$$

for $x \in (A \cup D)^c$, whenever the right-hand side exists.
Note that (3.16) implies that

$$\frac{d}{d\lambda} h_{A,D}^{\lambda=0}(x) = \mathbb{E}_x \tau_A \mathbb{1}_{\tau_A < \tau_D} \tag{3.17}$$

Differentiating the defining equation of $h_{A,D}^\lambda$ then implies that the function

$$w_{A,D}(x) = \begin{cases} \mathbb{E}_x \tau_A \mathbb{1}_{\tau_A < \tau_D}, & x \in (A \cup D)^c \\ 0, & x \in A \cup D \end{cases} \tag{3.18}$$

solves the inhomogeneous Dirichlet problem (to simplify notation, we set from now on $h_{A,D} \equiv h_{A;D}^0$, etc.)

$$\begin{aligned} Lw_{A,D}(x) &= h_{A,D}(x), & x \in (A \cup D)^c \\ w_{A,D}(x) &= 0, & x \in A \cup D \end{aligned} \tag{3.19}$$

Therefore, the mean hitting time in A of the process killed in D can be represented in terms of the Green's function as

$$\mathbb{E}_x \tau_A \mathbb{1}_{\tau_A < \tau_D} = \sum_{y \in (A \cup D)^c} G_{(A \cup D)^c}(x, y) h_{A,D}(y) \tag{3.20}$$

Note that in the particular case when $D = \emptyset$, we get the familiar Dirichlet problem

$$\begin{aligned} L w_A(x) &= 1, \quad x \in A^c \\ w_A(x) &= 0, \quad x \in A \end{aligned} \tag{3.21}$$

and the representation

$$\mathbb{E}_x \tau_A = \sum_{y \in A^c} G_{A^c}(x, y) \tag{3.22}$$

The full beauty of all this comes out when combining (3.6) with (3.20), resp. (3.22). Then, using Fubini's theorem,

$$\begin{aligned} \mathbb{Q}(z) \mathbb{E}_z \tau_A e_{z,A}(z) &= \sum_{y \in A^c} \mathbb{Q}(y) G_{A^c}(y, z) e_{z,A}(z) \\ &= \sum_{y \in A^c} \mathbb{Q}(y) h_{z,A}(y) \end{aligned} \tag{3.23}$$

and

$$\mathbb{Q}(z) \mathbb{E}_z \tau_A \mathbb{1}_{\tau_A < \tau_D} e_{z, A \cup D}(z) = \sum_{y \in (A \cup D)^c} \mathbb{Q}(y) h_{z, A \cup D}(y) h_{A,D}(y) \tag{3.24}$$

(3.24) yield directly formulae for mean hitting times in terms of capacities and equilibrium potentials.

Indeed (3.6) yield a formula for the Green function

$$G_{D^x}^\lambda(x, y) = \frac{h_{y,D}^\lambda(x)}{e_{y,D}^\lambda(y)} = \frac{\mathbb{Q}(y) h_{y,D}^\lambda(x)}{\mathrm{cap}_y^\lambda(D)} = \frac{\mathbb{Q}(y) h_{x,D}^\lambda(y)}{\mathrm{cap}_x^\lambda(D)} \tag{3.25}$$

which will play a key rôle in the sequel.

Remark. Equations (3.23)-(3.25) rely explicitly on the discrete structure on the state space, or more precisely that for any $x \in \Gamma$, $\mathbb{Q}(x) > 0$. In the case of continuous state space, such formulas do not hold in the strict sense, or are not useful, but suitable "integral versions", involving integrals over suitably chosen small neighborhoods of e.g. the points z in (3.23) are still valid, and can be used to more or less the same effect as the exact relations in the discrete case. This entails, however, some extra technical difficulties. In these notes we will therefore restrict our attention to the discrete case, where the principle ideas can be explained without being obscured by technicalities.

4. Metastability in Terms of Capacities and Mean Hitting Times

We now use the observations made in Section 2 to derive the desired alternative characterization of metastability in terms of potential theoretic quantities, namely capacities.

Definition 4.1 Assume that Γ is a discrete set. Then a family of Markov processes X_t^ϵ is metastable with respect to the set of points $\mathcal{M} \subset \Gamma$, if

$$\frac{\sup_{x \notin \mathcal{M}} \mathbb{Q}(x)/\mathrm{cap}_x(\mathcal{M})}{\inf_{x \in \mathcal{M}} \mathbb{Q}(x)/\mathrm{cap}_x(\mathcal{M}\backslash x)} \le \rho(\epsilon) \tag{4.1}$$

for some $\rho(\epsilon)$ that tends to zero as $\epsilon \downarrow 0$.

We will see that this definition is essentially equivalent to the provisional definition given in Section 2, at least in situations where the latter is reasonable. Definition 4.1 has far reaching consequences. As a first step we will show that it implies the mean values of transition times from minima to other minima can be computed very precisely.

Let $x \in \mathcal{M}$, $x \notin J \subset \mathcal{M}$. We want to compute $\mathbb{E}_x \tau_J$. This computation will be based on the formula (3.22), which yields that

$$\mathbb{E}_x \tau_J = \frac{\mathbb{Q}(x)}{\mathrm{cap}_x(J)} \sum_{y \in J^c} \frac{\mathbb{Q}(y)}{\mathbb{Q}(x)} h_{x,J\backslash x}(y) \tag{4.2}$$

Thus we have only to control expressions like

$$\sum_{y \in \mathcal{M}^c} \frac{\mathbb{Q}(y)}{\mathbb{Q}(x)} h_{x,\mathcal{M}\backslash x}(y) \tag{4.3}$$

The analysis of such sums requires some knowledge of the equilibrium potential. One of the cornerstones of our approach is the observation that the equilibrium potential, too, can be estimated in terms of capacities, and that in many cases, these estimates yield very sharp results. The reason underlying this fact is that it turns out that in metastable systems, we tend to have a dichotomy of the type: *Either* the equilibrium potential $h_{x,J}(y)$ is close to zero or to one, *or* the invariant measure $\mathbb{Q}(y)$ is very small.

Renewal Estimates. The estimation of the equilibrium through capacities is based on a renewal argument, that in the case of discrete state space is very simple.

Lemma 4.1 Let $A, D \subset \Gamma$ be disjoint sets, and let $x \notin A \cup D$. Then

$$h_{A,D}(x) \leq \frac{\text{cap}_x(A)}{\text{cap}_x(D)} \tag{4.4}$$

Proof. In the discrete case, this result is extremely easy to prove. Just use the simple observation that going from x to A the process either does or does not re-visit x, or does so a first time, one gets

$$h_{A,D}(x) = \mathbb{P}_x[\tau_A < \tau_D] = \mathbb{P}_x[\tau_A < \tau_{D \cup x}] + \mathbb{P}_x[\tau_x < \tau_{A \cup D}]\mathbb{P}_x[\tau_A < \tau_D] \tag{4.5}$$

Hence

$$h_{A,D}(x) \leq \frac{\mathbb{P}_x[\tau_A < \tau_{D \cup x}]}{1 - \mathbb{P}_x[\tau_x < \tau_{A \cup D}]} = \frac{\mathbb{P}_x[\tau_A < \tau_{D \cup x}]}{\mathbb{P}_x[\tau_{A \cup D} < \tau_x]} \leq \frac{\mathbb{P}_x[\tau_A < \tau_x]}{\mathbb{P}_x[\tau_D < \tau_x]} = \frac{\text{cap}_x(A)}{\text{cap}_x(D)} \tag{4.6}$$

Since even large probabilities do not exceed 1, the lemma is proven. □

Remark. Note that the power of Lemma 4.1 is more than doubled by judicious use of the elementary fact that $h_{A,D}(x) = 1 - h_{D,A}(x)$.

Remark. The bound (4.5) can easily be improved to $h_{A,D}(x) \leq$ $\min\left(\frac{\text{cap}_x(A)}{\text{cap}_x(D \cup A)}, 1\right)$, but this is seldom very useful.

Ultrametricity. An important fact that allows to obtain general results under our Definition of metastability is the fact that it implies approximate ultrametricity of capacities. This has been noted in [9].

Lemma 4.2 Assume that $x, y \in \Gamma$, $D \subset \Gamma$. Then, if for $0 < \delta < \frac{1}{2}$, $\text{cap}_y(D) \leq \delta\text{cap}_y(x)$, then

$$\frac{1 - 2\delta}{1 - \delta} \leq \frac{\text{cap}_x(D)}{\text{cap}_y(D)} \leq \frac{1}{1 - \delta} \tag{4.7}$$

Proof. The proof of this lemma given in [9] is probabilistic and uses splitting and renewal ideas. It should be possible to prove this result with purely potential theoretic arguments, but I have not worked this out. □

Lemma 4.2 has the following immediate corollary, which is the version of the ultrametric triangle inequality we are looking for:

Corollary 4.1 Let $x, y, z \in \mathcal{M}$. Then

$$\text{cap}_x(y) \geq \frac{1}{3} \min\left(\text{cap}_x(z), \text{cap}_y(z)\right) \tag{4.8}$$

In the sequel it will be useful to have the notion of a "valley" or "attractor" of a point in $\mathcal{M}$. We set for $x \in \mathcal{M}$,

$$A(x) \equiv \left\{ z \in \Gamma \,|\, \mathbb{P}_z[\tau_x = \tau_\mathcal{M}] = \sup_{y \in \mathcal{M}} \mathbb{P}_z[\tau_y = \tau_\mathcal{M}] \right\} \tag{4.9}$$

Note that valleys may overlap, but from Lemma 4.2 it follows easily that the intersection has a vanishing invariant mass. The notion of a valley in the case of a diffusion process coincides with the intuitive notion.

The following simple corollary will be most useful:

Corollary 4.2 Let $m \in \mathcal{M}$, $y \in A(m)$, and $J \subset \mathcal{M} \backslash m$. Then either

$$\frac{1}{2} \leq \frac{\mathrm{cap}_m(J)}{\mathrm{cap}_y(J)} \leq \frac{3}{2} \tag{4.10}$$

or

$$\frac{\mathbb{Q}(y)}{\mathbb{Q}(m)} \leq 3|\mathcal{M}| \frac{\mathbb{Q}(y)}{\mathrm{cap}_y(\mathcal{M})} \frac{\mathrm{cap}_m(J)}{\mathbb{Q}(m)} \tag{4.11}$$

Proof. Lemma 4.2 implies that *if* $\mathrm{cap}_m(y) \geq 3\mathrm{cap}_m(J)$, then (4.10) holds. Otherwise,

$$\frac{\mathbb{Q}(y)}{\mathbb{Q}(m)} \leq 3 \frac{\mathbb{Q}(y)}{\mathrm{cap}_y(m)} \frac{\mathrm{cap}_m(J)}{\mathbb{Q}(m)} \tag{4.12}$$

Since $y \in A(m)$,

$$\mathrm{cap}_y(\mathcal{M}) \leq \sum_{z \in \mathcal{M}} \mathrm{cap}_y(z) \leq |\mathcal{M}| \sup_{z \in \mathcal{M}} \mathrm{cap}_y(z) = |\mathcal{M}|\mathrm{cap}_y(m) \tag{4.13}$$

which yields (4.11). $\qquad\square$

We want to use this corollary in order to estimate the summands in the sum (4.2). We will set $\inf_y \mathbb{Q}(y)^{-1}\mathrm{cap}_y(\mathcal{M}) = a_\epsilon$.

Lemma 4.3 Let $x \in \mathcal{M}$ and $J \subset \mathcal{M}$ with $x \notin J$. Then:
(i) If $x = m$, either

$$h_{x,J}(y) \geq 1 - \frac{3}{2}|\mathcal{M}|\frac{\mathrm{cap}_m(J)}{\mathrm{cap}_y(\mathcal{M})} \tag{4.14}$$

or

$$\frac{\mathbb{Q}(y)}{\mathbb{Q}(x)} \leq 3|\mathcal{M}|a_\epsilon^{-1}\frac{\mathrm{cap}_m(J)}{\mathbb{Q}(m)} \tag{4.15}$$

(ii) If $m \in J$, then

$$\mathbb{Q}(y)h_{x,J}(y) \leq \frac{3}{2}|\mathcal{M}|a_\epsilon^{-1}\mathrm{cap}_m(x) \tag{4.16}$$

(iii) If $m \notin J \cup x$, then either

$$h_{x,J}(y) \leq 3 \frac{\text{cap}_m(x)}{\text{cap}_m(J)} \tag{4.17}$$

and

$$h_{x,J}(y) \geq 1 - 3 \frac{\text{cap}_m(J)}{\text{cap}_m(x)} \tag{4.18}$$

or

$$Q(y) \leq 3|\mathcal{M}|a_\epsilon^{-1} \max(\text{cap}_m(J), \text{cap}_m(x)) \tag{4.19}$$

Proof. We make use of the fact that by Lemma 4.1,

$$0 \leq h_{x,J}(y) \leq \frac{\text{cap}_y(x)}{\text{cap}_y(J)} \tag{4.20}$$

and

$$1 \geq h_{x,J}(y) \geq 1 - \frac{\text{cap}_y(J)}{\text{cap}_y(x)} \tag{4.21}$$

In case (i), we anticipate that only (4.21) will be useful. To get the first dichotomy, we use Corollary 4.2 to replace the numerator by $\text{cap}_m(J)$. To get the second assertion, note simply that

$$\frac{\text{cap}_m(J)}{\text{cap}_y(m)} \leq \frac{Q(y)\text{cap}_m(J)}{\text{cap}_y(x)Q(m)} \frac{Q(m)}{Q(y)} \tag{4.22}$$

and rewrite this inequality for $\frac{Q(y)}{Q(m)}$.

In case (ii), we use (4.20) and apply Corollary 4.2 to $\text{cap}_y(x)$.

In case (iii), we admit both possibilities and apply the corollary to both the numerators and the denominators. $\square$

Remark. Case (iii) in the preceding lemma is special in as much as it will not always give sharp estimates, namely whenever $\text{cap}_m(J) \sim \text{cap}_m(y)$. If this situation occurs, and the corresponding terms contribute to leading order, we cannot get sharp estimates with the tools we are exploiting here, and better estimates on the equilibrium potential will be needed.

Mean Times. Let us now apply this lemma to the computation of the sum (4.3) (we ignore the fact that the sets $A(m)$ may not be disjoint, as the overlaps give no significant contribution).

$$\sum_{y \in \mathcal{M}^c} \frac{Q(y)}{Q(x)} h_{x,J}(y) = \sum_{m \in \mathcal{M}} \frac{Q(m)}{Q(x)} \sum_{y \in A(m) \setminus m} \frac{Q(y)}{Q(m)} h_{x,J}(y)$$

$$\tag{4.23}$$

$$\equiv \sum_{m \in \mathcal{M}} \frac{Q(m)}{Q(x)} L(m)$$

(we ignore the fact that the sets $A(m)$ may not be disjoint, as the overlaps give no significant contribution).

We now estimate the terms $L(m)$ with the help of Lemma 4.3.

Lemma 4.4 With the notation introduced above and the assumptions of Lemma 4.3, we have that

(i) If $m = x$

$$L(x) \leq \frac{\mathbb{Q}(A(x))}{\mathbb{Q}(x)} \tag{4.24}$$

and

$$L(x) \geq \frac{\mathbb{Q}(A(x))}{\mathbb{Q}(x)} \left(1 - 6C|\mathcal{M}|a_\epsilon^{-1} \frac{\mathrm{cap}_m(J)}{\mathbb{Q}(A(m))} \right) \tag{4.25}$$

(ii) If $m \in J$, then

$$L(m) \leq Ca_\epsilon^{-1}|\mathcal{M}| \frac{\mathrm{cap}_m(x)}{\mathbb{Q}(m)} |\{y \in A(m) : \mathbb{Q}(y) \geq a_\epsilon^{-1}|\mathcal{M}|\mathrm{cap}_m(x)\}| \tag{4.26}$$

for some constant C independent of ϵ.

(iii) If $m \notin J \cup x$, then

$$L(m) \leq \frac{\mathbb{Q}(A(m))}{\mathbb{Q}(m)} \tag{4.27}$$

Moreover,

(iii.1) if $\mathrm{cap}_m(J) \leq \frac{1}{3}\mathrm{cap}_m(x)$, then

$$L(m) \geq \frac{\mathbb{Q}(A(m))}{\mathbb{Q}(m)} \left(1 - 3\frac{\mathrm{cap}_m(J)}{\mathrm{cap}_m(x)} \right) \left(1 - C|\mathcal{M}|a_\epsilon^{-1} \frac{\mathrm{cap}_m(x)}{\mathbb{Q}(A(m))} \right) \tag{4.28}$$

and

(iii.2) if $\mathrm{cap}_m(J) \geq \frac{1}{3}\mathrm{cap}_m(x)$, then

$$L(m) \leq \frac{\mathbb{Q}(A(m))}{\mathbb{Q}(m)} - C|\mathcal{M}|a_\epsilon^{-1} \frac{\mathrm{cap}_m(x)}{\mathbb{Q}(m)} \tag{4.29}$$

Proof. The proof of this lemma is rather straightforward and will be left as an exercise. Just note that to get (4.27) involves an optimal choice of δ in the application of Lemma 4.3. □

Remark. The statement of Lemma 4.4 looks a little complicated due to the rather explicit error terms. Ignoring all small factors, its statement boils down to:

(i) There is always the term

$$L(x) \approx \frac{\mathbb{Q}(A(x))}{\mathbb{Q}(x)}$$

(ii) If $m \in J$, roughly,

$$L(m) \leq \frac{\mathrm{cap}_m(x)}{\mathbb{Q}(m)}$$

(iii) If $m \notin J \cup x$, it is always true that

$$L(m) \leq \frac{\mathbb{Q}(A(m))}{\mathbb{Q}(m)}$$

i.e. roughly of order one. This bound is achieved, if

(iii.1) $\mathrm{cap}_m(J) \ll \mathrm{cap}_m(x)$, whereas in the opposite case
(iii.2) $\mathrm{cap}_m(J) >> \mathrm{cap}_m(x)$,

$$L(m) \leq \max \left(\frac{\mathrm{cap}_m(x)}{\mathbb{Q}(m)}, \frac{\mathrm{cap}_m(x)}{\mathrm{cap}_m(J)} \right)$$

Of course these arguments use that quantities like $\frac{\mathbb{Q}(m)}{\mathbb{Q}(A(m))}$ are not too small, i.e. that the most massive points in a metastable set have a reasonably large mass (compared to say, $\rho(\epsilon)$). If this condition is violated, the idea to represent metastable sets by single points is clearly mislead. We will discuss later what has to be done in such cases.

Taking into account that the $L(m)$ appear with the prefactor $\mathbb{Q}(m)/\mathbb{Q}(x)$ in the expression (4.23), we see that contributions from case (ii) are always sub-dominant; in particular, when $J = \mathcal{M} \backslash x$, the term $m = x$ gives always the main contribution. The terms from case (iii) have a chance to contribute only if $\mathbb{Q}(m) \geq \mathbb{Q}(x)$. If that is the case, and we are in sub-case (iii.1), they indeed contribute, and potentially dominate the sum, whereas in sub-case (iii.2) they never contribute, just as in case (ii).

From Lemma 4.3 we can now derive precise formulae for the mean arrival times in a variety of special cases. In particular,

Theorem 4.1 *Let $x \in \mathcal{M}$ and $J \subset \mathcal{M} \backslash x$ be such a that for all $m \notin J \cup x$ either $\mathbb{Q}(m) \ll \mathbb{Q}(x)$ or $\mathrm{cap}_m(J) m \mathrm{cap}_m(x)$, then*

$$\mathbb{E}_x \tau_J = \frac{\mathbb{Q}(A(x))}{\mathrm{cap}_x(J)} (1 + o(1)) \tag{4.30}$$

Proof. The proof of this result is straightforward from (4.2), (4.23) and Lemma 4.3.   □

Remark. In much the same way one can compute conditional mean times such as $\mathbb{E}_x[\tau_J | \tau_J \leq \tau_I]$. Formulae are given in [8, 9] and we will not go into these issues any further here.

Finally we want to compute the mean time to reach $\mathcal{M}$ starting from a general point.

Lemma 4.5 Let $z \notin \mathcal{M}$. Then

$$\mathbb{E}_z \tau_{\mathcal{M}} \le a_\epsilon^{-2} \left(|\{y : \mathbb{Q}(y) \ge \mathbb{Q}(z)|\} + C \right) \tag{4.31}$$

Proof. Using Lemma 4.1, we get that

$$
\begin{aligned}
\mathbb{E}_z \tau_{\mathcal{M}} \quad &\le \quad \frac{\mathbb{Q}(z)}{\mathrm{cap}_z(\mathcal{M})} \sum_{y \in \mathcal{M}^c} \frac{\mathbb{Q}(y)}{\mathbb{Q}(z)} \max \left(1, \frac{\mathrm{cap}_y(z)}{\mathrm{cap}_y(\mathcal{M})} \right) \\[2mm]
&= \quad \frac{\mathbb{Q}(z)}{\mathrm{cap}_z(\mathcal{M})} \sum_{y \in \mathcal{M}^c} \frac{\mathbb{Q}(y)}{\mathbb{Q}(z)} \max \left(1, \frac{\mathbb{P}_y[\tau_z < \tau_y]}{\mathbb{P}_y[\tau_{\mathcal{M}} < \tau_y]} \right) \\[2mm]
&\le \quad \sup_{y \in \mathcal{M}^c} \left(\frac{\mathbb{Q}(y)}{\mathrm{cap}_y(\mathcal{M})} \right)^2 \sum_{y \in \mathcal{M}^c} \max \left(\frac{\mathbb{Q}(y)}{\mathbb{Q}(z)}, \mathbb{P}_z[\tau_y < \tau_z] \right) \tag{4.32} \\[2mm]
&\le \quad \sup_{y \in \mathcal{M}^c} \left(\frac{\mathbb{Q}(y)}{\mathrm{cap}_y(\mathcal{M})} \right)^2 \left(\sum_{y : \mathbb{Q}(y) \le \mathbb{Q}(z)} \frac{\mathbb{Q}(y)}{\mathbb{Q}(z)} + \sum_{y : \mathbb{Q}(y) > \mathbb{Q}(z)} 1 \right) \\[2mm]
&\le \quad \sup_{y \in \mathcal{M}^c} \left(\frac{\mathbb{Q}(y)}{\mathrm{cap}_y(\mathcal{M})} \right)^2 (C + |\{y : \mathbb{Q}(y) > \mathbb{Q}(z)\}|)
\end{aligned}
$$

which proves the lemma.   □

Remark. If Γ is finite (resp. not growing to fast with ϵ), the above estimate combined with Theorem 4.1 shows that the two definitions of metastability we have given in terms of mean times rep. capacities are equivalent. On the other hand, in the case of infinite state space Γ, we cannot expect the supremum over $\mathbb{E}_z \tau_{\mathcal{M}}$ to be finite, which shows that our first definition was somewhat naive. We will later see that this definition can rectified in the context of spectral estimates.

5. Metastability and Spectral Theory

We now turn to the characterisation of metastability through spectral data. The connection between metastable behaviour and the existence of small eigenvalues of the generator of the Markov process has been realised for a very long time. Some key references are [17, 18, 19, 31, 33, 36, 43, 46, 49, 56, 60, 61]

We will show that Definition 4.1 implies that the spectrum of $1 - P$ decomposes into a cluster of $|\mathcal{M}|$ very small real eigenvalues that are separated by a gap from the rest of the spectrum.

A Priori Estimates. The first step of our analysis consists in showing that the matrix $(1 - P)^{\mathcal{M}}$ that has Dirichlet conditions in all the points of $\mathcal{M}$ has a minimal eigenvalue that is not smaller than $O(a_\epsilon)$.

Lemma 5.1 Let λ^0 denote the infimum of the spectrum of $L_\epsilon^\mathcal{M}$. Then

$$\lambda^0 \geq \frac{1}{\sup_{z \in \Omega} \mathbb{E}_z \tau_\mathcal{M}} \tag{5.1}$$

Proof. This result is a classical result of Donsker and Varadhan [26] (in the diffusion setting; see [9] for a simple proof in the discrete case). □

In the case when Γ is a finite set, (5.1) together with the estimate of Lemma 4.5 will yield a sufficiently good estimate. If $|\Gamma| = \infty$, the supremum on the right may be infinite and the estimate becomes useless. However, it is easy to modify the proof of Lemma 5.1 to yield an improvement.

Lemma 5.2 Let λ^0 denote the infimum of the spectrum of $(1 - P)^\mathcal{M}$ and denote by ϕ the corresponding eigenfunction . Let $D \subset \Gamma$ be any compact set, Then

$$\lambda^0 \geq \frac{1}{\sup_{z \in D} \mathbb{E}_z \tau_{\mathcal{M}_\epsilon}} \left(1 - \sum_{y \in D^c} \mathbb{Q}(y)|\phi(y)|^2 \right) \tag{5.2}$$

Moreover, for any $\delta > 0$, there exists D finite such that

$$\lambda^0 \geq \frac{1}{\sup_{z \in D} \mathbb{E}_z \tau_{\mathcal{M}_\epsilon}} (1 - \delta) \tag{5.3}$$

Proof. Let $w(x)$ denote the solution of the Dirichlet problem

$$\begin{aligned} (1 - P)w(x) &= 1, \quad x \in \Gamma \backslash \mathcal{M} \\ w(x) &= 1, \quad x \in \mathcal{M} \end{aligned} \tag{5.4}$$

Recall that $w(x) = \mathbb{E}_x \tau_{\mathcal{M}_\epsilon}$. Using that for any $C > 0$, $ab \leq \frac{1}{2}(Ca^2 + b^2/C)$ with $ab = \phi(x)\phi(y)$ and $C = w(y)/w(x)$, one shows readily that

$$\sum_{x \in \Gamma \backslash \mathcal{M}} \mathbb{Q}(x)\phi(x)(1 - P)\phi(x) \geq \sum_{x \in \Gamma \backslash \mathcal{M}} \mathbb{Q}(x)\frac{\phi(x)}{w(x)}(1 - P)w(x)\phi(x)$$

$$= \sum_{x \in \Gamma \backslash \mathcal{M}} \mathbb{Q}(x)\frac{\phi(x)}{w(x)}\phi(x)$$

$$\geq \frac{1}{\sup_{x \in D} w(x)} \sum_{x \in D \cap \Gamma \backslash \mathcal{M}} \phi^2(x)$$

$$\tag{5.5}$$

Choosing ϕ as the normalized eigenfunction with maximal eigenvalue yields (5.2).

We now claim that for any $\gamma > 0$,

$$\sum_{y \in \Gamma} \mathbb{Q}(y)^{-\gamma} |\phi(y)|^2 < C_\gamma < \infty \qquad (5.6)$$

This clearly implies (5.7). The estimate (5.6) follows from a standard Combes-Thomas estimate for the ground-state eigenfunction, ϕ. It is convenient to introduce $v(y) \equiv \mathbb{Q}(y)^{1/2} \phi(y)$, which is the corresponding ground state eigenfunction of the operator

$$H_\epsilon \equiv \mathbb{Q}(y)^{1/2}(1 - P)\mathbb{Q}(y)^{-1/2} \qquad (5.7)$$

which is a symmetric operator on $\ell^2(\Gamma)$. By a standard computation,

$$t(\alpha)[u] \equiv \sum_{y \in \Gamma} u^*(y)\mathbb{Q}(y)^{-i\alpha} H_\epsilon \mathbb{Q}(y)^{i\alpha} u(y) \qquad (5.8)$$

defines a closed sectoral form (in the sense of Kato [38]), which is analytic in the strip $|\Im \alpha| < 1/2$. An adaptation of the Combes-Thomas estimate (see e.g. [53]) then implies that v satisfies

$$\sum_{y \in \mathbb{Q}} \mathbb{Q}(y)^{(1-\gamma)} |v(y)|^2 < C_\gamma < \infty \qquad (5.9)$$

which is equivalent to (5.6). This completes the proof of the lemma.   □

If we combine this result with the estimate from Lemma 4.5, we obtain the following proposition.

Proposition 5.1 *Let λ^0 denote the principal eigenvalue of the operator $(1 - P)^\mathcal{M}$. Then there exists a constant $C > 0$, independent of ϵ, such that for all ϵ small enough,*

$$\lambda^0 \geq Ca_\epsilon^{-2} \qquad (5.10)$$

Remark. Proposition 5.1 links the fast time scale to the smallest eigenvalue of the Dirichlet operator, as should be expected. Note that the relation is not very precise. We will soon derive a much more precise relation between times and eigenvalues for the cluster of small eigenvalues.

Characterization of Small Eigenvalues. We will now obtain a representation formula for all eigenvalues that are smaller than λ^0. It is clear that there will be precisely $|\mathcal{M}|$ such eigenvalues. This representation was first exploited in [9], but already in 1973 Wentzell put forward very similar ideas (in the case of general Markov processes). As will become clear (hopefully more so then on the original paper [9], this is extremely simple in the context of discrete processes (see [11] for the more difficult continuous case.

The basic idea is to use the fact that the solution of the Dirichlet problem

$$(L - \lambda)f(x) = 0, \quad x \notin \mathcal{M}$$
$$f(x) = \phi_x, \quad x \in \mathcal{M} \tag{5.11}$$

already solves the eigenvalue equation $L\phi(x) = \lambda\phi(x)$ everywhere except possibly on $\mathcal{M}$. The question if whether an appropriate choice of boundary conditions and the right choice of the value of λ will actually lead to a solution. This is indeed the case.

Lemma 5.3 Assume that $\lambda < \lambda^0$ is an eigenvalue of L and $\phi(x)$ is the corresponding eigenfunction. Then the unique solution of (5.11) with $\phi_x = \phi(x)$, $x \in \mathcal{M}$, satisfies $f(y) = \phi(y)$, for all $y \in \Gamma$.

Proof. Since $\phi(x) = f(x)$ on $\mathcal{M}$, we have that $(f - \phi)(x)$ solves the Dirichlet problem

$$(L - \lambda)^{\mathcal{M}}(f - \phi)(x) = 0, \quad x \notin \mathcal{M}$$
$$(f - \phi)(x) = 0, \quad x \in \mathcal{M} \tag{5.12}$$

But since λ is not in the spectrum of $L^{S_k}_\epsilon$, (5.12) has a unique solution, which is identically zero, so that $f(x) = \phi(x)$ on Γ, which proves the lemma. □

From the lemma we conclude that we can find all eigenfunctions corresponding to eigenvalues smaller than λ^0 among the solutions of the Dirichlet problems (5.11).

Let now f be a solution of (5.11) with $\lambda < \lambda^0$. Clearly, f is an eigenfunction with eigenvalue λ, if

$$(L - \lambda)f(x) = 0, \quad x \in \mathcal{M} \tag{5.13}$$

Thus we need to compute the left-hand side of (5.13). Now $f(y)$ can be represented in terms of the equilibrium potentials $h^\lambda_{x,\mathcal{M}\setminus x} \equiv h^\lambda_x(y)$ defined in (3.3) as $f(y) = \sum_{x\in\mathcal{M}} \phi_x h^\lambda_x(y)$. Thus

$$(L - \lambda)f(x) = \sum_{z\in\mathcal{M}} \phi_z(L - \lambda)h^\lambda_z = \sum_{z\in\mathcal{M}} \phi_z e^\lambda_{z,\mathcal{M}\setminus z}(x) \tag{5.14}$$

where we used (3.7). Let us denote by $\mathcal{E}_\mathcal{M}(\lambda)$ the $|\mathcal{M}| \times |\mathcal{M}|$- matrix with elements

$$(\mathcal{E}_\mathcal{M}(\lambda))_{xy} \equiv e^\lambda_{z,\mathcal{M}\setminus z}(x) \tag{5.15}$$

We can then conclude that:

Lemma 5.4 A number $\lambda < \lambda^0$ is an eigenvalue of the matrix $L = (1 - P)$ if and only if

$$\det \mathcal{E}_\mathcal{M}(\lambda) = 0 \tag{5.16}$$

Anticipating that we are interested in small λ, we want to re-write the matrix $\mathcal{E}_{\mathcal{M}}$ in a more convenient form. To do so let us set

$$h_x^\lambda(y) \equiv h_x(y) + \psi_x^\lambda(y) \tag{5.17}$$

where $h_x(y) \equiv h_x^0(y)$ and consequently $\psi_x^\lambda(y$ solves the inhomogeneous Dirichlet problem

$$\begin{aligned}(L - \lambda)\psi_x^\lambda(y) &= \lambda h_x(y), \quad y \in \Gamma \backslash \mathcal{M} \\ \psi_x^\lambda(y) &= 0, \quad y \in \mathcal{M}\end{aligned} \tag{5.18}$$

Lemma 5.5

$$\begin{aligned}(\mathcal{E}_{\mathcal{M}}(\lambda))_{xz} = \mathbb{Q}(x)^{-1} &\left(\frac{1}{2} \sum_{y \neq y'} \mathbb{Q}(y')p(y', y)[h_z(y') - h_z(y)][h_x(y') \right. \\ &\left. -h_x(y)] - \lambda \sum_y \mathbb{Q}(y) \left(h_z(y)h_x(y) + h_x(y)\psi_z^\lambda(y) \right) \right)\end{aligned} \tag{5.19}$$

Proof. Note that

$$\begin{aligned}(L - \lambda)h_z^\lambda(x) &= (L - \lambda)h_z(x) + (L - \lambda)\psi_z^\lambda(x) \\ &= Lh_z(x) - \lambda h_z(x) + (L - \lambda)\psi_z^\lambda(x)\end{aligned} \tag{5.20}$$

Now by adding a huge zero,

$$Lh_z(x) = \mathbb{Q}(x)^{-1} \sum_{y' \in \Gamma} \mathbb{Q}(y')h_x(y')Lh_z(y')$$

$$= \mathbb{Q}(x)^{-1}\frac{1}{2} \sum_{y,y' \in \Gamma} \mathbb{Q}(y')p(y', y))[h_z(y') - h_z(y)][h_x(y') - h_x(y)] \tag{5.21}$$

Similarly,

$$(L - \lambda)\psi_z^\lambda(x) =$$

$$\mathbb{Q}(x)^{-1} \sum_{y' \in \Gamma} \left(\mathbb{Q}(y')h_x(y')(L - \lambda)\psi_z^\lambda(y') - \lambda \mathbb{1}_{y' \neq x}h_x(y')h_z(y') \right) \tag{5.22}$$

Since $\psi_z^\lambda(y) = 0$ whenever $y \in \mathcal{M}$, and $Lh_x(y)$ vanishes whenever $y \notin \mathcal{M}$, using the symmetry of L, we get that the right-hand side of (5.22) is equal to

$$-\lambda \mathbb{Q}(x)^{-1} \sum_{y' \in \Gamma} \left(\mathbb{Q}(y')h_x(y')(\psi_z^\lambda(y') + \mathbb{1}_{y' \neq x}h_x(y')h_z(y')) \right) \tag{5.23}$$

Adding the left-over term $-\lambda h_z(x) = -\lambda h_x(x)h_z(x)$ from (5.1) to (5.22), we arrive at (5.19).   □

Remark. Note that we get an alternative probabilistic interpretation of $Lh_z(x)$ as

$$Lh_z(x) = \begin{cases} -\mathbb{P}_z[\tau_x \leq \tau_{\mathcal{M}}], & \text{if} \quad z \neq x \\ \mathbb{P}_x[\tau_{\mathcal{M}\backslash x} < \tau_x] & \text{if} \quad z = x \end{cases} \tag{5.24}$$

Note that the off-diagonal quantities can in turn be expressed via the renewal equations as

$$\mathbb{P}_z[\tau_x \leq \tau_{\mathcal{M}}] = \mathbb{P}_z[\tau_x < \tau_{\mathcal{M}\backslash x}]\mathbb{P}_z[\tau_{\mathcal{M}\backslash z} < \tau_z] = h_{x,\mathcal{M}\backslash x}(z)e_{z,\mathcal{M}\backslash z}(z) \tag{5.25}$$

We will see in Section 6 that the equilibrium potential $h_{x,\mathcal{M}\backslash x}(z)$ can be estimated along with the capacities rather well.

We are now in a position to relate the small eigenvalues of $(1 - P)$ to the eigenvalues of the classical capacity matrix. Let us denote by $\|f\|_2$ the ℓ^2-norm with respect to the measure $\mathbb{Q}$, i.e. $\|f\|_2^2 = \sum_y \mathbb{Q}(y)f(y)^2$.

Theorem 5.1 If $\lambda < \lambda^0$ is an eigenvalue of L, then there exists an eigenvalue μ of the $|\mathcal{M}| \times |\mathcal{M}|$-matrix K whose matrix elements are given by

$$K_{zx} = \frac{\frac{1}{2} \sum_{y \neq y'} \mathbb{Q}(y')p(y', y)[h_z(y') - h_z(y)][h_x(y') - h_x(y)]}{\|h_z\|_2\|h_x\|_2} \tag{5.26}$$

such that $\lambda = \mu(1 + O(\rho(\epsilon)))$.

Proof. The proof will rely on the following general fact.

Lemma 5.6 Let A be a finite dimensional self-adjoint matrix. Let $B(\lambda)$ a Lipshitz continuous family of bounded operators on the same space that satisfies the bound $\|B(\lambda)\| \leq \delta + \lambda C$, and $\|B(\lambda) - B(\lambda')\| \leq C\|\lambda - \lambda'\|$ for $0 \leq \delta \ll 1$, and $0 \leq C < \infty$. Assume that A has k eigenvalues $\lambda_1, \ldots, \lambda_k$ in an interval $[0, a]$ with $a < \delta/C$. Then

(i) Any solution λ_i' of the equation

$$\det(A - \lambda(\mathbb{1} + B(\lambda))) = 0 \tag{5.27}$$

satisfies $|\lambda_i' - \lambda_i| \leq 4\delta\lambda_i$, for some $i = 1, \ldots, k$.

(ii) There exists $\delta_0 > 0$ and $a_0 > 0$ such that for all $\delta < \delta_0$, and $a < a_0$, Equation 5.27 has exactly k solutions $l_1', \ldots, \lambda_k'$.

(iii) If the eigenvalue λ_i is simple and isolated with $\min_{j \neq i} |\lambda_i - \lambda_j| \geq 2\delta\lambda_i$, then, if λ_i' is a solution of 5.27 with $|\lambda_i' - \lambda_i| \leq 4\delta\lambda_i$, ther exists a unique solution c of the equation

$$(A - \lambda_i'(\mathbb{1}_i' + B(\lambda_i')))c = 0 \tag{5.28}$$

Moreover, if c_i denotes the normalized eigenfunction of A with eigenvalue λ_i, then

$$\|c - c_i\|_2 \leq 2\lambda_i' \qquad (5.29)$$

A proof of this lemma can be found in the appendix of [11].

First we divide the row x of the matrix $\mathcal{E}(\lambda)$ by $\|h_x\|_2$ and multiply the columns by $\|h_z\|_2$ to obtain a matrix $\mathcal{G}(\lambda)$ with identical determinant that can be written as

$$\mathcal{G}(\lambda) = \mathcal{K} - \lambda\mathbb{1} - \lambda B(\lambda) \qquad (5.30)$$

where

$$B(\lambda)_{zx} = \frac{\sum\limits_y \mathbb{Q}(y)h_z(y)h_x(y)\left(\mathbb{1}_{x\neq y} + \psi_x^\lambda(y)/h_x(y)\right)}{\|h_z\|_2\|h_x\|_2} \qquad (5.31)$$

Note that $\mathcal{G}(\lambda)$ is symmetric. We must estimate the operator norm of $B(\lambda)$.

We will use the corresponding standard estimator $\left(\sum\limits_{x,z\in\mathcal{M}} B_{zx}^2\right)^{1/2}$.

We first deal with the off-diagonal elements that have no additional λ or other small factor in front of them.

Lemma 5.7 There is a constant $C < \infty$ such that

$$\max_{x\neq z\in\mathcal{M}} \frac{\sum\limits_{y\in\Gamma} \mathbb{Q}(y)h_x(y)h_z(y)}{\|h_x\|_2\|h_z\|_2} \leq Ca_\epsilon^{-1} \max_{m\in\mathcal{M}} \mathbb{Q}(m)^{-1}\mathrm{cap}_m(\mathcal{M}\backslash m) \leq \rho(\epsilon) \qquad (5.32)$$

Proof. Note first by the estimate (4.4) the equilibrium potentials $h_x(y)$ are essentially equal to one on $A(x)$. Thus the denominator in (5.32) is bounded from below by

$$\sqrt{\sum\limits_{y\in A(x)} \mathbb{Q}(y)h_x^2(y) \sum\limits_{y\in A(y)} \mathbb{Q}(y)h_z^2(y)h_i^2(y)} \geq \sqrt{\mathbb{Q}(A(x))\mathbb{Q}(A(z))} \qquad (5.33)$$

To bound the numerators, we will use Lemma 4.3 in the special situation when $J = \mathcal{M}\backslash x$.

Lemma 5.8 For any $x \neq z \in \mathcal{M}$,

$$\sum\limits_{y\in\Gamma} \mathbb{Q}(y)h_x(y)h_z(y) \leq C\rho(\epsilon)\sqrt{\mathbb{Q}(x)\mathbb{Q}(z)} \qquad (5.34)$$

Proof. By (ii) of Lemma 4.4, if $y \in A(m)$, then

(i) If $m = z$, either

$$Q(y) \leq \frac{3}{2}Q(x)a_\epsilon^{-1}|\mathcal{M}|\frac{\mathrm{cap}_m(x)}{Q(x)} \tag{5.35}$$

or

$$Q(y)h_x(y)h_z(y) \leq \frac{3}{2}Q(x)a_\epsilon^{-1}|\mathcal{M}|\frac{\mathrm{cap}_m(x)}{Q(x)} \tag{5.36}$$

(ii) If $m = x$,

$$Q(y) \leq \frac{3}{2}Q(z)a_\epsilon^{-1}|\mathcal{M}|\frac{\mathrm{cap}_m(z)}{Q(z)} \tag{5.37}$$

or

$$Q(y)h_x(y)h_z(y) \leq \frac{3}{2}Q(z)a_\epsilon^{-1}|\mathcal{M}|\frac{\mathrm{cap}_m(z)}{Q(z)} \tag{5.38}$$

(iii) Let $m \notin \{x, z\}$, and assume w.r.g. that $\mathrm{cap}_m(x) \geq \mathrm{cap}_m(z)$. Then, if $\mathrm{cap}_m(y) > 3\mathrm{cap}_m(z)$, already

$$Q(y)\sqrt{h_x(y)h_z(y)} \leq \frac{3}{2}\sqrt{Q(x)Q(z)}a_\epsilon^{-1}|\mathcal{M}|\sqrt{\frac{\mathrm{cap}_m(x)\mathrm{cap}_m(z)}{Q(x)Q(z)}} \tag{5.39}$$

while otherwise

$$Q(y) \leq 3Q(y)\frac{\mathrm{cap}_m(z)}{\mathrm{cap}_y(m)} \leq 3a_\epsilon^{-1}|\mathcal{M}|\sqrt{\mathrm{cap}_m(z)\mathrm{cap}_m(x)} \tag{5.40}$$

Summing over y yields e.g. in case (i)

$$\sum_{y \in A(m)} Q(y)h_x(y)h_y(y) \leq C|\{y \in A(m) : Q(y) \geq 32a_\epsilon^{-1}|\mathcal{M}|\mathrm{cap}_m(x)\}|$$
$$\times a_\epsilon^{-1}|\mathcal{M}|\mathrm{cap}_m(x) \tag{5.41}$$

and in case (ii) the same expression with x replaced by z. The case (iii) is concluded in the same way.

This implies the statement of the lemma.   □

Remark. Note that the estimates in the proof of Lemma 5.7 also imply that

$$Q(A(x))(1 - O(\rho(\epsilon)))\sum_y Q(y)h_x(y)^2 \leq Q(A(x))(1 + O(\rho(\epsilon))) \tag{5.42}$$

The remaining contribution to the matrix elements of $B(\lambda)$ are of order λ, and thus the crudest estimates will suffice:

Lemma 5.9 If λ^0 denotes the principal eigenvalue of the operator L with Dirichlet boundary conditions in $\mathcal{M}$, then

$$\left| \sum_{y \in \Gamma} \mathbb{Q}(y) \left(h_z(y) \psi_x^\lambda(y) \right) \right| \leq \frac{\lambda}{(\lambda^0 - \lambda)} \|h_z\|_2 \|h_x\|_2 \qquad (5.43)$$

Proof. Recall that ψ_x^λ solves the Dirichlet problem (5.18). But the Dirichlet operator $L^\mathcal{M} - \lambda$ is invertible for $\lambda < \lambda^0$ and is bounded as an operator on $\ell^2(\Gamma, \mathbb{Q})$ by $1/(\lambda^0 - \lambda)$. Thus

$$\|\psi_x^\lambda\|_2^2 \leq \left(\frac{\lambda}{\lambda^0 - \lambda} \right)^2 \|h_x\|_2^2 \qquad (5.44)$$

The assertion of the lemma now follows from the Cauchy-Schwartz inequality.   □

As a consequence of the preceding lemmata, we see that the matrix $B(\lambda)$ is indeed bounded in norm by

$$\|B(\lambda)\| \leq C\rho(\epsilon) + c\frac{\lambda}{\lambda^0 - \lambda} \qquad (5.45)$$

The theorem follows from Lemma 5.6.   □

The computation of the eigenvalues of the capacity matrix is now in principle a finite, though in general not trivial problem. The main difficulty is of course the computation of the capacities and induction coefficients. Capacities can be estimated quite efficiently, as we will see in the next section, the off-diagonal terms however, pose in general a more serious problem, although in many practical cases exact symmetries may be very helpful. On the other hand, a particularly nice situation arises when *no* symmetries are present.

In fact we will prove the following theorem.

Theorem 5.2 *Assume that there exists $x \in \mathcal{M}$ such that for some $\delta \ll 1$*

$$\frac{\mathrm{cap}_x(\mathcal{M} \backslash x)}{\|h_x\|_2^2} \geq \delta \max_{z \in \mathcal{M} \backslash x} \frac{\mathrm{cap}_z(\mathcal{M} \backslash z)}{\|h_z\|_2^2} \qquad (5.46)$$

Then the largest eigenvalue of L is given by

$$\lambda_x = \frac{\mathrm{cap}_x(\mathcal{M} \backslash x)}{\|h_x\|_2^2}(1 + O(\delta)) \qquad (5.47)$$

and all other eigenvalues of L satisfy

$$\lambda \leq C\delta\lambda_x \qquad (5.48)$$

Moreover, the eigenvector, ϕ, corresponding to the largest eigenvalues normalized s.t. $\phi_x = 1$ satisfies $\phi_z \leq C\delta$, for $z \neq x$.

Proof. This is a simple perturbation argument. Note that we can write

$$K = \hat{K} + \check{K} \tag{5.49}$$

where $\hat{K}_{uv} = K_{xx}\delta_{xv}\delta_{xv}$. Now we estimate the norm of $\check{K}$.
By the Cauchy-Schwartz inequality,

$$\left| \tfrac{1}{2} \sum_{y,y'} \mathbb{Q}(y')p(y',y)[h_x(y') - h_x(y)][h_z(y') - h_z(y)] \right| \tag{5.50}$$

$$\leq \sqrt{\mathrm{cap}_x(\mathcal{M}\backslash x)\mathrm{cap}_z(\mathcal{M}\backslash z)}$$

Thus

$$\left| \frac{q_z}{q_x} K_{zx}^2 \right| \leq K_{xx}K_{zz} \tag{5.51}$$

Whence by assumption,

$$\|\check{K}\| \leq K_{xx}\sqrt{\delta|\mathcal{M}| + \delta^2|\mathcal{M}|^2} \tag{5.52}$$

Since obviously $\hat{K}$ has one eigenvalue K_{xx} with the obvious eigenvector and all other eigenvalues are zero, the announced result follows from standard perturbation theory.   □

Theorem 5.2 has the following simple corollary, that allows in many situations a complete characterization of the small eigenvalues of L.

Corollary 5.1 *Assume that we can construct a sequence of metastable sets $\mathcal{M}_k \supset \mathcal{M}_{k-1} \supset \ldots \supset \mathcal{M}_2 \supset \mathcal{M}_1 = x_0$, such that for any i, $\mathcal{M}_i\backslash\mathcal{M}_{i-1} = x_i$ is a single point, and that each $\mathcal{M}_i$ satisfies the assumptions of Theorem 5.2. Then L has k eigenvalues*

$$\lambda_i = \frac{\mathrm{cap}_{x_i}(\mathcal{M}_{i-1})}{\mathbb{Q}(A(x_i))}(1 + O(\delta)) \tag{5.53}$$

The corresponding normalized eigenfunction is given by

$$\psi_i(y) = \frac{h_{x_i}(y)}{\|h_{x_i}\|_2} + O(\delta) \tag{5.54}$$

6. Computation of Capacities

We have seen so far that in metastable dynamics we can largely reduce the computation of key quantities to the computation of *capacities*. The usefulness of all this thus depends on how well we can compute capacities.

While clearly the universality of our approach ends here, and model specific properties have to enter the game, it is rather surprising to what extent precise computations of capacities are possible in a multitude of specific systems.

6.1. GENERAL PRINCIPLES

The key to success is the variational representation of capacities through the Dirichlet principle, i.e. Eq. (3.11). The Dirichlet principle immediately yields two avenues towards bounds:

- Upper bounds via judiciously chosen test functions
- Lower bounds via monotonicity of the Dirichlet form in the transition probabilities via simplified processes.

These two principles are well-known and give rise to the so-called "Rayleigh's short-cut rules" in the language of electric networks (see e.g. [25] and references wherein). In the context of metastable systems, the usefulness of these principle can be enhanced by an iterative method.

The key idea of iteration is to get first of all control of the minimizer in the Dirichlet principle, i.e. the equilibrium potential. In metastable systems, when we are interested in computing e.g. the capacity $\text{cap}_{B_x}(B_y)$ where B_x, B_y represent two metastable sets, our first goal will always be to identify domains where $h_{B_x,B_y}(z)$ is close to zero or close to one. This is done with the help of the renewal estimate of Lemma 4.1. While with looks cyclic at first glance (we need to know the capacities in order to estimate the equilibrium potential, which we want to use in order to estimate capacities....) it yields a tool to enhance "poor" bounds in order to get good ones. Thus the first step in the program is to get a first estimate on capacities of the form $\text{cap}_z(B)$ for arbitrary z, B.

(i) Choose a roughly ok looking test function for the upper bound.
(ii) Dramatically simplify the state space of the process to obtain a system that can be solved exactly for the lower bound. In most examples, this leads to choosing a one-dimensional or quasi-one-dimensional system.
(iii) Insert the resulting bounds in (4.4) to obtain bounds on $h_{B_x,B_y}(z)$.

Using this bound we can now identify the set

- $D_x \equiv \{z : h_{B_x,B_y}(z) < \delta\}$
- $D_y \equiv \{z : h_{B_x,B_y}(z) > 1 - \delta\}$ for $\delta \ll 0$ suitably chosen.
  If the complement of the set $D_x \cup D_y$ contains no further metastable set, we define
- $I \equiv \{z \in (D_x \cup D_y)^c : \mathbb{Q}(z) < \rho \sup_{w \in (D_x \cup D_y)^c} \mathbb{Q}(w)\}$ for $\rho \ll 1$ conveniently chosen.

Let us denote by $S \equiv (D_x \cup D_y \cup I)^c$.

The idea is that the set I will be irrelevant for the value of the capacity, no matter what value $h_{B_x,B_y}(z)$ takes where, and that the sets D_x and D_y give no contribution to the capacity to leading order. The only problem is thus to find the equilibrium potential, or a reasonably good approximation to it on the set S. We return to this problem shortly. Of course this idea can only make sense if the sets D_x and D_y can be connected through S. If that is not the case, we will have to analyse the set $(D_x \cup D_y)^c$ more carefully.

$D_x \cup D_y$ contains further metastable sets, say w, then it will be possible to identify domains D_w on which $h_{B_x,B_y}(z)$ takes on a constant values c_w (to be determined later). Note that this can be done again with the help of the renewal bounds; The starting point (in the discrete case) is of course the observation that

$$
\begin{aligned}
h_{B_x,B_y}(z) &= \mathbb{P}_z[\tau_{B_x} < \tau_{B_y}] \\
&= \mathbb{P}_z[\tau_{B_x} < \tau_{B_y}, \tau_w < \tau_{B_x}] + \mathbb{P}_z[\tau_{B_x} < \tau_{B_y}, \tau_w \geq \tau_{B_x}] \\
&= \mathbb{P}_z[\tau_w < \tau_{B_x \cup B_y}]\mathbb{P}_w[\tau_{B_x} < \tau_{B_y}] \\
&= \mathbb{P}_z[\tau_w < \tau_{B_x \cup B_y}]c_w
\end{aligned}
\tag{6.1}
$$

The problem, to be solved with the help of (4.4) and the a priori bounds on capacities is thus to determine the set of points for which $\mathbb{P}_z[\tau_w < \tau_{B_x \cup B_y}] > 1 - \delta$.

Once with is done, we proceed as in the former case, but increasing the set $D_x \cup D_y$ in their definition of I to $D \equiv D_x \cup D_y \cup D_{w_1} \cup \ldots \cup D_{W_k}$ if k such sets can be identified. It should now be the case that the set $D_x \cup D_y \cup D_{w_1} \cup \ldots \cup D_{W_k} \cup S$ is connected. The remaining problem consists then in the determination of the equilibrium potential on the set S and of the values c_{w_i}.

At this stage we can then obtain upper and lower bounds in terms of variational problems that involve only the sets S; to what extend these problems then can be solved depends on the problem at hand.

Upper Bound. To obtain the upper bound, we choose a test-function h^+ with the properties that

$$
\begin{aligned}
h^+(z) &= 1, & z \in D_x \\
h^+(z) &= 0, & z \in D_y \\
h^+(z) &= c_{w_i}, & z \in D_{w_i}
\end{aligned}
\tag{6.2}
$$

where the constants c_{w_i} are determined only later. On I, the function h^+ can be chosen essentially arbitrarily, while on S, we chose h^+ such that it optimizes the restriction of the Dirichlet form to S with boundary conditions implied by (6.2) on $\partial S \cap \partial D$. Finally, the constants c_{w_i} are determined by minimizing the result as a function of these constants.

Lower Bound. For the lower bound, we use that if h^* denotes the true minimizer, then

$$\Phi(h^*) \geq \Phi_{\mathcal{S}}(h^*) \tag{6.3}$$

where $\Phi_{\mathcal{S}}$ is the restriction of the Dirichlet form to the subset $\mathcal{S}$, i.e.

$$\Phi_{\mathcal{S}}(h) = \frac{1}{2} \sum_{\substack{x \vee y \in \mathcal{S} \\ x \wedge y \in \mathcal{S} \cup D}} \mathbb{Q}(x) p(x,y) [h(x) - h(y)]^2 \tag{6.4}$$

Finally we minorise $\Phi_{\mathcal{S}}(h^*)$ by taking the infimum over all h on $\mathcal{S}$, with boundary conditions imposed by what we know a priori about the equilibrium potential. In particular we know that these boundary conditions are close to constants on the different components of D. Of course we do not really know the constants c_{w_i}, but taking the infimum over these, we surely are on the safe side.

Thus, if we can show that the minimizers in the lower bound differ little from the minimizers with constant boundary conditions, we get upper and lower bounds that coincide up to small error terms. Of course in general, it may remain difficult to actually compute these minimizers. However, the problem is greatly reduced in complexity with respect to the original problem, and in many instances this problem can be solved quite explicitly (see [8, 12]).

7. What to do when Points are too Small?

In the previous chapters we have mainly relied on the fact that by removing individual points from state space we already lifted the spectrum of the generator beyond the small eigenvalues corresponding to metastable transitions, or, in other terms, the fact that a set of points is reached in times much smaller that metastable transition times. This is not the case when the cardinality of phase space is too large, or even uncountable. Obvious examples are diffusion processes and spin systems at finite temperature.

Such situations require always some coarse-graining of state space. Metastable sets are then no longer collections of points, but collections of disjoint subsets. This coarse graining causes problems. One concerns the use of the fundamental relation (3.6) that cannot immediately be used to obtain a formula for the Green's function à la (3.25), if A cannot be chosen a single point. The solution must then be to find admissible sets A on those boundary $G_\Omega(x, y)$ is constant or varies very little. Of course such a procedure requires some a priori knowledge about the Green's function. A very similar problem arises in the use of renewal arguments to derive Lemma 4.1.

So far, we have no general rule for how to proceed in such situations. In the case of diffusion processes, it has turned out that general elliptic regularity theory (based on Harnack and Hölder inequalities) allows to obtain all desired results by replacing points with suitable ϵ-dependent balls around points. The details can be found in [10, 11], and we will not go into these problems.

A second method proved useful in the context of certain mean field spin systems and was successfully used in [8, 1] is "lumping" [40]. As this method is the basis of some of the estimates that are crucial in the analysis of ageing in the Random Energy Model (REM) that we will discuss later, we present this in some detail here, even though it is rather extensively presented in [8, 1].

Lumping. Let us suppose we have an ergodic reversible Markov chain those invariant measure is constant on the level set of some function $m : \Gamma \to \Sigma$. Let us further assume that we are interested only in events that can be expressed in terms of m. Of course the idea will always be that Σ should be a much smaller space than Γ. In such a situation it may appear natural to define metastable sets in terms of subsets of Σ rather than of Γ.

Example 7.1 Curie-Weiss Model. The simplest example of this type is furnished by the Curie-Weiss model. Here $\Gamma = \{-1, 1\}^N$, $\Sigma = \{-1, -1 + 2/N, \ldots, 1-2/N, 1\}$ and $m(x) = N^{-1} \sum_{i=1}^{N} x_i$. The invariant measure is given by

$$Q(x) = Q_{\beta,N} \equiv \frac{e^{\beta N m(x)^2}}{Z_\beta} \tag{7.1}$$

and the transition probabilities are e.g. (for $x \neq y$)

$$p(x, y) = N^{-1} e^{-N[m(y)-m(x)]_+} \mathbb{1}_{\|x-y\|_2=2} \tag{7.2}$$

where $[\cdot]_+$ denote the positive part of $\cdot$. If $\beta > 1$, the measure $Q_{\beta,N}$ concentrates (for large N) near the points $\pm m^*(\beta)$ where $m^*(\beta) = \tanh(\beta m^*(\beta)) > 0$. Thus it would be natural to assume that a metastable set for our Markov chain could consist of the two subsets $M_\pm \equiv \{x \in \Gamma : m(x) \approx \pm m^*(\beta)$.

It is instructive to review our basic notions in this context. We see that the definition of metastability given in Section 4 may now not be very appropriate since it would involve ratios of quantities such as $\mathbb{P}_x[\tau_{M_+ \cup M_-} < \tau_x]$ and $\mathbb{P}_x[\tau_{M_+} < \tau_x]$ which might tent to be close to one simply because of entropic reasons it is very difficult for a process starting in x to ever return to that point before an exponentially long time. Looking back at our tentative definition in Section 2, this now seems to be more promising, as we still expect the mean times for arrival in $M_- \cup M_+$ to be much shorter than than transition times between M_- and M_+. The question is whether and when we can re-express such mean times in terms of capacities?

To understand this issue, recall that by (3.23), we have that (e.g.)

$$\mathbb{E}_x \tau_{M_+} = \frac{\sum\limits_{y \notin M_+ \cup x} \mathbb{Q}(y) h_{x,M_+}(y)}{\mathrm{cap}_x(M_+)} \tag{7.3}$$

In fact, this formula is quite annoying since again the denominator is difficult to evaluate due to the fact that we anticipate some hard to control local behaviour of the equilibrium potential $h_{x,M_+}(y)$ near x.

In fact, it would be highly desirable if we could obtain a formula where only capacities of "fat" sets, e.g. sets that can be described in the form $\{x : m(x) \in A \subset [-1,1]\}$ enter. to do so let us go back to the original form of (3.6). If we multiply equation (3.22) on both side by $e_{M(x),M_+}(y)$, where $M(x) \equiv \{y : m(y) = m(x)\}$ and then sum over $M(x)$, we get

$$\sum\limits_{y \in M(x)} \mathbb{Q}(x) \mathbb{E}_x \tau_{M_+} e_{M(x),M_+}(y) = h_{x,M_+}(y) \sum\limits_{x \in M(x)} \sum\limits_{y \notin M_+ \cup x} \mathbb{Q}(y) \tag{7.4}$$

Let us first consider the right-hand side. Since $\mathbb{Q}(y)$ depends only on $m(y)$, we can write

$$\sum\limits_{x \in M(x)} \sum\limits_{y \notin M_+ \cup x} \mathbb{Q}(y) h_{x,M_+}(y) = \sum\limits_{y \notin M_+ \cup x} \mathbb{Q}(y) h_{M(x),M_+}(y) \tag{7.5}$$

Now if we knew that $h_{M(x),M_+}(y) = g_{M(x),M_-}(m(y))$ depended only on $m(y)$, then this would become simply

$$\sum\limits_{m \neq m^*} \mathbb{Q}(M(m)) g_{x,M_+}(m) \tag{7.6}$$

On the other hand, if $\mathbb{E}_x \tau_{M_+} = f_{m(x)}(M_+)$ was a function of $m(x)$ only, the left hand side of (7.4) would reduce to

$$\mathrm{cap}_{M(x)}(M_+) \mathbb{E}_x \tau_{M_+} \tag{7.7}$$

and we would have the nice formula

$$\mathbb{E}_x \tau_{M_+} = \frac{1}{\mathrm{cap}_{M(x)}(M_+)} \sum\limits_{m \neq m^*} \mathbb{Q}(M(m)) g_{M(x),M_+}(m) \tag{7.8}$$

In fact, it is easy to see that in our simple model, both properties are in fact verified. The reason for this is that the transition probabilities verify the property that for any $x, x' \in \Gamma$ such that $m(x) = m(x')$ and $m' \in [0,1]$,

$$\sum\limits_{y:m(y)=m'} p(x,y) = \sum\limits_{y:m(y)=m'} p(x',y) \tag{7.9}$$

In fact it is an old result due to Burke and Rosenblatt [13] that condition (7.9) is necessary and sufficient for the fact that the image process $m(t) \equiv m(x(t))$ is a Markov chain on $m(\Gamma)$ with transition rates given by

$$r(m, m') = \sum_{y:m(y)=m'} p(x, y) \tag{7.10}$$

where x is any point such that $m(x) = m$. We refer to the chain $m(t)$ as the *lumped chain*. Note also that in this case we have that

$$g_{M(x),M_+}(m) = \tilde{h}_{m(x),m^*}(m) \tag{7.11}$$

where $\tilde{h}$ is the equilibrium potential for the lumped Markov chain $m(t)$. Also,

$$\mathrm{cap}_{M(x)}(M_+) = \widetilde{\mathrm{cap}}_{m(x)}(m^*) \tag{7.12}$$

where $\widetilde{\mathrm{cap}}$ is the capacity for the lumped chain, and formula (7.8) can be derived directly in the context of the lumped chain. Thus, the study of the metastability problem in the high-dimensional Curie-Weiss model can be reduced readily to the study of a one-dimensional discrete problem.

The Curie-Weiss model is a particularly simple incident of the lumping technique. In general, if we are given some map m, the process $m(t) = m(x(t))$ will not be a Markov chain. In some cases it is however possible to construct a map into some higher-dimensional space that verifies property (7.9). In the context of spin systems when $\Gamma = \{-1, 1\}^N$ (and similar constructions work when $\{-1, 1\}$ is replaced by a general finite set), there is a natural class of such maps that proof often helpful.

Theorem 7.1 *Assume that according to some rule there is a partition of $\Lambda \equiv \{1, \ldots, N\}$ into k subsets $\Lambda_1, \ldots, \Lambda_k$. denote by m_i the maps*

$$m_i(x) = \frac{1}{|\Lambda_i|} \sum_{i \in \Lambda_i} x_i \tag{7.13}$$

and let m denote the k-dimensional vector $(m_1, \ldots, m_k)$. Assume that $\mathbb{Q}(x)$ depends only in $m(x)$ and that the Markov chain $x(t)$ has transition rates that are of the form

$$p(x, x') = f(m(x), m(x')) \mathbb{1}_{\|x-x'\|_2^2 \le 2} \tag{7.14}$$

Then $m(t) \equiv m(x(t))$ is a Markov chain on $m(\Gamma)$ with transition rates given by equation (7.10).

Proof. Consider two possible values of m and m' that may be connected by a simple transition, i.e. changing the sign of one component of $x \in M(m)$. Note that this has to happen in one of the boxes Λ_i, and changes the value of one component of m, namely m_i by plus or minus $2/|\Lambda_i|$. Suppose then that $m'_i = m_i + 2/|\Lambda_i|$. By (7.14),

$$\sum_{x':m(x')=m'} p(x,x') = f(m(x),m(x')) \sum_{j\in\Lambda_i} \mathbb{1}_{x_i=-1} \tag{7.15}$$

But $\sum_{j\in\Lambda_i} \mathbb{1}_{x_i=-1} = |\Lambda_i|(1 - m_i(x))/2$ depends only on $m(x)$, which proves our case. $\qquad\square$

Remark. Lumping is a useful tool to treat some random mean field models, such as the random field Curie-Weiss model [8] and the Hopfield model (in the Hopfield model, the construction can be found in the context of large deviation theory in [44, 34]).

8. Simple Random Walk on the Hypercube

To illustrate the lumping procedure and to show what it can achieve, we turn to the ordinary random walk on $\{-1,1\}^N$. This will provide a preparation for the treatment of the Random Energy Model. This model has been studied in the past, mainly in view of convergence to equilibrium, see e.g. [41, 23, 55]. Problems that are more closely related to our questions were studied in [14, 20]. This section summarizes results obtained in [2, 4].

We consider a Markov chain $\sigma(t)$ in discrete time on $\mathcal{S}_N \equiv \{-1,1\}^N$ with transition probabilities

$$p(\sigma,\sigma') = 1/N, \quad \text{if} \quad \|\sigma - \sigma'\|_2^2 = 2 \tag{8.1}$$

and zero else. We will be interested in hitting probabilities on a certain subset (of moderate cardinality) $\mathcal{M} \subset \mathcal{S}_N$.

Proposition 8.1 *Set* $M = |\mathcal{M}|$, $d = 2^M$. *There exists a constant* $c > 0$, *such that*

i) For all $\eta \in \mathcal{M}$ *and all* $\sigma \in \Gamma\backslash\mathcal{M}$,

$$\left|\mathbb{P}_\sigma\left(\tau_\eta < \tau_{\mathcal{M}\backslash\eta}\right) - \tfrac{1}{M}\right| \leq \tfrac{c}{N} \tag{8.2}$$

ii) For all $\eta \in \mathcal{M}$ *and* $\bar\eta \in \mathcal{M}$ *with* $\eta \neq \bar\eta$,

$$\left|\mathbb{P}_{\bar\eta}\left(\tau_\eta < \tau_{\mathcal{M}\backslash\{\eta,\bar\eta\}}\right) - \tfrac{1}{M-1}\right| \leq \tfrac{c}{N} \tag{8.3}$$

iii) For all $\sigma \notin \mathcal{M}$,

$$\frac{M}{M+1}\left(1 - \frac{c}{N}\right) \leq \mathbb{P}_\sigma\left(\tau_\mathcal{M} < \tau_\sigma\right) \leq \frac{M}{M+1} \tag{8.4}$$

iv) For all $\sigma \notin \mathcal{M}$ and all $\bar{\sigma} \notin \mathcal{M} \cup \sigma),$

$$\left|\mathbb{P}_{\bar{\sigma}}\left(\tau_\sigma \leq \tau_\mathcal{M}\right) - \frac{1}{M+1}\right| \leq \frac{c}{N} \tag{8.5}$$

Proof. The key tool of the proof of this proposition is the construction of a lumped chain in the sense explained above. In constructing such a chain, we must take care that the events whose probabilities we are computing are mapped one-to-one into the reduced state space. To construct such a mapping, we consider a collection i of vectors $\xi^1, \ldots, \xi^{|I|}$ as a $|I| \times N$ matrix ξ those rows are the vectors ξ^μ. We will denote by $\xi_i \in \{-1, 1\}^M$ the column vectors of this matrix.

Next, let $\{e_1, \ldots, e_k, \ldots, e_d\}$ be an arbitrarily chosen labeling of all $d = 2^{|I|}$ elements of $\mathcal{S}_M$. Then ξ induces a partition of Λ into d disjoint (possibly empty) subsets, $\Lambda_k(I)$,

$$\Lambda_k(I) = \{i \in \Lambda \mid \xi_i = e_k\} \tag{8.6}$$

This is the partitioning we will use for the construction of the lumped chain according to the construction given above. We will write

$$\mathcal{P}_I(\Lambda) = \{\Lambda_k(I), 1 \leq k \leq d\} \tag{8.7}$$

m_I, that maps the elements of $\mathcal{S}_N$ into d-dimensional vectors,

$$m_I(\sigma) = \left(m_I^1(\sigma), \ldots, m_I^k(\sigma), \ldots, m_I^d(\sigma)\right), \quad \sigma \in \mathcal{S}_N \tag{8.8}$$

where, for all $k \in \{1, \ldots, d\}$,

$$m_I^k(\sigma) = \frac{1}{|\Lambda_k(I)|} \sum_{i \in \Lambda_k(I)} \sigma_i \tag{8.9}$$

A few elementary properties of m_I are listed in the lemma below.

Lemma 8.1   i) The range of m_I, $m_{N,d}(I) \equiv m_I(\mathcal{S}_N)$, is a discrete subset of the d-dimensional cube $[-1, 1]^d$ and may be described as follows. Let $\{u_k\}_{k=1}^d$ be the canonical basis of $\mathbb{R}^d$. Then,

$$x \in m_{N,d}(I) \iff x = \sum_{k=1}^d \frac{n_k}{|\Lambda_k(I)|} u_k \tag{8.10}$$

where, for all $1 \leq k \leq d$, $|n_k| \leq |\Lambda_k(I)|$ has the same parity as $|\Lambda_k(I)|$.

ii)

$$|\{\sigma \in \mathcal{S}_N \mid m_I(\sigma) = x\}| = \prod_{k=1}^{d} \left(\frac{|\Lambda_k(I)|}{|\Lambda_k(I)|^{\frac{1+x_k}{2}}} \right), \quad \forall x \in m_{N,d}(I) \quad (8.11)$$

In particular, the restriction of m_I to I is a one-to-one mapping from I onto $m_I(I)$.

iii) The elements of I are mapped onto corners of $[-1, 1]^d$: for all $\sigma \in I$

$$m_I(\sigma) = (\sigma_{i_1}, \ldots, \sigma_{i_k} \ldots, \sigma_{i_d}), \quad \text{for any choice of indices} \quad i_k \in \Lambda_k(I) \tag{8.12}$$

iv) Let $\sigma \in \mathcal{S}_N$ be such that $\inf_{\eta \in I \setminus \sigma} \|\sigma - \eta\|_2 \geq \sqrt{\varepsilon N}$ for some $\varepsilon > 0$. Set $x \equiv m_I(\sigma)$ and $\mathcal{I} \equiv m_I(I)$. Then

$$\inf_{y \in \mathcal{I} \setminus x} \|x - y\|_2 \geq \frac{\varepsilon N}{2\sqrt{d} \max_k |\Lambda_k(I)|} \tag{8.13}$$

Proof of Lemma 8.1. Assertions i), ii), and iii) result from elementary observations. To prove assertion iv) note that for any $\eta \in I \setminus \sigma$, setting $y \equiv m_I(\eta)$ and using (8.12), we have:

$$\varepsilon N \leq \sum_{i=1}^{N} (\sigma_i - \eta_i)^2 = \sum_{k=1}^{d} \sum_{i \in \Lambda_k} (\sigma_i - y_k)^2$$
$$= 2 \sum_{k=1}^{d} |\Lambda_k(I)|(1 - y_k x_k) \leq 2 \max_k |\Lambda_k(I)|(y, y - x) \tag{8.14}$$

where we used in the last line that $1 - y_k x_k = y_k(y_k - x_k)$. But $(y, y - x) \leq \|y\|_2 \|y - x\|_2 = \sqrt{d} \|y - x\|_2$, so that

$$\|x - y\|_2 \geq \frac{\varepsilon N}{2\sqrt{d} \max_k |\Lambda_k(I)|} \tag{8.15}$$

which, together with assertion ii) yields (8.13). $\square$

Note in particular that $\{\sigma_N(t)\}$ is reversible w.r.t. the measure

$$\mu_N(\sigma) = 2^{-N}, \quad \sigma \in \mathcal{S}_N \tag{8.16}$$

We will denote by $\{X_{I,N}(t)\}_{t \in \mathbb{N}}$ and call the *I-lumped chain* or the *lumped chain induced by I*, the chain defined through

$$X_{I,N}(t) \equiv m_I(\sigma_N(t)), \quad \forall t \in \mathbb{N} \tag{8.17}$$

To $m_{N,d}(I)$ we associate an undirected graph, $\mathcal{G}(m_{N,d}(I)) = (V(m_{N,d}(I)), E(m_{N,d}(I)))$, with set of vertices $V(m_{N,d}(I)) = m_{N,d}(I)$ and set of edges:

$$E(m_{N,d}(I))$$
$$= \left\{ (x, x') \in m_{N,d}(I) \mid \exists_{k \in \{1,\dots,d\}}, \exists_{s \in \{-1,1\}} : x' - x = s \frac{2}{|\Lambda_k(I)|} u_k \right\}$$
$$(8.18)$$

The properties of $\{X_{I,N}(t)\}$ are summarized in the lemma below.

Lemma 8.2 Given any subset $I \in \mathcal{S}_N$:

i) The process $\{X_{I,N}(t)\}$ is Markovian no matter how the initial distribution π° of $\{\sigma_N(t)\}$ is chosen.

ii) Set $\mathbb{Q}_N = \mu_N m_\xi^{-1}$. Then $\mathbb{Q}_N$ is the unique reversible invariant measure for the chain $\{X_{I,N}(t)\}$. In explicit form, the density of $\mathbb{Q}_N$ reads:

$$\mathbb{Q}_N(x) = \frac{1}{2^N} |\{\sigma \in \mathcal{S}_N \mid m_I(\sigma) = x\}|, \quad \forall x \in m_{N,d}(I) \qquad (8.19)$$

iii) The transition probabilities $r_N(\cdot, \cdot)$ of $\{X_{I,N}(t)\}$

$$r_N(x, x') = \begin{cases} \frac{|\Lambda_k(I)|}{N} \frac{1 - s x_k}{2} & \text{if } (x, x') \in E(m_{N,d}(I))) \\ & \text{and } x' - x = s \frac{2}{|\Lambda_k(I)|} u_k \quad (8.20) \\ 0, & \text{otherwise} \end{cases}$$

Proof. These results follow from the general Theorem 7.1 and explicit calculations. $\qquad \square$

8.1. MAIN INGREDIENTS OF THE PROOF OF PROPOSITION 8.1

Observe that the entropy produced by the lumping procedure gives rise through (8.19) to a potential, $F_N(x) \equiv -\frac{1}{N} \ln \mathbb{Q}_N(x)$. It moreover follows from assertions ii) and iii) of Lemma 8.1 that this potential is convex and takes on its global minimum at 0 and its global maximum at the corners of the cube $[-1, 1]^d$. Thus the key idea in all the computations will be that the potential will have the tendency to drive the lumped process quickly to zero, before it does anything else; in other words, with overwhelming probability all the events we are interested in are realized in such a way that the process passes through zero.

Note. The diligent reader may have observed that depending on the particular properties of the vectors ξ^μ, the point 0 may or may not be in the range of m_I. To avoid notational complications, in the sequel 0 will be understood to stand for one of the points in $m_I(\mathcal{S}_N)$ closest to zero.
    The next two Lemmata quantify this statement.

Lemma 8.3 Let $x \in m_I(I)$ and $y \in m_I(S_N)\backslash\{x\}$ Then

$$\mathbb{R}_y^\circ(\tau_x < \tau_0) \leq \frac{c}{N} \tag{8.21}$$

Lemma 8.4 There exists a constant $c > 0$ such that, for all N large enough,

$$\mathbb{R}_x^\circ(\tau_0 < \tau_x) \geq \left(1 - \frac{c}{N}\right)^2, \quad \text{for all} \quad x \in m_I(I) \tag{8.22}$$

A consequence of the previous two Lemmata will be that the process starting from 0 hits the set of corners of the hypercube $[-1, 1]^d$ with essentially uniform probability, more precisely:

Lemma 8.5 For all $J \subseteq m_I(I)$, $x \in J$,

$$\frac{\vartheta}{|J|} \leq \mathbb{R}_0^\circ(\tau_x \leq \tau_J) \leq \frac{1}{\vartheta|J|}, \tag{8.23}$$

where

$$\vartheta = \left(1 - \frac{c}{N}\right)^2 \tag{8.24}$$

The basis for the proofs of the preceding Lemmata is the following a priori estimate.

Lemma 8.6 There is a constant $c > 0$ such that, if $\mathbb{Q}(y) \leq e^{-\delta N}$,

$$\mathbb{R}_y^\circ[\tau_0 < \tau_y] \geq c\delta^2 \tag{8.25}$$

while otherwise

$$\mathbb{R}_y^\circ[\tau_0 < \tau_y] \geq cN^{-1} \tag{8.26}$$

Proof. Our general strategy for getting a priori lower bounds on such capacity type probabilities is the use of dramatically simplified chains. An L-steps path ω on $m_{N,d}(I)$, beginning at x and ending at y is defined as sequence of L sites $\omega = (\omega_0, \omega_1, \ldots, \omega_L)$, with $\omega_0 = x$, $\omega_L = y$, and $\omega_l = (\omega_l^k)_{k=1,\ldots,d} \in V(m_{N,d}(I))$ for all $1 \leq l \leq L$, that satisfies:

$$(\omega_l, \omega_{l-1}) \in E(m_{N,d}(I)), \quad \text{for all} \quad l = 1, \ldots, L \tag{8.27}$$

(We may also write $|\omega| = L$ to denote the length of ω.) If ω is such a path with $\omega_0 = y$ and $\omega_L = 0$, it is clear that

$$\text{cap}_y(0) \equiv \inf_{h:h(0)=1,h(y)=0} \Phi(h)$$

$$\geq \inf_{h:h(0)=1,h(y)=0} \frac{1}{2} \sum_{z,z' \in \omega} \mathbb{Q}(z) r_N(z, z')(h(z) - h(z'))^2 \tag{8.28}$$

This lower bound corresponds to a simple one-dimensional problem involving the process restricted to the path ω and its solution is of course well known. In fact, if we enumerate sites of the path ω by $y = \omega_0, \ldots, \omega_L = 0$, (8.28) yields

$$\text{cap}_y(0) \geq \frac{1}{\displaystyle\sum_{k=0}^{L-1} \frac{1}{\mathbb{Q}(\omega_k)r_N(\omega_k,\omega_{k+1})}} \tag{8.29}$$

We optimize this bound by choosing the path ω in a more or less optimal way. Assume w.r.g. that $|\Lambda_1|y_1^2 \geq |\Lambda_2|y_2^2 \geq \ldots > |\Lambda_d|y_d^2$ and that $y_\mu \geq 0$, for all μ. Then our path will consist of a sequence of straight pieces along the coordinate axis, starting with the first and ending with the last one.

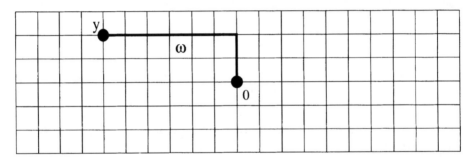

This allows a very explicit representation of the denominator in (8.29) in the form

$$\sum_{\mu=1}^{d} 2^N \prod_{k=1}^{\mu-1} \left(\frac{|\Lambda_k|}{|\Lambda_k|/2}\right)^{-1} \prod_{k=\mu+1}^{d} \left(\frac{|\Lambda_k|}{|\Lambda_k|\frac{1+y_k}{2}}\right)^{-1}$$

$$\sum_{n=1}^{[y_\mu|\Lambda_\mu|/2]} \left(\frac{|\Lambda_\mu|}{|\Lambda_\mu|\frac{1+y_k}{2}-n}\right)^{-1} \frac{N}{|\Lambda_\mu|} \frac{2}{1+y_\mu-2n/|\Lambda_\mu|}$$

$$= \mathbb{Q}(y)^{-1} \sum_{\mu=1}^{d} \prod_{k=1}^{\mu-1} \frac{\left(\frac{|\Lambda_k|}{|\Lambda_k|\frac{1+y_k}{2}}\right)}{\left(\frac{|\Lambda_k|}{|\Lambda_k|/2}\right)} \tag{8.30}$$

$$\sum_{n=1}^{[y_\mu|\Lambda_\mu|/2]} \frac{\left(\frac{|\Lambda_\mu|}{|\Lambda_\mu|\frac{1+y_\mu}{2}}\right)}{\left(\frac{|\Lambda_\mu|}{|\Lambda_\mu|\frac{1+y_\mu}{2}-n}\right)} \frac{N}{|\Lambda_\mu|} \frac{2}{1+y_\mu-2n/|\Lambda_\mu|}$$

From this formula one can in derive reasonably good bounds for all possible choices of y. Here our concern is only the situation when y is somewhat far from 0. In fact, assume that for some $\delta > 0$, there exists no k such that $|\Lambda_k|y_k^2 \geq \delta N$. Then it is easy to see that $\mathbb{Q}(y) \geq e^{-c\delta N}$. Otherwise, let by convention $|\Lambda_1|y_1^2 > \delta N$. Note that this implies in particular $\frac{|\Lambda_1|}{N} > \delta$. Then all terms in the sum (8.30) with $\mu > 1$ are bounded by $e^{-\delta N}$ while the term with $\mu = 1$ is bounded by

$$\frac{1}{\mathbb{Q}(y)} \sum_{n=1}^{[y_1|\Lambda_1|/2]} \frac{\binom{|\Lambda_1|}{|\Lambda_1|\frac{1+y_1}{2}}}{\binom{|\Lambda_1|}{|\Lambda_1|\frac{1+y_1}{2}-n}} \delta^{-1} \leq C\delta^{-2} \tag{8.31}$$

The bound (8.26) is trivial. $\qquad\qquad\qquad\qquad\qquad\qquad\qquad\square$

Proof of Lemma 8.3. Note first that

$$\mathbb{R}_y^\circ (\tau_x < \tau_0) = \frac{\mathbb{R}_y^\circ (\tau_x < \tau_{y\cup 0})}{\mathbb{R}_y^\circ (\tau_{x\cup 0} < \tau_y)} \tag{8.32}$$

Let us first consider the case when

$$2^{-N}\mathbb{Q}(y)^{-1} \leq \delta N^{-1} \tag{8.33}$$

By reversibility its numerator may be rewritten as

$$\mathbb{R}_y^\circ (\tau_x < \tau_{y\cup 0}) = \frac{\mathbb{Q}(x)}{\mathbb{Q}(y)} \mathbb{R}_x^\circ (\tau_y < \tau_{x\cup 0}) \leq 2^{-N}/\mathbb{Q}(y) \tag{8.34}$$

Then by Lemma 8.6, the denominator of (8.32) obeys the bound

$$\mathbb{R}_y^\circ (\tau_{x\cup 0} < \tau_y) \geq cN^{-1} \tag{8.35}$$

if $\mathbb{Q}(y) > e^{-\delta N}$ (with, say,. $\delta = 0.1$). Thus in this case we get that

$$\mathbb{R}_y^\circ (\tau_x < \tau_0) \leq 2^{-N} e^{\delta N} N \tag{8.36}$$

which is exponentially small. On the other hand, if $\mathbb{Q}(y) < e^{-\delta N}$ then

$$\mathbb{R}_y^\circ (\tau_{x\cup 0} < \tau_y) \geq c\delta^{-2} \tag{8.37}$$

and we are done if (8.33) is satisfied.

Otherwise (8.37) always holds, and we can use that (if $y \neq x$),

$$\begin{aligned}
\mathbb{R}_y^\circ (\tau_x < \tau_0) &= \sum_{y'} p(y,y') \mathbb{R}_{y'}^\circ (\tau_x < \tau_0) \\
&\leq C \sum_{y'} p(y,y') \mathbb{Q}(y')^{-1} 2^{-N} \\
&= \sum_{\mu=1:|y_\mu|\neq 1}^{N} \sum_{\pm} \frac{|\Lambda_\mu(I)|}{N} \frac{1\pm y_\mu}{2} 2^{-N} \mathbb{Q}(y)^{-1} \\
&\quad + \sum_{\mu=1:y_\mu=\pm 1}^{N} \frac{|\Lambda_\mu(I)|}{N} \frac{1\pm y_\mu}{2} 2^{-N} \mathbb{Q}(y \mp 2u_\mu/|\Lambda_\mu|)^{-1}
\end{aligned} \tag{8.38}$$

Using the explicit representation of $\mathbb{Q}$ in (8.19) shows that all terms in the sums are smaller than cN^{-1}, which concludes the proof of the lemma. $\square$

Remark. Note that we have striven to obtain only the crudest uniform upper bound, and this cannot be improved. of course we can get much sharper bounds as functions of y.

Proof of Lemma 8.4. To prove Lemma 8.4, we use that $\mathbb{R}_x^\circ(\tau_0 < \tau_x) = \mathbb{Q}(x)^{-1}\mathrm{cap}_0(x)$, while

$$\mathrm{cap}_0(x) = \frac{1}{2}\sum_{z,z'}\mathbb{Q}(z)r_N(z,z')[h^*(z) - h^*(z')]^2$$

$$\geq \sum_{k=1}^{d}\mathbb{Q}(x)r_N(x, x - 2u_k/|\Lambda_k|)\left[h^*(x) - h^*(x - 2u_k/|\Lambda_k|)\right]^2$$

$$(8.39)$$

where $h^*(z) = \mathbb{R}_z^\circ(\tau_0 < \tau_x)$ if $z \notin \{x, 0\}$, and $h^*(x) = 0$. Thus

$$\mathrm{cap}_0(x) \geq \sum_{k=1}^{d}\mathbb{Q}(x)r_N(x, x - 2u_k/|\Lambda_k|)\left(1 - cN^{-1}\right)^2 = \mathbb{Q}(x)\left(1 - cN^{-1}\right)^2$$

$$(8.40)$$

This yields the claimed estimate. $\square$

Proof of Lemma 8.5. Again using renewal,

$$\mathbb{R}_0^\circ(\tau_x \leq \tau_J) = \frac{\mathbb{R}_0^\circ(\tau_x \leq \tau_{J\cup 0})}{\mathbb{R}_0^\circ(\tau_J < \tau_0)} = \frac{\mathbb{R}_0^\circ(\tau_x \leq \tau_{J\cup 0})}{\sum_{y\in J}\mathbb{R}_0^\circ(\tau_y \leq \tau_{J\cup 0})} \tag{8.41}$$

so that we are left to bound a term of the form $\mathbb{R}_0^\circ(\tau_y \leq \tau_{J\cup 0})$, $y \in J$. To do so observe that

$$\mathbb{R}_0^\circ(\tau_y \leq \tau_{J\cup 0}) = \mathbb{R}_0^\circ(\tau_y < \tau_0) - \mathbb{R}_0^\circ\left(\tau_{J\setminus y} < \tau_y < \tau_0\right) \tag{8.42}$$

and that

$$\mathbb{R}_0^\circ\left(\tau_{J\setminus y} < \tau_y < \tau_0\right) = \sum_{z\in J\setminus y}\mathbb{R}_0^\circ(\tau_z \leq \tau_{J\cup 0})\mathbb{R}_z^\circ(\tau_y < \tau_0) \tag{8.43}$$

By assumption, the probabilities $\mathbb{R}_z^\circ(\tau_y < \tau_0)$ in the r.h.s. above obey the bound (8.21) of Lemma 8.3. Thus

$$\mathbb{R}_0^\circ\left(\tau_{J\setminus y} < \tau_y < \tau_0^\circ\right) \leq \frac{c}{N}\sum_{z\in J\setminus y}\mathbb{R}_0^\circ(\tau_z \leq \tau_{J\cup 0})$$

$$(8.44)$$

$$\leq \frac{c}{N}\mathbb{R}_0^\circ(\tau_J < \tau_0)$$

From (8.42) and (8.44) we deduce that

$$\mathbb{R}_0^\circ(\tau_y < \tau_0) - \frac{c}{N}\mathbb{R}_0^\circ(\tau_J < \tau_0) \leq \mathbb{R}_0^\circ(\tau_y \leq \tau_{J\cup0}) \leq \mathbb{R}_0^\circ(\tau_y < \tau_0) \qquad (8.45)$$

and, summing over $y \in J$,

$$\sum_{y\in J}\mathbb{R}_0^\circ(\tau_y \leq \tau_0) - |J|\frac{c}{N}\mathbb{R}_0^\circ(\tau_J < \tau_0) \leq \mathbb{R}_0^\circ(\tau_J \leq \tau_0) \leq \sum_{y\in J}\mathbb{R}_0^\circ(\tau_y < \tau_0) \tag{8.46}$$

Inserting the bounds (8.45) and (8.46) into (8.41), and using that

$$\frac{\mathbb{R}_0^\circ(\tau_J \leq \tau_0)}{\sum_{y\in J}\mathbb{R}_0^\circ(\tau_y < \tau_0)} \leq 1 \qquad (8.47)$$

we arrive at:

$$R - \frac{c}{N} \leq \mathbb{R}_0^\circ(\tau_x \leq \tau_J) \leq R\left(\frac{1}{1 - |J|\frac{c}{N}}\right) \qquad (8.48)$$

where

$$R \equiv \frac{\mathbb{R}_0^\circ(\tau_x \leq \tau_0)}{\sum_{y\in J}\mathbb{R}_0^\circ(\tau_y \leq \tau_0)} \qquad (8.49)$$

To estimate the above ratio we use first that, by reversibility,

$$R = \frac{\mathbb{Q}_N(x)\mathbb{R}_x^\circ(\tau_0 \leq \tau_x)}{\sum_{y\in J}\mathbb{Q}_N(y)\mathbb{R}_y^\circ(\tau_0 \leq \tau_y)} \qquad (8.50)$$

and next that, by Lemma 8.4,

$$\vartheta\overline{R} \leq \mathbb{R}_0^\circ(\tau_x \leq \tau_J) \leq \frac{\overline{R}}{\vartheta} \qquad (8.51)$$

where ϑ is defined in (8.24) and

$$\overline{R} \equiv \frac{\mathbb{Q}_N(x)}{\sum_{y\in J}\mathbb{Q}_N(y)} \qquad (8.52)$$

Now since $J \subseteq m_I(I)$, and since $\mathbb{Q}_N(y) = 2^{-N}$ for all $y \in m_I(I)$,

$$\overline{R} = \frac{1}{|J|} \qquad (8.53)$$

and (8.23) is proven. □

8.2. PROOF OF PROPOSITION 8.1

We are now ready to prove Proposition 8.1.

Notation. The following notation will be used throughout: $T = m_{\mathcal{M}}(\mathcal{M})$, $y = m_{\mathcal{M}}(\sigma)$, $x = m_{\mathcal{M}}(\eta)$ and $\bar{x} = m_{\mathcal{M}}(\bar{\eta})$.

Proof of Proposition 8.1, i), and ii). Firstly

$$\mathbb{P}_\sigma\left(\tau_\eta < \tau_{\mathcal{M}\backslash\eta}\right) = \mathbb{R}_y^\circ\left(\tau_x < \tau_{T\backslash x}\right) \tag{8.54}$$

Defining

$$R_1 \equiv \mathbb{R}_y^\circ\left(\{\tau_x < \tau_{T\backslash x}\} \cap \{\tau_0 < \tau_x\}\right)$$
$$R_2 \equiv \mathbb{R}_y^\circ\left(\{\tau_x < \tau_{T\backslash x}\} \cap \{\tau_x < \tau_0\}\right) \tag{8.55}$$

$\mathbb{R}_y^\circ\left(\tau_x < \tau_{T\backslash x}\right)$ may be decomposed as

$$\mathbb{R}_y^\circ\left(\tau_x < \tau_{T\backslash x}\right) = R_1 + R_2 \tag{8.56}$$

Obviously

$$0 \le R_2 \le \mathbb{R}_y^\circ\left(\tau_x < \tau_0\right) \le \frac{c}{N} \tag{8.57}$$

while

$$R_1 = \mathbb{R}_y^\circ\left(\tau_0 < \tau_x < \tau_{T\backslash x}\right)$$
$$= \mathbb{R}_y^\circ\left(\tau_0 < \tau_T\right)\mathbb{R}_0^\circ\left(\tau_x < \tau_{T\backslash x}\right) \tag{8.58}$$
$$= \left[1 - \mathbb{R}_y^\circ\left(\tau_T < \tau_0\right)\right]\mathbb{R}_0^\circ\left(\tau_x < \tau_{T\backslash x}\right)$$

which, together with the bound

$$\mathbb{R}_y^\circ\left(\tau_T < \tau_0\right) \le \sum_{x'\in T}\mathbb{R}_y^\circ\left(\tau_{x'} < \tau_0\right) \tag{8.59}$$

yields

$$\mathbb{R}_0^\circ\left(\tau_x < \tau_{T\backslash x}\right)\left[1 - M\sup_{x'\in T}\mathbb{R}_y^\circ\left(\tau_{x'} < \tau_0\right)\right] \le R_1 \le \mathbb{R}_0^\circ\left(\tau_x < \tau_{T\backslash x}\right) \tag{8.60}$$

We are thus left to bound the quantities $\sup_{x'\in T}\mathbb{R}_y^\circ\left(\tau_{x'} < \tau_0\right)$ and $\mathbb{R}_0^\circ\left(\tau_x < \tau_{T\backslash x}\right)$, which will be done by means of, respectively, Lemma 8.3 and Lemma 8.5.

$$\left|\mathbb{R}_0^\circ\left(\tau_x < \tau_{T\backslash x}\right) - \frac{1}{M}\right| \le \frac{c}{N} \tag{8.61}$$

for some constant $c_0 > 0$.

Collecting the previous estimates we obtain that, for large enough N,

$$\left| \mathbb{R}_y^\circ \left(\tau_x < \tau_{T \setminus x} \right) - \frac{1}{M} \right| \leq \frac{c}{N} \tag{8.62}$$

for some constant $c > 0$. This yields the claim of assertion i). The proof of assertion ii) is very similar to that of assertion i).   □

In order to study the probabilities appearing in (iii) and (iv) we must construct the lumped chain based on the vectors from $\mathcal{M}$ and σ. Otherwise, there is little difference. The following notation will be used throughout: $I \equiv \mathcal{M} \cup \sigma$, $\mathcal{I} \equiv m_I(I)$, $y \equiv m_I(\sigma)$, and $\bar{y} \equiv m_I(\bar{\sigma})$. It will moreover be assumed that $w \in \mathcal{E}_N$.

Proof of Proposition 8.2, iii) and iv). With the notation introduced above

$$\mathbb{P}_\sigma \left(\tau_{\mathcal{M}} < \tau_\sigma \right) = \mathbb{R}_y^\circ \left(\tau_{\mathcal{I} \setminus y} < \tau_y \right) \tag{8.63}$$

and

$$\mathbb{P}_{\bar{\sigma}} \left(\tau_\sigma \leq \tau_{\mathcal{M}} \right) = \mathbb{R}_{\bar{y}}^\circ \left(\tau_y < \tau_{\mathcal{I} \setminus y} \right) \tag{8.64}$$

Let us first consider the capacity-like quantity (8.63). To prove an upper bound, we will chose as a test function in the formula for the capacity $\mathrm{cap}_{\mathcal{I} \setminus y}(y)$ the function

$$h(z) = \begin{cases} \frac{1}{M+1}, & \text{if} \quad z \notin \mathcal{I} \\ 1, & \text{if} \quad z \in \mathcal{I} \setminus y \\ 0, & \text{if} \quad z = y \end{cases} \tag{8.65}$$

This gives that

$$\mathbb{R}_y^\circ \left(\tau_{\mathcal{I} \setminus y} < \tau_y \right) \leq Q(\mathcal{I} \setminus y)^{-1} \left(Q(y) \sum_{k=1}^d r_N(y, y - 2 \, \mathrm{sign} \, (y_k) u_k / |\Lambda_k|) \frac{1}{(M+1)^2} \right.$$

$$\left. + \sum_{x \in \mathcal{I} \setminus y} Q(x) \sum_{k=1}^d r_N(x, x - 2 \, \mathrm{sign} \, (x_k) u_k / |\Lambda_k|) \left(1 - \frac{1}{M+1} \right)^2 \right)$$

$$= \frac{M}{M+1} \tag{8.66}$$

To get the corresponding lower bound we have only to show that in fact

$$\mathbb{R}_z \left(\tau_{\mathcal{I} \setminus y} < \tau_y \right) = \frac{1}{M+1} (1 + O(1/N)) \tag{8.67}$$

for $z \notin \mathcal{I}$. But Lemma 8.5,

$$\mathbb{R}_0 \left(\tau_{\mathcal{I} \backslash y} < \tau_y \right) = \frac{1}{M+1} \left(1 + O(1/N) \right) \tag{8.68}$$

while

$$\mathbb{R}_z \left(\tau_0 < \tau_{\mathcal{I}} \right) \geq 1 - \frac{c}{N} \tag{8.69}$$

Combining these facts in the by now familiar way, we see that (8.67) holds, and hence assertion (iii) of the proposition is proven. Assertion (iv) is then proven in just the same way as assertion (i).   □

9. Dynamics of the REM

We will now see show that the results from the last section allow us to analyse the dynamics of a much more complicated model, namely the random energy model.

The REM is a very simple, but instructive model for spin glasses that was introduced by B. Derrida [De1,De2] in the '80ies and that has been studied extensively since then [27, 32, 52, 54]. The configuration space again the hypercube $\mathcal{S}_N = \{-1, 1\}^N$. On an abstract probability space $(\Omega, \mathcal{F}, P)$ we define the family of i.i.d. standard normal random variables $\{X_\sigma\}_{\sigma \in \mathcal{S}_N}$. We set $E_\sigma \equiv [X_\sigma]_+ \equiv (X_\sigma \vee 0)$. We define a random (Gibbs) probability measure on $\mathcal{S}_N$, $\mu_{\beta,N}$, by setting

$$\mu_{\beta,N}(\sigma) \equiv \frac{e^{\beta \sqrt{N} E_\sigma}}{Z_{\beta,N}} \tag{9.1}$$

where $Z_{\beta,N}$ is the normalizing partition function. For our purposes it is enough to know that if $\beta > \sqrt{2 \ln 2}$, then the Gibbs measure is asymptotically concentrated on the random set of vertices σ for which E_σ is maximal. I.e. if

$$E_{\sigma(1)} \geq E_{\sigma(2)} \geq E_{\sigma(3)} \geq \ldots \geq E_{\sigma(2^N)} \tag{9.2}$$

(note that this ordering depends of course on N) then for any finite k,

$$\lim_{k \uparrow \infty} \lim_{N \uparrow \infty} \mu_{\beta,N} \left(\{\sigma^{(1)}, \ldots, \sigma^{(k)}\} \right) = 1, \quad \text{a.s.} \tag{9.3}$$

This fact suggests that for any Markov chain that is reversible with respect to the Gibbs measure (9.1), the states $\sigma^{(1)}, \ldots, \sigma^{(k)}$ are good candidates for metastable states. It will be important to have a precise understanding of their respective energy values. This information is contained in a classical result of extreme value theory that states that:

Proposition 9.1 *Define*

$$u_N(x) \equiv \sqrt{2N \ln 2} + \frac{x}{\sqrt{2N \ln 2}} - \frac{1}{2} \frac{\ln(N \ln 2) + \ln 4\pi}{\sqrt{2N \ln 2}} \tag{9.4}$$

and define the point process

$$\mathcal{P}_N \equiv \sum_{\sigma \in \{-1,1\}^N} \delta_{u_N^{-1}(X_\sigma)} \tag{9.5}$$

Moreover, let $\mathcal{P}$ denote the Poisson point process on $\mathbb{R}$ with intensity measure P. Then,

$$\mathcal{P}_N \overset{\mathcal{D}}{\to} \mathcal{P} \tag{9.6}$$

Let us now define a particular Glauber dynamics for this model. We will construct a Markov chain $\sigma(t)$ with state space $\mathcal{S}_N$ and discrete time $t \in \mathbb{N}$ by prescribing transition probabilities $p_N(\sigma, \eta) = \mathbb{P}[\sigma(t+1) = \eta | \sigma(t) = \sigma]$ by

$$p_N(\sigma, \eta) = \begin{cases} \frac{1}{N} e^{-\beta \sqrt{N} E_\sigma}, & \text{if} & \|\sigma - \eta\|_2 = \sqrt{2} \\ 1 - e^{-\beta \sqrt{N} E_\sigma}, & \text{if} & \sigma = \eta \\ 0, & \text{otherwise} \end{cases} \tag{9.7}$$

Note that the dynamics is also random, i.e. the law of the Markov chain is a measure valued random variable on Ω that takes values in the space of Markov measures on the path space $\mathcal{S}_N^{\mathbb{N}}$. We will mostly take a pointwise point of view, i.e. we consider the dynamics for a given fixed realization of the disorder parameter $\omega \in \Omega$.

One may now think of the sets

$$T_N(E) \equiv \{\sigma \in \mathcal{S}_N | E_\sigma \geq u_N(E)\} \tag{9.8}$$

with $E \in \mathbb{R}$ as candidates for sets of metastable states, as they are the "deepest minima". The difficulty here is, however, that there is no good separation between the states in $T_N(E)$ and those outside: in fact the difference in depth is only of the order $N^{-1/2}$ between the most shallow minimum within $T_N(E)$ and the deepest one without. This will lead us to ask rather different questions when before, but as we will see, our tools are still of good use. Note that this is also related to the fact, observed by Fontes, Isopi, Kohayakawa, and Picco [30], that the phase transition in this model is not visible in terms of the behaviour of the spectral gap of the generator (see also [47] for an analysis of the dynamical phase transition in this model).

9.1. AGEING

In systems like the REM, physicist have discovered a novel concept charac-
terizing long term dynamics that is called "ageing" (we refer to the reviews
[6, 16] for an overview and further references. [5] contains a short review
from a mathematical perspective). This phenomenon is typically characterized in terms of the behaviour of an autocorrelations function, that could
for instance be taken as

$$C_N(t,s) = \frac{1}{N} \sum_{i=1}^{N} \sigma_i(t)\sigma_i(t+s) \tag{9.9}$$

Ageing is then said to occur whenever the $C_N(t,s)$ does not become independent of t as both s and t tend to infinity. In fact in many cases of ageing
dynamics it turns out that asymptotically, the correlation function tends
to a limit that is a function of s/t only.

The idea that the long-time dynamics of the model can be described
effectively in terms of metastable transitions between the states in $T_N(E)$
gave rise to the ad hoc definition of an effective "trap model" that should
effectively represent this dynamics. This model introduced by Bouchaud
and Dean [7] can easily be described as follows: the state space is reduced
to M points, representing the elements of $T_N(E)$. Each of the states is
assigned a positive random energy E_k which is taken to be exponentially
distributed with rate one, which is justified from the Poisson convergence
result Proposition 9.1. Of course this does not represent the actual energy
of the state, but the properly rescaled one, more precisely $\sqrt{N}u_N^{-1}(X_\sigma)$.

The dynamics is now a continuous time Markov chain $Y(t)$ taking values
in $S_M \equiv \{1,\ldots,M\}$. If the process is in state k, it waits an exponentially
distributed time with mean proportional to $e^{E_k \alpha}$ where $\alpha = \beta/\beta_c$, and
then jumps with equal probability in one of the other states $k' \in S_M$. This
process is then analyzed using essentially techniques from *renewal* theory.
The essential point is that if one starts the process from the uniform distribution, it is possible to show that if one only considers the times, T_i, at
which the process changes its state, then the counting process, $c(t)$, that
counts the number of these jumps in the time interval $(0,t]$ is a classical
renewal (counting) process; moreover, as $n \uparrow \infty$, this renewal process converges to a renewal process with a *deterministic* law for the renewal time
with a heavy-tailed distribution (in the sense that the mean is infinite[1])
whose density is proportional to $t^{-1-1/\alpha}$. It is the emergence of such *non-
Markovian* limit processes that is ultimately responsible for all the ageing

---

[1]This is clearly due to the fact that the average of the waiting time $e^{\alpha E_i}$ over the
disorder is infinite.

phenomena observed in the abundant literature on this and related models. The correlation function to be studied in the trap model is then simply the probability, $\Pi_N(t, s)$, that no jump occurs in a time interval $[t, t + s]$. One sees easily that this quantity satisfies a renewal equation

$$\Pi_N(t, s) = 1 - F_N(s + t) + \int_0^t \Pi_N(t - u, s)dF_N(u) \tag{9.10}$$

where F_N is is just the mean waiting time distribution, i.e. $F_N(t) = \sum_{i=1}^N \left(1 - e^{-t/\tau_i}\right)$ where $\tau_i \equiv \tau_0 \exp(E_i/\alpha)$ and E_i are exponentially distributed random variables. The key point is that F_N converges a.s. to the deterministic function

$$F_\infty(t) \equiv \alpha \int_1^\infty dx e^{-t/x} x^{-1-\alpha} \tag{9.11}$$

and consequently Π_N converges to the solution Π_∞ of the renewal equation

$$\Pi_\infty(t, s) = 1 - F_\infty(s + t) + \int_0^t \Pi_\infty(t - u, s)dF_\infty(u) \tag{9.12}$$

The particular behaviour of the solution is due to the fact that the kernel F_∞ of the equation is singular in the sense that the mean renewal time is infinite. It is, however, not hard to analyse the asymptotics of the solution using Laplace transform methods. It turns out that to leading order (as $t, s \uparrow \infty$),

$$\Pi_\infty(t, s) \sim \frac{1}{\pi \mathrm{cosec}(\alpha\pi)} \int_{s/t}^\infty dx \frac{1}{(1 + x)x^\alpha} \equiv H_0(s/t) \tag{9.13}$$

i.e. is indeed a function of s/t.

The purpose of the analysis presented in [1, 2, 3] is to justify the predictions of this trap model in a rigorous way. We will briefly review the main aspects of this analysis.

9.2. JUSTIFYING THE TRAP MODEL

Three assumptions entering in the definition of the trap model that need to be justified: 1) the uniform distribution of the jump distribution, 2) the distribution of the random mean exit time, and 3) the exponential distribution of the transition times. We will show that the first two assumptions can be derived. The last assumption will in fact not hold true strictly speaking, which will be the cause of a lot of trouble.

To prove the uniformity of the distribution of the jumps over $T_N(E)$, we have to show that

$$\mathbb{P}_\eta[\tau_{\eta'} = \tau_{T_N(E)}] \sim \frac{1}{|T_N(E)|} \tag{9.14}$$

for any $\eta \neq \eta' \in T_N(E)$. The nice thing is that this follows from Proposition 8.1 due to the simple fact that

$$\mathbb{P}_\eta[\tau_{\eta'} = \tau_{T_N(E)}] = e^{-\beta\sqrt{N}E_\eta}\mathbb{P}_\eta^\circ[\tau_{\eta'} = \tau_{T_N(E)}] \tag{9.15}$$

where $\mathbb{P}^\circ$ denotes the law of the simple random walk studied in Section 7. The reason for this is very simple: the probability in our REM process to jump to any neighboring site, *conditioned to jump* is the same as in the ordinary random walk, while $e^{-\beta\sqrt{N}E_\eta}$ is the probability not to move away from η.

So this was easy (but recall that we had to work a little in Section 7!). Next we turn to the mean values $\mathbb{E}_\eta\tau_{T_N(E)\backslash\eta}$. Not surprisingly, we will draw on our formulas for mean hitting times of Section 4. In fact, (4.2) reads here

$$\mathbb{E}_\eta\tau_{T_N(E)\backslash\eta} = \frac{1}{\text{cap}_\eta(T_N(E)\backslash\eta)} \sum_{\sigma\notin T_N(E)\backslash\eta} \mu_{\beta,N}(\sigma)h_{\eta,T_N(E)\backslash\eta}(\sigma)$$

$$= \frac{1}{\mu_{\beta,N}(\eta)\mathbb{P}_\eta[\tau_{T_N(E)} < \tau_\eta]} \left(\mu_{\beta,N}(\eta) + \sum_{\sigma\notin T_N(E)} \mu_{N,\beta}(\sigma)\mathbb{P}_\sigma[\tau_\eta < \tau_{T_N(E)\backslash\eta}] \right)$$

$$= \frac{1}{e^{\beta\sqrt{N}E_\eta}\mathbb{P}_\eta[\tau_{T_N(E)} < \tau_\eta]} \left(e^{\beta\sqrt{N}\eta} + \sum_{\sigma\notin T_N(E)} e^{\beta\sqrt{N}E_\sigma}\mathbb{P}_\sigma[\tau_\eta < \tau_{T_N(E)\backslash\eta}] \right) \tag{9.16}$$

as in (9.15) we have that

$$e^{\beta\sqrt{N}E_\eta}\mathbb{P}_\eta[\tau_{T_N(E)} < \tau_\eta] = \mathbb{P}_\eta^\circ[\tau_{T_N(E)} < \tau_\eta] \tag{9.17}$$

while

$$\mathbb{P}_x[\tau_\eta < \tau_{T_N(E)\backslash\eta}] = \mathbb{P}_\sigma^\circ[\tau_\eta < \tau_{T_N(E)\backslash\eta}] \tag{9.18}$$

This allows to express our mean time entirely in terms of probabilities computed in the simple random walk:

$$\mathbb{E}_\eta\tau_{T_N(E)\backslash\eta}$$

$$= \frac{1}{\mathbb{P}_\eta^\circ[\tau_{T_N(E)} < \tau_\eta]} \left(e^{\beta\sqrt{N}E_\eta} + \sum_{\sigma\notin T_N(E)} e^{\beta\sqrt{N}E_\sigma}\mathbb{P}_\sigma^\circ[\tau_\eta < \tau_{T_N(E)\backslash\eta}] \right) \tag{9.19}$$

and those have all been computed in Proposition 8.1. With $|T_N(E)| \equiv M$, this gives

$$\mathbb{E}_\eta \tau_{T_N(E) \backslash \eta} = \frac{1}{1 - \frac{1}{M}} \left(e^{\beta \sqrt{N} E_\eta} + \sum_{\sigma \notin T_N(E)} e^{\beta \sqrt{N} E_\sigma} \frac{1}{M} \right) (1 + O(1/N))$$

$$= \frac{e^{\beta \sqrt{N} E_\eta}}{1 - \frac{1}{M}} \left(1 + M^{-1} e^{-\beta \sqrt{N} E_\eta} Z_{\beta,N}(T_N(E)^c) \right) (1 + O(1/N)) \tag{9.20}$$

where

$$Z_{\beta,N}(T_N(E)^c) \equiv \sum_{\sigma \notin T_N(E)} e^{\beta \sqrt{N} E_\sigma} \tag{9.21}$$

Thus we are reduced to computing a single, purely equilibrium quantity, the restricted partition function $Z_{\beta,N}(T_N(E)^c)$. Note that the two terms in the bracket correspond rather naturally to the time it takes the process to leave for the first time η and to that time it takes then to travel from η to $T_N(E)$. The restricted partition function $Z_{\beta,N}(T_N(E)^c)$ was studied in the context of analysing the equilibrium measure of the REM, and we have very good control over it. In fact,

Lemma 9.1 The partition function $Z_{\beta,N}(T_N(E)^c)$ can be written as

$$Z_{\beta,N}(T_N(E)^c) = \frac{e^{(\alpha-1)E + \beta\sqrt{N}u_N(0)}}{\alpha - 1} \left(1 + \mathcal{V}_{N,E} e^{E/2} \frac{\alpha - 1}{\sqrt{2\alpha - 1}} \right) \tag{9.22}$$

where $\mathcal{V}_{N,E}$ is a random variable of mean zero and variance one, all of those moments are finite.

Using this information, we can express our mean time as follows:

Lemma 9.2 With the preceding notation

$$\mathbb{E}_\eta \tau_{T_N(E) \backslash \eta} = \frac{e^{\beta\sqrt{N}u_N(0) + \alpha u_N^{-1}(E_\eta)}}{1 - \frac{1}{M}}$$

$$\left(1 + \frac{e^{-\alpha u_N^{-1}(E_\eta) + (\alpha-1)E}}{M(\alpha - 1)} \left(1 + \mathcal{V}_{N,E} e^{E/2} \frac{\alpha - 1}{\sqrt{2\alpha - 1}} \right) \right) (1 + O(1/N)) \tag{9.23}$$

Notice that the second term in the bracket tends to zero as $E \downarrow -\infty$ for any "fixed" η, but this convergence is not uniform. Note that $M \sim e^{-E}$. Modulo this non-uniformity, this result does however support the assumption in the trap model that modulo an overall factor $(\exp(\beta\sqrt{N}u_N(0)))$, the mean exit time from η converges to a random variable of the form $\exp(\sigma e_\eta)$ where e_η is exponentially distributed with mean one.

Next we should investigate the exponential distribution of these times. Our usual way is to look at Laplace transforms. To do so, we have to get some control on the spectrum. What corresponds to our previous a priori estimates is now the bound on the maximal mean time it takes to hit $T_N(E)$. Form the analysis above we deduce readily

Lemma 9.3 With the notation from above, let

$$\widehat{\Theta}(E) \equiv (1 - \tfrac{1}{|T(E)|})^{-1} e^{\beta\sqrt{N}u_N(0)+\alpha E}$$

$$\left[1 + \frac{e^{-E}}{|T(E)|(\alpha-1)}\left(1 + \mathcal{V}e^{E/2}\frac{\alpha-1}{\sqrt{2\alpha-1}}\right)\right](1 + O(1/N)) \tag{9.24}$$

Then

$$\Theta(E) \equiv \max_{\sigma\in S_N} \mathbb{E}_\sigma \tau_{T(E)} \leq \widehat{\Theta}(E) \tag{9.25}$$

Using this a priori estimate, by using renewal equations and Taylor expansions exclusively for the ensuing Laplace transforms of times that terminate on arrival at $T_N(E)$, we can then proof the rather detailed estimate on $G^\sigma_{T_N(E)\backslash\sigma}(u) \equiv \mathbb{E}_\sigma e^{u\tau_{T_N(E)\backslash\sigma}}$:

Theorem 9.1 For any $\sigma \in T(E)$, the Laplace transform $G^\sigma_{T(E)\backslash\sigma}(u)$ can be written as

$$G^\sigma_{T(E)\backslash\sigma}(u) = \frac{a_\sigma}{1 - (1 - e^{-u})\mathbb{E}_\sigma \tau_{T(E)\backslash\sigma} b_\sigma} + R_\sigma(u) \tag{9.26}$$

where

$$a_\sigma = 1 + O(\widehat{\Theta}(E)/\mathbb{E}_\sigma \tau_{T(E)\backslash\sigma}) \tag{9.27}$$

$$b_\sigma = 1 + O(\widehat{\Theta}(E)/\mathbb{E}_\sigma \tau_{T(E)\backslash\sigma}) \tag{9.28}$$

and $R_\sigma(u)$ is analytic in the half-plane $\Re(u) < 1/\widehat{\Theta}(E)$, periodic with period 2π in the imaginary direction, and satisfies
(i) for all $|u| \leq a/\widehat{\Theta}(E)$,

$$|R_\sigma(u)| \leq C(a)\left(e^{-\beta\sqrt{N}E_\sigma}\widehat{\Theta}(E)\right)^2 \tag{9.29}$$

and
(ii) for all u with $\Re(u) < (1-\epsilon)\widehat{\Theta}(E)$ and $|1 - e^{-u}| \geq 2\epsilon^{-1}e^{-\beta\sqrt{N}E_\sigma}$

$$|R_\sigma(u)| \leq 2\frac{e^{-\beta\sqrt{N}E_\sigma}}{|1 - e^{-u}|(1 - \Re(u)\widehat{\Theta}(E))} \tag{9.30}$$

Moreover,

$$a_\sigma + R_\sigma(0) = 1 \tag{9.31}$$

This proposition allows in fact to prove very good estimates on the distribution function of $\tau_{T(E)\backslash\sigma}$. Note first that if

$$\mathcal{L}(u) \equiv \sum_{n=0}^{\infty} e^{un} \mathbb{P}_\sigma[\tau_{T(E)\backslash\sigma} > n] \tag{9.32}$$

then

$$\mathcal{L}(u) = \frac{G^\sigma_{T(E)\backslash\sigma}(u) - 1}{e^u - 1} \tag{9.33}$$

Corollary 9.1 *With the notation of Theorem 9.1, for any $\epsilon > 0$ and for any positive integer $n \in \mathbb{N}$,*

$$\mathbb{P}_\sigma[\tau_{T(E)\backslash\sigma} = n] = \frac{a_\sigma}{\mathbb{E}_\sigma \tau_{T(E)\backslash\sigma} b_\sigma} e^{-n/\mathbb{E}_\sigma \tau_{T(E)\backslash\sigma} b_\sigma}$$

$$+ O\left(e^{-n(1-\epsilon)/\widehat{\Theta}(E)} e^{-\beta\sqrt{N}E_\sigma} \epsilon^{-1} \ln\left(\widehat{\Theta}(E)\epsilon\right)\right) \tag{9.34}$$

and (for $n > 0$)

$$\mathbb{P}_\sigma[\tau_{T(E)\backslash\sigma} > n] = a_\sigma e^{-n/\mathbb{E}_\sigma \tau_{T(E)\backslash\sigma} b_\sigma}$$

$$+ O\left(e^{-n(1-\epsilon)/\widehat{\Theta}(E)} e^{-\beta\sqrt{N}E_\sigma} \widehat{\Theta}(E)\epsilon^{-1}\right) \tag{9.35}$$

For the proofs of these assertions, see [9].

We see that the deviations from the exponential distribution are substantial, when E_σ is close to E. This means that the assumptions leading to the trap model are not fulfilled, and that we cannot expect that the trap model is a true "limit" of our dynamics. On the other hand, when E_σ is fixed and $E \downarrow -\infty$, the distribution of the exit time converges indeed to the exponential distribution. Thus, we may still hope that with regard to the long-term asymptotics, the trap model yields the correct predictions.

9.3. THE RENEWAL EQUATIONS

The first step now is to define a correlation function that has a good chance to resemble the one used in the trap model. A good choice seems the probability that the process does not jump from a state in the top to another state in the top during a time interval of the form $[n, n+m]$. To define this precisely, we introduce the following random times. For any $k \in \mathbb{N}$, let k_- denote the last time before k at which the process has visited the top, i.e.

$$k_- \equiv \sup\{l < k \mid \sigma(l) \in T_N(E)\} \tag{9.36}$$

Then set

$$\Pi(n, m, N, E) \equiv \mathbb{P}\left[\forall_{k \in [n+1, n+m]}\, \sigma(k) \notin T_N(E)\backslash\sigma(k_-)\right] \tag{9.37}$$

To be as close as possible to Bouchaud, the natural choice is the uniform distribution on $T_N(E)$ that we will denote by π_E. However, we will also need to introduce the respective functions with starting point in an arbitrary state σ. Thus we set

$$\Pi_\sigma(m, n, N, E) \equiv \mathbb{P}\left[\forall_{k \in [n+1, n+m]}\, \sigma(k) \notin T_N(E)\backslash\sigma(k_-) \mid \sigma(0) = \sigma\right] \tag{9.38}$$

and

$$\Pi(m, n, N, E) \equiv \frac{1}{|T_N(E)|} \sum_{\sigma \in T_N(E)} \Pi_\sigma(m, n, N, E) \tag{9.39}$$

We will also use vector notation and write $\underline{\Pi}(n, m, N, E)$ for the M dimensional vector with components $\Pi_\sigma(n, m, N, E)$, $\sigma \in T_N(E)$.

Note that it is easy to derive a renewal equation for the quantities (9.38). Just observe that event in the probability in (9.38) occurs either

(i) $\sigma(k) \notin T(E)\backslash\sigma$, for all $k \in [0, n+m]$, or
(ii) there is $0 < l \le n$, s.t. $l = \inf\{k \le n \mid \sigma(k) \in T(E)\backslash\sigma\}$, and $\forall_{k \in [n+1, n+m]}\, \sigma(k) \notin T_N(E)\backslash\sigma(k_-)$.

Since this decomposition is disjoint, it implies system of renewal equations:

$$\Pi_\sigma(m, n, E) = \mathbb{P}_\sigma[\tau_{T(E)\backslash\sigma} > m + n]$$

$$+ \sum_{k=1}^{n} \sum_{\sigma' \in T(E)\backslash\sigma} \mathbb{P}_\sigma[\tau_{T(E)\backslash\sigma} = k, X_k = \sigma', X_l \notin T(E)\backslash X_{l_-}, \forall n \le l \le m+n]$$

$$= \mathbb{P}_\sigma[\tau_{T(E)\backslash\sigma} > m + n]$$

$$+ \sum_{k=1}^{n} \sum_{\sigma' \in T(E)\backslash\sigma} \mathbb{P}_\sigma[\tau_{\sigma'} = \tau_{T(E)\backslash\sigma} = k]\Pi_{\sigma'}(m, n - k, E)$$

$$\tag{9.40}$$

Now it would be nice to be able to transform this into a single equation for Π by summing over σ. This would work if we had the relation

$$\mathbb{P}_\sigma[\tau_{\sigma'} = \tau_{T(E)\backslash\sigma} = k] = \frac{\pi_E(\sigma')}{1 - \pi_E(\sigma)}\mathbb{P}_\sigma[\tau_{T(E)\backslash\sigma} = k] \tag{9.41}$$

sadly this is not true, and we cannot even proof a reasonable approximate version of this. All we can show, that this relation holds averaged over k. This leaves us for the time being with no alternative but to analyse the full system of equations (9.40).

The method of choice for doing this are Laplace transforms.

We set

$$\Pi_\sigma^*(m, u, E) \equiv \sum_{n=0}^{\infty} e^{nu} \Pi_\sigma(m, n, E) \tag{9.42}$$

for $u \in \mathbb{C}$ whenever this sum converges. Let us define

$$F_\sigma^*(m, u) \equiv \sum_{n=0}^{\infty} e^{nu} \mathbb{P}_\sigma[\tau_{T(E)\backslash\sigma} > m + n] \tag{9.43}$$

Then it follows from (9.40) that for any $\sigma \in T(E)$,

$$\Pi_\sigma^*(m, u, E) = F_\sigma^*(m, u) + \sum_{\sigma' \in T(E)\backslash\sigma} G_{\sigma', T(E)\backslash\sigma}^\sigma(u) \Pi_{\sigma'}^*(m, u, E) \tag{9.44}$$

Let us denote by $K_E^*(u)$ the $|T(E)| \times |T(E)|$ matrix with elements[2]

$$(K_E^*(u))_{\sigma,\sigma'} \equiv \begin{cases} G_{\sigma', T(E)\backslash\sigma}^\sigma(u) & \text{if} \quad \sigma \neq \sigma' \\ 0, & \text{if} \quad \sigma = \sigma' \end{cases} \tag{9.45}$$

Then clearly the solution of equation (9.44) can be written as

$$\underline{\Pi}^*(m, u, E) = \left([\mathbb{1} - K_E^*(u)]^{-1} K_E^*(u) + \mathbb{1} \right) \underline{F}^*(m, u) \tag{9.46}$$

where $\underline{\Pi}^*$ and $\underline{F}^*$ denote the vectors with components Π_σ^*, and F_σ^*.

The matrix

$$M_E^*(u) \equiv [\mathbb{1} - K_E^*(u)]^{-1} K_E^*(u) \tag{9.47}$$

is known as the Laplace transform of the resolvent of the system of renewal equations.

Our task is to compute the inverse Laplace transform of the right hand side of (9.46). This requires estimates in the complex u-plane. Basically, there are two difficulties:

i) Inversion of the matrix $\mathbb{1} - K_E^*(u)$. This is in general quite difficult, and we will not be able to do this for all values of u. However, we are greatly helped by the fact that at $u = 0$, the matrix $K_E^*(0)$ has a very simple form in that it has almost constant columns. Thus the vector

[2]We will often write $K_{\sigma,\sigma'}^*(u)$ instead of $(K_E^*(u))_{\sigma,\sigma'}$ whenever no confusion is possible

1 is an eigenvector with eigenvalue zero, and all other eigenvalues are close to zero. This property can be carried over perturbatively to small values of $|u|$. This will allow us to compute on a small neighborhood of the origin the inverse to leading orders in $1/u$, which will be responsible for the leading long time behaviour of the inverse Laplace transform.

ii) For the use of the Laplace inversion formula, we need reasonable control of the Laplace transform also away from the origin. We expect that these not to give important contributions, but this requires both a judicious choice of the integration contour in the complex plane, and some bounds on the integrands on these contours. Let us recall that

$$\Pi(n, m, E) = \frac{1}{2\pi i} \int_{-i\pi}^{i} \pi du\, e^{-un} \left[(1, M_E^*(u)\underline{F}^*(m, u)) + (1, F^*(m, u)) \right]$$
(9.48)

We will deform the integration path to the contour $\mathcal{C}$ given as follows: consisting of the three parts

$$\mathcal{A} \equiv \left\{ u \in \mathbb{C} : \Re z = 1/2,\ |\Im z| \in [1/\sqrt{2\kappa}, \pi\widehat{\Theta}] \right\}$$
(9.49)

$$\mathcal{B} \equiv \left\{ u \in \mathbb{C} : \Re z \in [1/\tilde{t}, 1/2],\ \Re z = \kappa|\Im z|^2 \right\}$$
(9.50)

and

$$\mathcal{D} \equiv \mathcal{D}_1 \cup \mathcal{D}_2$$
(9.51)

where

$$\mathcal{D}_1 \equiv \left\{ u \in \mathbb{C} : |z| = 1/t,\ \Re z < c|\Im z|^2 \right\}$$

$$\mathcal{D}_2 \equiv \left\{ u \in \mathbb{C} : \Re z \in [\sqrt{1/(4\kappa^2) + 1/t^2} - 1/(2\kappa), 1/\tilde{t}],\ \Re z = \kappa|\Im z|^2 \right\}$$
(9.52)

Here t and κ are positive parameters that are assumed to be chosen such that $\mathcal{C}$ lies in the domain of validity of certain estimates. In what follows, t must be thought of as very large compared with one. At this stage no constraint is imposed on the parameter $\tilde{t}$; it will be chosen as $\tilde{t} = t^\eta$, for suitable $0 < \eta < 1$, later. For future reference let us define the points:

$$z_A = r_A + i s_A, \quad z_B = r_B + i s_B, \quad z_D = r_D + i s_D$$
$$r_A = 1/2, \qquad r_B = 1/\tilde{t}, \qquad r_D = \sqrt{1/(4\kappa^2) + 1/t^2} - 1/(2\kappa)$$
$$s_A = 1/\sqrt{2\kappa}, \qquad s_B = 1/\sqrt{\kappa \tilde{t}}, \qquad s_D = \left(\left(\sqrt{1 + (2\kappa/t)^2} - 1\right)/2\kappa^2\right)^{1/2}$$

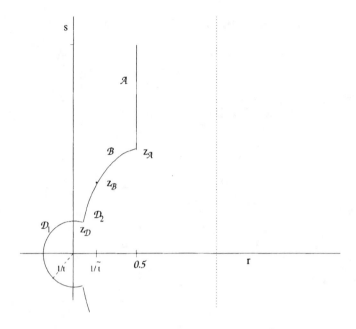

The contour C in the variables r and s

The main contribution of the integral will come from the integral along $\mathcal{D}$.

Our main result is then the following

Theorem 9.2 *Let $\beta > \sqrt{2\ln 2}$. Then there is a sequence $c_N \sim \exp(\beta\sqrt{N}u_N(E))$ such that for any $\epsilon > 0$*

$$\lim_{t,s\uparrow\infty}\ \lim_{E\downarrow-\infty}\ \lim_{N\uparrow\infty} P\left[\left|\frac{\Pi([c_Ns],[c_Nt],N,E)}{\Pi_\infty(s,t)} - 1\right| > \epsilon\right] = 0 \qquad (9.53)$$

where $\Pi_\infty(s,t)$ is the limiting correlation function of the trap model.

In the sequel I will indicate the main steps of the necessary estimates, while for details I have to refer to [3].

9.4. UNIFORM ESTIMATES ON M_E^*

The first thing we want to do is to show that the resolvent is small as soon as we go away from zero along our path C. It is almost a miracle that we are able to do this, because for this we need to show that the norm of $K_E^*(u)$ is smaller than 1. But at $u = 0$, K_E^* has an eigenvalue 1, so we are in a delicate situation where we really need to get quite precise estimates. In particular, we need to use a norm that gives the right values 1 at zero.

Fortunately, operator norm in $L_\infty(\mathbb{C}^M)$, which is given as

$$\|K\| \equiv \max_{\sigma \in T(E)} \sum_{\sigma' \in T(E)} |K_{\sigma,\sigma'}| \tag{9.54}$$

has the right property and is well suited for our matrices. First, its value is one at $u = 0$. Second, we can show that it decays along the imaginary axis up to the value $\pm i\pi$ at rate $v\bar{\Theta}^{-1}$. More precisely, we have

Lemma 9.4 Let $v \in [-\pi, \pi]$. Then (for N large enough),

$$\|K_E^*(iv)\| \leq \frac{1}{\sqrt{2(1 - \cos v)\bar{\Theta}^2\left(1 - O(\bar{\Theta}^{-1})\right) + 1 - \frac{4}{M-1}\left(1 + O(d/N)\right)}} \tag{9.55}$$

The proof of the Lemma is somewhat tedious, but essentially makes use of the simple idea to use renewal to represent

$$K_{\sigma,\sigma'}^*(u) = \frac{G_{\sigma',T(E)}^\sigma(u)}{1 - G_{\sigma,T(E)}^\sigma(u)} \tag{9.56}$$

Here the numerator is always small, if u is purely imaginary. On the other hand, the Laplace transform in the denominator is dominated by the process realising the event to go from σ to σ in a single step. This part is easily computed and gives, if $u = iv$, $1 - e^{iv}p(\sigma, \sigma)$. This yields roughly the behaviour

$$\|K_E^*(iv)\| \sim \frac{\mathbb{P}_\sigma[\tau_{T(E)\backslash\sigma} < \tau_{T(E)}]}{|1 - e^{iv}|} \tag{9.57}$$

which explains the estimate given in the lemma.

The estimate (9.55) can now be extended a little bit off the imaginary axis in the positive real direction. This uses simply Taylor expansions about $u = iv$; of course, the trick is to use again the representation (9.56) and to exploit the fact that the functions appearing in the numerator and denominator are analytic in the the half-space $\Re u < 1/\bar{\Theta}$, well beyond the first pole of K_E^* itself.

To state these estimates, we need some notation: First, let

$$z = \hat{\Theta}(E)u \tag{9.58}$$

The real and imaginary parts of z will always be called r and s:

$$z = r + is \tag{9.59}$$

Thus

$$\begin{aligned} r &= \hat{\Theta}(E)w \\ s &= \hat{\Theta}(E)v \end{aligned} \tag{9.60}$$

Definition 9.1 $0 < C_1, C_2 < \infty$, and $0 < \gamma < 1$ be numerical constants. With the above notation we define the sets:

$$D_1(C_1) \equiv \left\{ u \in \mathbb{C} : \sqrt{r^2 + s^2} \geq C_1/\sqrt{M} \right\}$$

$$D_2(C_2, \gamma) \equiv \left\{ u \in \mathbb{C} : 0 \leq r < \min\left(\frac{\gamma s^2}{C_2\sqrt{1+s^2}}, 1 - \gamma \right), \ v \in [-\pi, \pi] \right\}$$

$$D_3 \equiv \{ u \in \mathbb{C} : -1 \leq r < 0, \ |s| < 1 \}$$

$$D_4 \equiv \{ u \in \mathbb{C} : |r| < 1, \ |s| < 1 \}$$

$$(9.61)$$

Lemma 9.5 There exist constants $0 < C, C' < \infty$ such that, for all $0 < \gamma < 1$ and all $u \in D_2(C', \gamma)$,

$$\|K_E^*(u)\| \leq$$

$$\frac{1 + C\gamma^{-1}r}{\sqrt{1 + \hat{\Theta}^2 2(1 - \cos v)(1 - O(\bar{\Theta}^{-1})) - \frac{4}{M-1}(1 + O(d/N)) - C'\gamma^{-1}r}}$$

$$(9.62)$$

Consequently, for all $0 < \gamma < 1$ there exists a constant $0 < L < \infty$ (depending on C, C' and γ) such that, for all $u \in D_1(4) \cap D_2(L, \gamma)$,

$$\|K_E^*(u)\| < 1 \qquad (9.63)$$

and

$$\|M_E^*(u)\| \leq$$

$$\frac{1 + C\gamma^{-1}r}{\sqrt{1 + \hat{\Theta}^2 2(1 - \cos v)(1 - O(\bar{\Theta}^{-1})) - 1 - \frac{4}{M-1}(1 + O(d/N)) - (C + C')\gamma^{-1}r}}$$

$$(9.64)$$

The last estimate of this kind we will need concerns the case when $|u|$ is very small and $w \leq 0$. Its derivation is very similar to that of the preceding ones.

Lemma 9.6 For M large enough,
(i) for all $u \in D_3$,

$$\|K_E^*(u)\| \leq \frac{1}{\sqrt{1 + r^2 + s^2} - \frac{5}{M}} \qquad (9.65)$$

(i) for all $u \in D_1(4) \cap D_3$, $\|K_E^*(u)\| < 1$ and

$$\|M_E^*(u)\| \leq \frac{1}{\sqrt{1 + r^2 + s^2} - 1 - \frac{5}{M}} \qquad (9.66)$$

The estimates in this subsection suffice to show that the the contributions from the integration excepting the little circle around the origin do not give a significant contribution. To extract the leading behaviour, we need of course far more precise control on our kernels in this domain.

9.5. PERTURBATIVE ESTIMATES FOR SMALL u

So far we have been able to avoid the problem of inverting the matrix $1 - K^*(u)$. Fortunately, for small u, this matrix is very close to a matrix with constant columns whose inverse we can of course compute easily. The idea is thus to use perturbation theory.

The basis is the expansion

$$K^*_{\sigma,\sigma'}(u) = \frac{1}{1-G^\sigma_{\sigma,T}(u)}\left(G^\sigma_{\sigma',T}(0) + u\frac{d}{du}G^\sigma_{\sigma',T}(0) + \frac{u^2}{2}\frac{d^2}{du^2}G^\sigma_{\sigma',T}(\tilde{u})\right)$$

$$= \frac{1}{1-G^\sigma_{\sigma,T}(u)}\left(\mathbb{P}[\tau^\sigma_{\sigma'} \leq \tau^\sigma_T] + u\mathbb{E}\tau^\sigma_{\sigma'}\mathbb{1}_{\{\tau^\sigma_{\sigma'}\leq\tau^\sigma_T\}} + \frac{u^2}{2}\frac{d^2}{du^2}G^\sigma_{\sigma',T}(\tilde{u})\right)$$

$$(9.67)$$

We define

$$\mathcal{K}^{*(0)}_{\sigma,\sigma'}(u) \equiv \frac{1}{1-G^\sigma_{\sigma,T}(u)}\left(\frac{1}{M}\mathbb{P}[\tau^\sigma_{T\setminus\sigma} < \tau^\sigma_T]\left(1 + u\mathbb{E}[\tau^\sigma_{T\setminus\sigma}|\tau^\sigma_{T\setminus\sigma} = \tau^\sigma_T]\right)\right),$$

$$\forall \sigma, \sigma' \in T(E)$$

$$(9.68)$$

as the leading part. We then prove a norm estimate, valid for

$$\{u \in \mathbb{C} \mid r \leq s^2/4\} \subseteq D_2(L,\gamma) \cap D_4 \qquad (9.69)$$

that states that

$$\left|K^*(u) - \mathcal{K}^{*(0)}(u)\right| \leq \frac{C\gamma^{-1}(s^2+r^2) + (1+\sqrt{s^2+r^2})O(1/(M-1))}{\sqrt{1+(s^2+r^2)/2 - 5/M}}$$

$$(9.70)$$

Now $\mathcal{K}^{*(0)}(u)$ has a unique non-zero eigenvalues

$$\lambda(u) \equiv \sum_{\sigma\in T}\mathcal{K}^{*(0)}_{\sigma,\sigma'}(u) \qquad (9.71)$$

The corresponding left eigenvector is proportional to $(1,1,\ldots,1)$. Based on this, we define

$$M^{*(0)}(u) \equiv [\mathbb{1} - \mathcal{K}^{*(0)}(u)]^{-1}\mathcal{K}^{*(0)}(u) \qquad (9.72)$$

and decompose the Laplace transform of the resolvent (defined in (9.47)) into

$$M^*(u) \equiv M^{*(0)}(u) + M^{*(1)}(u) \qquad (9.73)$$

It is a simple matter to see that

Lemma 9.7 Set

$$
\begin{aligned}
R(u) &\equiv [\mathbb{1} - \mathcal{K}^{*(0)}(u)]^{-1} \\
\rho(u) &\equiv \max\left(|1 - \lambda(u)|^{-1}, 1\right)
\end{aligned}
\tag{9.74}
$$

Then,

$$
M^{*(1)}(u) = R(u)\mathcal{K}^{*(1)}(u)R(u)\frac{1}{\mathbb{1} - R(u)\mathcal{K}^{*(1)}(u)}
\tag{9.75}
$$

and, if $\|R(u)\mathcal{K}^{*(1)}(u)\| < 1$,

$$
\|M^{*(1)}(u)\| \leq \frac{\|\mathcal{K}^{*(1)}(u)\|\rho(u)^2}{1 - \|\mathcal{K}^{*(1)}(u)\|\rho(u)}
\tag{9.76}
$$

It will turn out that this estimate is good enough to show that $M^{*(1)}(u)$ can be neglected. The main remaining issue is now to compute the eigenvalue $\lambda(u)$. Explicitly, we have

$$
1 - \lambda(u) = \frac{1}{|T|}\sum_{\sigma \in T}\left[1 - \frac{G^{\sigma}_{T\setminus\sigma,T}(0)}{1 - G^{\sigma}_{\sigma,T}(u)}\left(1 + u\mathbb{E}[\tau^{\sigma}_{T\setminus\sigma}|\tau^{\sigma}_{T\setminus\sigma} = \tau^{\sigma}_{T}]\right)\right]
\tag{9.77}
$$

This is still quite a complicated expression, and we need to do some more simplification. The key is the following lemma.

Lemma 9.8 Recall that $u = z/\widehat{\Theta}(E)$ and set

$$
z_{\sigma} \equiv \left(1 - \tfrac{1}{M}\right)e^{-\beta\sqrt{N}E_{\sigma}}\widehat{\Theta}(E)
\tag{9.78}
$$

If u belongs to the set

$$
D_{\delta} \equiv \left\{u \in \mathbb{C} \mid r < s^2/4, |z| \leq \delta\right\}, \quad 0 < \delta < 1
\tag{9.79}
$$

then, for N large enough,

$$
\left|1 - \frac{G^{\sigma}_{T\setminus\sigma,T}(0)}{1 - G^{\sigma}_{\sigma,T}(u)}\left(1 + u\mathbb{E}[\tau^{\sigma}_{T\setminus\sigma}|\tau^{\sigma}_{T\setminus\sigma} = \tau^{\sigma}_{T}]\right) - \frac{z}{z - z_{\sigma}}\right| \leq C(\delta)|z|
\tag{9.80}
$$

for some constant $0 < C(\delta) < \infty$ that only depends on δ.

The point of this lemma is that it shows that the summands in (9.77) can be replaced by $z/(z - z_{\sigma})$, since this is dominating the error of order $|z|$. The nice thing is that the sum over these leading terms can be expressed as integrals with respect to our Poisson point process. Also, we see that these are now Laplace transforms of exponentially distributed random variables.

Thus, we are approaching something the looks like the trap model. Note, however, that the proof of Lemma 9.8 is quite involved.

9.6. POISSON CONVERGENCE AND SELF AVERAGING

The next step is now to represent the sum over the $z/(z-z_\sigma)$ as an integral with respect to the point process $\sum_\sigma \delta_{u_N^{-1}(E_\sigma)}$. Then we want to use to facts: first, that this point process converges to a Poisson point process, and second that the integral over that Poisson point process converges, as we take the cut of E to minus infinity, to a deterministic integral. Both facts are of course well known.

Let us write

$$z_\sigma = (1 - 1/M)e^{-\beta\sqrt{N}E_\sigma}\hat{\Theta}(E) \equiv \frac{1}{e^{\alpha(u_N^{-1}(E_\sigma)-E)}\tau_{E,N}} \tag{9.81}$$

It will also be convenient to define

$$\mathcal{N}^*_{N,E} \equiv \sum_{\sigma\in\{-1,1\}^N} \delta_{\exp\{\alpha(-E+u_N^{-1}(E_\sigma))\}} = \sum_{\sigma\in\{-1,1\}^N} \delta_{1/(z_\sigma\tau_{N,E})} \tag{9.82}$$

It is easy to see that this process converges weakly to the Poisson point process $\mathcal{N}^*_E$ on $[1,\infty)$ with intensity measure $\alpha^{-1}e^E x^{-1-1/\alpha}dx$. On the other hand, one can show without difficulty that

$$\lim_{E\downarrow-\infty}\lim_{N\uparrow\infty}\tau_{N,E} = 1 - 1/\alpha \equiv \tau_\infty, \quad \text{in Probability.} \tag{9.83}$$

The following Lemma yields all we need to analyse sums like (9.77).

Lemma 9.9 Let g be a bounded continuous function on $\mathbb{R}^+$, such that $\left|\int_0^\infty \frac{dx}{x^{1+1/\alpha}}g(x)\right| < +\infty$, and let X_N be a family of positive random variables that converge in distribution to the positive random variable X. Then for any $b > 0$,

(i) $\int_b^\infty \mathcal{N}^*_{N,E}(dx)g(xX_N)$ converges, as $N \uparrow \infty$, to the random variable $\int_b^\infty \mathcal{N}^*_E(dx)g(xX)$.

(ii) If X_E is a family of random variables such that, as $E \downarrow -\infty$, $X_E \to a \in \mathbb{R}^+$ almost surely, then

$$\lim_{E\downarrow-\infty} e^{+E}\int_1^\infty \mathcal{N}^*_E(dx)g(xX_E) = \alpha^{-1}\int_1^\infty \frac{dx}{x^{1+1/\alpha}}g(xa), \quad \text{a.s.}$$
$$\tag{9.84}$$

(iii) If g is a complex valued function on $\mathbb{C}$, and if for some domain $B \subset \mathbb{C}$, for all $x \in \mathbb{R}^+$, $z \in B$, $g(zx)$ is bounded, and for all $z \in B$,

$$\left|\int_0^\infty \frac{dx}{x^{1+1/\alpha}}g(zx)\right| < \infty \tag{9.85}$$

holds, then

$$\lim_{E\downarrow-\infty} P\left[\limsup_{N\uparrow\infty} \left| e^E \int_1^\infty \mathcal{N}_E^*(dx)g(zxX_E) - \right. \right.$$

$$\left. \left. (az)^{1/\alpha}\alpha^{-1}\int_{az}^\infty \frac{dx}{x^{1+1/\alpha}}g(x)\right| > \epsilon \right] = 0 \tag{9.86}$$

Applying this lemma to the sum in (9.77) yields the

Corollary 9.2 *Uniformly in* $\Re(z) < \max(|\Im(z)|, 1/2)$,

$$\lim_{E\downarrow-\infty}\lim_{N\uparrow\infty} (1 - \lambda(u))$$

$$= \alpha^{-1}\int_1^\infty \frac{dx}{x^{1+1/\alpha}}\frac{xz\tau_\infty}{xz\tau_\infty - 1} + O(|z|)$$

$$= (-z\tau_\infty)^{1/\alpha}\pi\mathrm{cosec}(\pi/\alpha) + O(|z|), \quad \text{in Probability.} \tag{9.87}$$

9.7. LAPLACE INVERSION

Let us now show how roughly how to go back to real space. The term that will give the main contribution is of the form

$$\left(\mathbb{1}, M^{*(0)}(u)\underline{F}^*(m,u)\right) = \frac{\lambda(u)}{1 - \lambda(u)}(\mathbb{1}, \underline{F}^*(m,u)) \equiv h_{N,E}(u,m) \tag{9.88}$$

It turns out the the limit of this expression has the following nice representation:

Proposition 9.2 *For u on* $\mathcal{C}$*, we have that*

$$\lim_{E\downarrow-\infty}\lim_{N\uparrow\infty} h_{N,E}(u,m) = H_0^*(s,z)(1 + O(|z|^{1-1/\alpha}, |z|^{1/\alpha})) + O(z^{-1/\alpha}e^{-s/\tau_\infty}) \tag{9.89}$$

where $H_0^*(s,u) \equiv \int_0^\infty dt e^{zt}\int_{s/t}^\infty \frac{dx}{x^{1/\alpha}(1+x)}$ *is the Laplace transform of the function* H_0 *defined in (9.13).*

Proof. As we already know $\lambda(u)$, the main work goes into the analysis of $(\mathbb{1}, \underline{F}^*(m,u))$. This goes largely parallel to the analysis of $M^*(u)$. It turns out that the leading contribution can be written in the form

$$(\mathbb{1}, \underline{F}^*(m,u)) \sim \frac{1}{|T(E)|}\sum_{\sigma\in T(E)} e^{-me^{-\sqrt{N}E_\sigma}}\frac{\widehat{\Theta}(E)}{z_\sigma - z} \tag{9.90}$$

and

Lemma 9.10 Set $s \equiv m/\widehat{\Theta}$. Then, uniformly on $\Re z < \max(\Im z, 1/2)$, and $\Re(u) \leq |\Im u|$,

$$\lim_{E\downarrow-\infty} \lim_{N\uparrow\infty} \frac{1}{|T(E)|} \sum_{\sigma\in T(E)} e^{-me^{-\beta\sqrt{N}E_\sigma}} \frac{1}{z_\sigma - z}$$

$$= (-z\tau_\infty)^{1/\alpha} \left(z^{-1}\pi\operatorname{cosec}(\pi/\alpha) - \int_0^\infty dt e^{zt} \int_0^{s/t} \frac{dx}{x^{1/\alpha}(1+x)} \right) \qquad (9.91)$$

$$+ O(e^{-s/\tau_\infty}) \quad \text{in Probability.}$$

Thus

$$\lim_{E\downarrow-\infty} \lim_{N\uparrow\infty} \frac{\lambda(u)}{1-\lambda(u)} \frac{1}{|T(E)|} \sum_{\sigma\in T(E)} e^{-me^{-\sqrt{N}E_\sigma}} \frac{\widehat{\Theta}(E)}{z_\sigma - z}$$

$$= u^{-1} - \frac{\int_0^\infty dt e^{ut} \int_0^{s/t} \frac{dx}{x^{1/\alpha}(1+x)}}{\pi\operatorname{cosec}(\pi/\alpha)} \left(1 + O(|z|^{1-1/\alpha}, |z|^{1/\alpha}) \right) \qquad (9.92)$$

$$+ O\left(z^{-1/\alpha} e^{-s/\tau_\infty} \right)$$

The leading term is readily identified as the Laplace transform of

$$H_0(s/t) \equiv 1 - \frac{\int_0^{s/t} \frac{dx}{x^{1/\alpha}(1+x)}}{\pi\operatorname{cosec}(\pi/\alpha)} \qquad (9.93)$$

which we recognise as precisely the function that appeared as the leading asymptotic contribution in the trap model.

It remains the rather painstaking task to show that indeed all other contributions can be neglected. The interested reader will find them in [3]. In any case the logic of these terms is simple: First off all, only the behaviour of a term near $u = 0$ matters, since the a priori estimates show that the contributions to the inverse Laplace transform from the path off zero are small. Then, whenever a term has a higher power in u than the leading term, its inverse Laplace transform decays faster in time. Since this is the case for all error terms we have produced, we can indeed conclude that our Main Theorem holds.

Acknowledgement. First of all I would like to thank the organisers of the II Workshop on Dynamics and Randomness, Alejandro Maass, Servet

Martínez, and Jaime San Martín for the kind invitation to lecture at the exiting meeting at Santiago de Chile and for inviting the preparation of these notes. Special thanks are due to Gladys Cavallone for the perfect organisation of all practical matters and, in particular, for dealing with my TeX-file. Needless to say, I'm deeply grateful to my collaborators on the topics of these notes. I can only hope that the presentation of their work I am attempting here does some justice to it. Last but not least, I thank Alessandra Faggionato for helping me to reduce the number of missprints in the manuscript.

Work supported in part by the DFG in the Priority programme 1033 "Interagierende stochastische Systeme von hoher Kompexität".

References

1. Ben Arous, G., Bovier, A. and Gayrard, V. (2002) Aging in the Random Energy Model, *Phys. Rev. Letts.*, **Vol.88**, pp. 087201.
2. Ben Arous, G., Bovier, A. and Gayrard, V. (2003) Glauber Dynamics of the Random Energy Model. 1. Metastable Motion on the Extreme States, *Commun. Math. Phys.*, **Vol.235**, pp. 379-425.
3. Ben Arous, G., Bovier, A. and Gayrard, V. (2003) Glauber Dynamics of the Random Energy Model. 2. Aging below the Critical Temperature, *Commun. Math. Phys.*, **Vol.236**, pp. 1-54.
4. Ben Arous, G., Bovier, A. and Gayrard, V. Random Walks on the Hypercube, in preparation.
5. Ben Arous, G. (2002) Aging and Spin Glasses, in *Proceedings of the International Congress of Mathematicians 2002*, Beijing, China, (Li, Ta, Tsien et al. Eds.), China: Higher Education Press, **Vol.3**, pp. 3–14.
6. Bouchaud, J.P., Cugliandolo, L., Kurchan, J., Mézard, M. (1998) Out-of-Equilibrium Dynamics in Spin-Glasses and other Glassy Systems, in *Spin-Glasses and Random Fields* (A.P. Young, Ed.), World Scientific, Singapore.
7. Bouchaud, J.P. and Dean, D. (1995) Aging on Parisi's Tree, *J. Phys. I*, France, **Vol.5**, pp. 265.
8. Bovier, A., Eckhoff, M., Gayrard, V. and Klein, M. (2001) Metastability in Stochastic Dynamics of Disordered Mean-Field Models, *Probab. Theor. Rel. Fields*, **Vol.119**, pp. 99–161.
9. Bovier, A., Eckhoff, M., Gayrard, V. and Klein, M. (2002) Metastability and Low-Lying Spectra in Reversible Markov Chains, *Commun. Math. Phys.*, **Vol.228**, pp. 219–255.
10. Bovier, A., Eckhoff, M., Gayrard, V. and Klein, M. (2002) Metastability in Reversible Diffusion Processes I. Sharp Asymptotics for Capacities and Exit Times, Preprint.
11. Bovier, A., Gayrard, V. and Klein, M. (2002) Metastability in Reversible Diffusion Processes II. Precise Asymptotics for Small Eigenvalues, Preprint.
12. Bovier, A. and Manzo, F. (2002) Metastability in Glauber Dynamics in the Low-Temperature Limit: beyond Exponential Asymptotics, *J. Statist. Phys.*, **Vol.107**, pp. 757–779.
13. Burke, C.J. and Rosenblatt, M. (1958) A Markovian Function of a Markov Chain, *Ann. Math. Statist.*, **Vol.29**, pp. 1112–1122.
14. Cassandro, M., Galves, A. and Picco, P. (1991) Dynamical Phase Transitions in Disordered Systems: the Study of a Random Walk Model, *Ann. Inst. H. Poincaré Phys. Thór.*, **Vol.55**, pp. 689–705.

15. Cugliandolo, L. (2002) Dynamics of Glassy Systems, Les Houches Lecture Notes, cond-mat/0210312.

16. Cugliandolo, L. and Kurchan, J. (1999) Thermal Properties of Slow Dynamics, *Physica*, **Vol.A 263**, pp. 242–253.

17. Davies, E.B. (1982) Metastable States of Symmetric Markov Semigroups. I. *Proc. Lond. Math. Soc. III, Ser.*, **Vol.45**, pp. 133–150.

18. Davies, E.B. (1982) Metastable States of Symmetric Markov Semigroups. II. *J. Lond. Math. Soc. II, Ser.*, **Vol.26**, pp. 541–556.

19. Davies, E.B. (1983) Spectral Properties of Metastable Markov Semigroups, *J. Funct. Anal.*, **Vol.52**, pp. 315–329.

20. den Hollander, W.T.F. and Shuler, K.E. (1992) Random Walks in a Random Field of Decaying Traps, *J. Statist. Phys.*, **Vol.67**, pp. 13–31.

21. Derrida, B. (1980) Random Energy Model: Limit of a Family of Disordered Models, *Phys. Rev. Letts.*, **Vol.45**, pp. 79–82.

22. Derrida, B. (1981) Random Energy Model: An Exactly Solvable Model of Disordered Systems, *Phys. Rev. B*, **Vol.24**, pp. 2613–2626.

23. Diaconis, P. (1988) Applications of Noncommutative Fourier Analysis to Probability Problems, *École d'Été de Probabilités de Saint-Flour XV–XVII*, 1985–87, Lecture Notes in Math., **Vol.1362**, Springer Verlag, Berlin, pp. 51–100.

24. Doob, J.L. (1984) *Classical Potential Theory and its Probabilistic Counterpart*, Grundlehren der Mathematischen Wissenschaften **Vol.262**, Springer Verlag, Berlin.

25. Doyle, P.G. and Snell, J.L. (1984) *Random Walks and Electrical Networks*, Carus Mathematical Monographs, **Vol.22**, Mathematical Association of America, Washington, DC.

26. Donsker, M.D. and Varadhan, S.R.S. (1976) On the Principal Eigenvalue of Second-Order Elliptic Differential Operators, *Comm. Pure Appl. Math.*, **Vol.29**, pp. 595–621.

27. Eisele, Th. (1983) On a Third Order Phase Transition, *Commun. Math. Phys.*, **Vol.90**, pp. 125–159.

28. Eyring, H. (1935) The Activated Complex in Chemical Reactions, *J. Chem. Phys.*, **Vol.3**, pp. 107–115.

29. Eyring, H. and Polanyi, M. (1931) *Z. Physik. Chemie.*, **Vol.B12**, pp. 279.

30. Fontes, L.R.G., Isopi, M., Kohayakawa, Y. and Picco, P. (2001) The Spectral Gap of the REM under Metropolis Dynamics, *Ann. Appl. Probab.*, **Vol.8**, pp. 917–943.

31. Freidlin, M.I. and Wentzell, A.D. (1984) *Random Perturbations of Dynamical Systems*, Springer Verlag, Berlin-Heidelberg-New York.

32. Galvez, A., Martínez, S. and Picco, P. (1989) Fluctuations in Derrida's Random Energy and Generalized Random Enery Models, *J. Stat. Phys.*, **Vol.54**, pp. 515–529.

33. Gaveau, B. and Schulman, L.S. (1998) Theory of Nonequilibrium First-Order Phase Transitions for Stochastic Dynamics, *J. Math. Phys.*, **Vol.39**, pp. 1517–1533.

34. Gayrard, V. (1992) Thermodynamic Limit of the q-State Potts-Hopfield Model with Infinitely many Patterns, *J. Statist. Phys.*, **Vol.68**, pp. 977–1011.

35. Glasstone, S., Laidler, K.J. and Eyring, H. (1941) *The Theory of Rate Processes*, McGraw-Hill, New York.

36. Holley, R.A., Kusuoka, S., and Stroock, S.W. (1989) Asymptotics of the Spectral Gap with Applications to the Theory of Simulated Annealing, *J. Funct. Anal.*, **Vol.83**, pp. 333–347.

37. Huisinga, W., Meyn, S. and Schütte, Ch. (2002) Phase Transitions and Metastability for Markovian and Molecular Systems, FU Berlin, Preprint.

38. Kato, T. (1976) *Perturbation Theory for Linear Operators*, Second edition, Grundlehren der Mathematischen Wissenschaften, Band 132, Springer-Verlag, Berlin-New York.

39. Kakutani, S. (1941) Markov Processes and the Dirichlet Problem, *Proc. Jap. Acad.*

Vol.21, pp. 227–233.

40. Kemeny, J.G. and Snell, J.L. (1960) *Finite Markov Chains*, D. van Nostrand Company, Princeton.

41. Kemperman, J.H.B. (1961) *The Passage Problem for a Stationary Markov Chain*, Statistical Research Monographs, Vol.I, The University of Chicago Press.

42. Kolokoltsov, V.N. (2000) *Semiclassical Analysis for Diffusions and Stochastic Processes*, Springer Verlag, Berlin.

43. Kolokoltsov, V.N. and Makarov, K.A. (1996) Asymptotic Spectral Analysis of a Small Diffusion Operator and the Life Times of the Corresponding Diffusion Process, *Russian J. Math. Phys.*, Vol.4, pp. 341–360.

44. Koch, H. and Piasko, J. (1989) Some Rigorous Results on the Hopfield Neural Network Model, *J. Statist. Phys.*, Vol.55, pp. 903–928.

45. Kramers, H.A. (1940) Brownian Motion in a Field of Force and the Diffusion Model of Chemical Reactions, *Physica*, Vol.7, pp. 284–304.

46. Mathieu, P. (1995) Spectra, Exit Times and Long Times Asymptotics in the Zero White Noise Limit, *Stoch. Rep.*, Vol.55, pp. 1–20.

47. Mathieu, P. and Picco, P. (2000) Convergence to Equilibrium for Finite Markov Processes with Application to the Random Energy Model, CPT-2000/P.39, Preprint.

48. Matthews, P. (1987) Mixing Rates for a Random Walk on the Cube, *SIAM J. Algebraic Discrete Methods*, Vol.8, pp. 746–752.

49. Miclo, L. (1995) Comportement de Spectres d'Opérateurs de Schrödinger à Basse Température, *Bull. Sci. Math.*, Vol.119, pp. 529–553.

50. Matkowsky, B.J. and Schuss, Z. (1979) The Exit Problem: a New Approach to Diffusion Across Potential Barriers, *SIAM J. Appl. Math.*, Vol.36, pp. 604–623.

51. Maier, R.S. and Stein, D.L. (1997) Limiting Exit Location Distributions in the Stochastic Exit Problem, *SIAM J. Appl. Math.*, Vol.57, pp. 752–790.

52. Olivieri, E. and Picco, P. (1991) On the Existence of Thermodynamics for the Random Energy Model, *Commun. Math. Phys.*, Vol.96, pp. 125–144.

53. Reed, M. and Simon, B. (1978) *Methods of Modern Mathematical Physics. IV. Analysis of Operators*, Academic Press, New York-London.

54. Ruelle, D. (1987) A Mathematical Reformulation of Derrida's REM and GREM, *Commun. Math. Phys.*, Vol.108, pp. 225–239.

55. Saloff-Coste, L. (1997) Lectures on Finite Markov Chains, *Lectures on Probability Theory and Statistics*, Saint-Flour, 1996, Lecture Notes in Math., Springer Verlag, Berlin, Vol.1665, pp. 301–413.

56. Scoppola, E. (1995) *Renormalization and Graph Methods for Markov Chains*, Advances in Dynamical Systems and Quantum Physics, Capri, 1993, World Sci. Publishing, River Edge, NJ, pp. 260–281.

57. Sznitman, A.-S. (1998) *Brownian Motion, Obstacles and Random Media*, Springer Monographs in Mathematics, Springer Verlag, Berlin.

58. Soardi, P.M. (1994) *Potential Theory on Infinite Networks*, LNM 1590, Springer Verlag, Berlin.

59. Talkner, P. (1987) Mean First Passage Times and the Lifetime of a Metastable State, *Z. Phys.*, Vol.B 68, pp. 201–207.

60. Wentzell, A.D. (1972) On the Asymptotic Behaviour of the Greatest Eigenvalue of a Second Order Elliptic Differential Operator with a Small Parameter in the Higher Derivatives, *Soviet Math. Docl.*, Vol.13, pp. 13–17.

61. Wentzell, A.D. (1973) Formulas for Eigenfunctions and Eigenmeasures that are Connected with a Markov Process, *Teor. Verojatnost. i Primenen.*, Vol.18, pp. 329.

62. Wigner, E.P. (1938) *Trans. Faraday Soc.* Vol.34, pp. 29.

ALGEBRAIC SYSTEMS OF GENERATING FUNCTIONS AND RETURN PROBABILITIES FOR RANDOM WALKS

STEVEN P. LALLEY

Department of Statistics
University of Chicago
5734 University Avenue, Chicago IL 60637, U.S.A.
`lalley@galton.uchicago.edu`

Abstract. A technique for analyzing the leading singularity of an infinite algebraic system of generating functions is developed. The generating functions are assumed to have positive coefficients, and to be interrelated by a system of holomorphic functional equations. It is shown that if the Jacobian operator is of Perron-Frobenius type at the lead singularity then the generating functions have a square-root type singularity. Standard Tauberian theorems then imply that the power series coefficients satisfy power laws with exponent 3/2. Examples of systems obeying the hypotheses include random walks on homogeneous trees and random walks on regular languages.

1. Introduction

The purpose of these lectures is to set out a method for analyzing the principal singularities of certain systems of generating functions $F_i(z)$ that are interrelated by functional equations of the form

$$F_i(z) = zQ_i(F_1(z), F_2(z), \ldots).\tag{1.1}$$

Such systems occur in a variety of combinatorial and probabilistic contexts, several of which are discussed below. Our primary interest in them stems from their occurrence in random walk problems, especially for random walks on homogeneous trees and tree like structures. We shall see that the asymptotic behavior of *returnprobabilities* for such random walks is governed by the leading singularity of systems of the form (1.1).

A. Maass et al. (eds.), Dynamics and Randomness II, 81–122.

1.1. THE LAGRANGE INVERSION FORMULA

The scalar form of the system (1.1), in which there is a single generating function $F(z)$ that satisfies a functional equation

$$F(z) = zQ(F(z)), \qquad (1.2)$$

has a history that dates to Lagrange (or perhaps even earlier). Lagrange discovered what is now known as the *Lagrange Inversion Formula*, which gives an explicit formula for the coefficients of the power series $F(z) = \sum_{n \geq 0} a_n z^n$ representing $F(z)$ in terms of the link function $Q(w)$ and its powers.

Theorem 1.1 *(Lagrange Inversion Formula) Let $Q(w) = \sum_{n \geq 0} b_n w^n$ be a formal power series in w with constant coefficient $b_0 \neq 0$. There is a unique formal power series $F(z) = \sum_{n \geq 0} a_n z^n$ that satisfies the functional equation (1.2). Its coefficients are given by*

$$a_n = n^{-1}[w^{n-1}]Q(w)^n. \qquad (1.3)$$

Here $[w^{n-1}]Q(w)^n$ denotes the coefficient of w^{n-1} in the power series expansion of the function Q^n.

Various proofs of Lagrange's theorem are known: see [24], [10], and [23] for some of the standard ones. A more intricate but also more interesting proof, based on a form of Spitzer's Combinatorial Lemma ([6],Ch. XII, Section 6), is given in [20]: this proof shows that the Lagrange Formula is intimately connected with the combinatorics of lattice paths. The essence of this proof is distilled in section 1.2 below.

   The validity of the Lagrange formula does not require convergence of the formal power series $Q(w)$. However, if Q has a nonzero radius of convergence, then so will $F(z)$. (*Exercise!*) In this event, the Cauchy Integral Formula may be used in conjunction with (1.2) to relate the coefficients a_n of $F(z)$ with the values of $Q(w)$ on any contour γ surrounding the origin such that $Q(w)/w$ is analytic everywhere inside γ except at $w = 0$:

$$a_n = \frac{1}{2\pi i n} \oint_\gamma \frac{Q(w)^n}{w^n} \, dw. \qquad (1.4)$$

This formula is useful for asymptotic calculations. Assume that the coefficients b_n of the power series $Q(w)$ are nonnegative (this will be the case in most combinatorial and probabilistic applications). If there exists $\rho > 0$ less than the radius of convergence of $Q(w)$ such that $\rho Q'(\rho) = Q(\rho)$, then $w = \rho$ is a *saddle point* for the integrand in (1.4), and so the saddle point method of asymptotic evaluation ([5], section 2.2) may be applied. In particular, unless $Q(w)$ is a linear function, the second derivative at the saddle point $w = \rho$ will be positive, and so

$$a_n \sim \frac{1}{\sigma R^n n^{3/2}}$$ (1.5)

for a suitable constant $\sigma > 0$, where

$$R = \rho/Q(\rho),$$ (1.6)

provided that $|Q(w)| < Q(\rho)$ for all $w \neq \rho$ such that $|w| = \rho$. (This last proviso will be satisfied unless the set of indices n such that $q_n \neq 0$ is contained in a coset of a proper subgroup of the integers.) In particular, it follows that R is the radius of convergence of the power series $F(z)$.

1.2. HITTING TIMES FOR LEFT-CONTINUOUS 1D RANDOM WALKS

Let $\{q_x\}_{x=0,1,\dots}$ be a probability distribution on the nonnegative integers, and let $\xi_1, \xi_2, \dots$ be independent, identically distributed random variables with distribution $\{q_x\}$. Define

$$S_n = \sum_{j=1}^{n}(\xi_j - 1) \, ;$$ (1.7)

the sequence S_n is a *left-continuous* random walk on the integers. Define τ to be the first time n that $S_n = -1$ (or ∞ if there is no such n). One may easily check (*Exercise!*) that the generating function $F(z) := Ez^\tau$ satisfies the functional equation(1.2), with

$$Q(w) = \sum_{m=0}^{\infty} q_m w^m.$$

The Lagrange Inversion Formula (1.3) translates as

$$P\{\tau = n\} = n^{-1}P\{S_n = -1\}.$$ (1.8)

Exercise. Give a direct proof of (1.8) using Spitzer's Combinatorial Lemma. A special case of Spitzer's Lemma may be stated as follows:

Lemma 1.1 Let $x_1, x_2, \dots, x_n$ be a sequence of integers ≥ -1 with sum -1. Then there is a unique cyclic permutation π of the integers $1, 2, \dots, n$ such that

$$\sum_{j=1}^{k} x_{\pi(j)} \geq 0 \ \forall \, k = 1, 2, \dots, n - 1.$$ (1.9)

The proof of Lemma 1.1 is another *Exercise*. The trick is to guess where the cycle must begin.

Now consider the asymptotic behavior of the probabilities $P\{\tau = n\}$. There are two trivial cases: (A) If $q_0 + q_1 = 1$, then the random walk S_n makes no jumps to the right: it stays at the starting point 0 for a geometrically-distributed number of steps, then jumps to -1, and so τ has a geometric-plus-one distribution. (B) If $q_0 = 0$ then the jumps of the random walk are all nonnegative, and so $\tau = \infty$. When $q_0 > 0$ and $q_0 + q_1 < 1$, the generating function $Q(w)$ is strictly convex. There are several possibilities, depending on whether the mean $Q'(1)$ of the distribution $\{q_x\}$ is greater, or less, or equal to 1:(C) If $Q'(1) > 1$ then there exists $0 < \rho < 1$ such that $\rho Q'(\rho) = Q(\rho)$; in this case $R = \rho/Q(\rho) > 1$. (D) If $Q'(1) < 1$, and if the radius of convergence of $Q(w)$ is infinite, then there exists $1 < \rho < \infty$ such that $\rho Q'(\rho) = Q(\rho)$, and once again $R = \rho/Q(\rho) > 1$. (E) If $Q'(1) = 1$ then the saddle point is at$\rho = 1$, and $R = 1$. In cases (C), (D), and (E), if the distribution $\{q_x\}$ is *nonlattice* (that is, not supported by a coset of a proper subgroup of $\mathbb{Z}$) then $|Q(w)| < Q(\rho)$ for all $w \neq \rho$ on the contour $|w| = \rho$, and thus, by relation(1.5),

$$P\{\tau = n\} \sim \frac{1}{\sigma R^n n^{3/2}}. \tag{1.10}$$

(Note: In case (E), if the radius of convergence of $Q(w)$ is 1 then the Laplace expansion of the integral (1.4) requires in addition that $Q''(1) < \infty$, that is, that the distribution $\{q_x\}$ has a finite second moment.) Finally, if $Q'(1) < 1$ but the radius of convergence of $Q(w)$ is finite, then there may be no saddle point, in which case the decay of $P\{\tau = n\}$ as $n \to \infty$ may be considerably more complicated.

1.3. SIZE OF A GALTON-WATSON TREE

Let N be the *size* of a Galton-Watson tree for which the offspring distribution is $\{q_x\}_{x=0,1,\ldots}$, that is, N is the total number of individuals (including the progenitor) ever born. This will be finite if and only if the mean offspring number $\mu = \sum x q_x$ is no greater than 1. The random variable N may be decomposed as a sum $1 + N_1 + N_2 + \cdots + N_Z$, where Z is the number of offspring of the progenitor and $N_1, N_2, \ldots$ are the sizes of the Galton-Watson trees engendered by the members of the first generation. These are conditionally independent, given Z, each with the same distribution as N. Therefore, the probability generating function $F(z) = Ez^N$ satisfies the functional equation (1.2), where $Q(w)$ is the probability generating function of the offspring distribution $\{q_x\}$. Because solutions to the Lagrange equation (1.2) are unique, it follows that the generating function $F(z)$ coincides with that of the random variable τ discussed in section

1.2 above; hence, the random variables N and τ have the same distribution. It makes an interesting *Exercise* to prove this directly, without using generating functions.

In the critical case, when $\mu = 1$, the solution ρ of the saddle point equation $\rho Q'(\rho) = Q(\rho)$ is at $\rho = 1$ (since $Q(1) = 1$) and $R = 1$. Consequently, by (1.10), if the offspring distribution is nonlattice then for a suitable constant $C > 0$,

$$P\{N = n\} \sim C/n^{3/2}. \tag{1.11}$$

1.4. MULTITYPE GALTON-WATSON TREES

Next, consider a *multitype* Galton-Watson tree with finitely many types $i = 1, 2, \ldots, I$. For each ordered type i, there is a probability distribution $q_i = \{q_i(m_1, m_2, \ldots, m_I)\}$ on the set $\mathbb{Z}_+^I$ of $I-$vectors of nonnegative integers that governs the numbers of offspring of different types produced by a type-i individual in one generation. Denote by N the size of the tree, that is, the number of individuals of *all* types ever born. Clearly, the distribution of N will depend on the type of the progenitor. Denote by $F_i(z)$ the probability generating function of N when the progenitor has type i; then by a one-step analysis entirely analogous to that used for unitype Galton-Watson trees above, one finds that

$$F_i(z) = zQ_i(F_1(z), F_2(z), \ldots, F_I(z)) \tag{1.12}$$

where $Q_i(w_1, w_2, \ldots, w_I)$ is the (multivariate) generating function of the distribution q_i. This system is evidently of type (1.1).

The first serious attempt to analyze *systems* of functional equations of type (1.1) was made by I. J. Good in the 1950s (see [9]). The primary impetus for Good's study seems to have been the problem we have just discussed – the size of the multitype Galton-Watson tree. Good discovered, among other things, a remarkable generalization of Lagrange's Inversion Formula that has itself engendered a substantial literature. Unfortunately, Good's formula involves determinants, and so, because of the cancellations in the determinants, asymptotic analysis using the saddle point method cannot be carried out in the same manner as in the scalar case. Thus, for the purpose of asymptotic analysis, Good's formula proves to be a wrong turn. In sections 3 and 5 below, we shall present a method for asymptotic analysis of the coefficients of power series $F_i(z)$ that are interrelated by functional equations of type(1.1). Before coming to this, however, we shall discuss the occurrence of such systems of functional equations in random walk problems.

2. Random Walks on Trees and Free Products

Let $\mathcal{T}^d$ be the infinite homogeneous tree of degree $d \geq 3$, with a vertex $\emptyset$ designated as the *root*. Assume that the edges of the tree are assigned *colors* from the set $[d] = \{1, 2, \ldots, d\}$ in such a way that every vertex of the tree is incident to exactly one edge of each color. Given a probability distribution $\{p_i\}_{i \in [d]}$ on the set of colors, a *nearest-neighbor random walk* on $\mathcal{T}^d$ may be performed as follows: At each step, choose a color i at random from the step distribution $\{p_i\}$, independently of all previous choices, and move across the unique edge incident to the current vertex with color i.

The reader will recognize that the random walk just described may also be described as a nearest-neighbor (right) random walk on the free product $\Gamma = \mathbb{Z}_2^{*d}$ of d copies of the two-element group $\mathbb{Z}_2$. (See [3] or [25] for a formal definition of the free product of a collection of groups.) The group elements are finite reduced words whose letters are elements of the set $[d]$ of colors, that is, words in which no color i appears twice consecutively. Note that there is a one-to-one correspondence between reduced words and vertices of $\mathcal{T}^d$: the letters of the word representing a given vertex indicate the unique path from the root $\emptyset$ to the vertex. Multiplication in the group consists of concatenation followed by reduction (successive elimination of adjacent matching pairs of letters). The location X_n of the random walker after n steps is given by

$$X_n = \xi_1, \xi_2, \ldots, \xi_n, \tag{2.1}$$

where $\xi, \xi_2, \ldots$ are i.i.d. with distribution $\{p_i\}$. One may also use equation (2.1) to define *non-nearest-neighbor* random walks: for such walks, the common distribution of the steps ξ_n is no longer restricted to words of length 1. We shall say that a random walk is *finite-range* if the step distribution $\{p_x\}$ is supported by a finite subset of Γ.

Our primary interest is in the asymptotic behavior of the transition probabilities of the random walk X_n, and in particular on the behavior of the return probabilities $P\{X_n = \emptyset\}$. The method of analysis that we shall develop applies generally to all finite-range random walks on $\mathcal{T}^d$, and even to a large class of infinite-range random walks. As we shall see, the method extends also to *countable* free products. However, to present the method in its simplest form, we shall for now restrict attention to the nearest-neighbor case. To avoid complications stemming from periodicity, we shall also assume that the step distribution attaches positive probability $p_\emptyset$ to the empty word $\emptyset$.

Remark. In the nearest-neighbor case, there is another approach to the analysis of transition probabilities: see[8] and [2]. This approach does not generalize to the non-nearest-neighbor case, unfortunately.

2.1. THE GREEN'S FUNCTION

The Green's function of a random walk on a discrete group Γ is defined to be the generating function of the return probabilities:

$$G(z) = \sum_{n=0}^{\infty} P\{X_n = \emptyset\}z^n. \qquad (2.2)$$

In this section, we shall establish some fundamental properties of the Green's function and introduce a denumerable family of auxiliary generating functions to which the Green's function is related by a system of algebraic functional equations.

For any element $x \in \Gamma$, define the generating functions

$$G_x(z) = \sum_{n=0}^{\infty} P\{X_n = x\}z^n \qquad \text{and} \qquad (2.3)$$

$$F_x(z) = \sum_{n=1}^{\infty} P\{\tau_x = n\}z^n, \qquad (2.4)$$

where

$$\tau_x = \min\{n \geq 0 : X_n = x\}. \qquad (2.5)$$

Note that $G = G_\emptyset$. Also, if the random walk is irreducible (that is, for any two states x, y there is at least one positive probability path leading from x to y) then the sums of the series (2.3) and (2.4) are strictly positive (possibly $+\infty$) for all positive arguments z. Because the coefficients of the power series defining the functions G_x and F_x are probabilities, each has radius of convergence at least one; moreover, the first-passage generating functions $F_x(z)$ are bounded in modulus by 1 for all $|z| \leq 1$. In fact, all of the the functions $G_x(z)$ have common radius of convergence R, as we shall see below.

The functions $G_x(z)$ can be expressed in terms of the first-passage generating functions $F_x(z)$ by an application of the Markov property. Conditioning on (i) the first step of the random walk, and then (ii) the value of the first-passage time τ_x, one obtains the relations

$$G(z) = 1 + p_\emptyset z G(z) + \sum_{x \neq \emptyset} p_x z F_{x^{-1}}(z) G(z) \qquad \text{and} \qquad (2.6)$$

$$G_x(z) = F_x(z) G(z) \ \forall x \neq \emptyset. \qquad (2.7)$$

The first of these may be solved for $G(z)$:

$$G(z) = \left\{ 1 - p_\emptyset z - \sum_{x \neq \emptyset} p_x z F_{x^{-1}}(z) \right\}^{-1} \tag{2.8}$$

2.2. THE LAGRANGIAN SYSTEM

The arguments used thus far have been completely general, and do not depend on the particular structure of the group Γ. Assume now that Γ is a free product (finite or denumerable) of copies of $\mathbb{Z}_2$, and that X_n is a nearest-neighbor random walk on Γ. Under this assumption, the first-passage generating functions, which by equation (2.6) determine the Green's function, are themselves inter related by a system of functional equations that derive from the Markov property. Because the random walk is nearest-neighbor, the only values of x that occur in the relation (2.8) are words with a single letter i; consequently, we shall consider only these.

For those random paths that ultimately visit the state i are three possibilities for the first step: X_1 could be $\emptyset$, it could be i, or it could be a one-letter word j for some $j \neq i$. In the last case, the path must first return to $\emptyset$ before visiting i, and so we may condition on the first time at which this happens. This leads to the equations

$$F_i(z) = z \left\{ p_i + p_\emptyset F_i(z) + \sum_{j \neq i} p_j F_j(z) F_i(z) \right\} \tag{2.9}$$

We shall refer to this system as the *Lagrangian system* of the random walk. Observe that it has the generic form (1.1). In addition, the equations may be iterated, by successive re-substitutions on the right sides. If the random walk is irreducible, as we shall assume henceforth, then for any two colors i, j, some iterate of the equation for $F_i(z)$ includes on the right side a term in which $F_j(z)$ occurs as a factor with a positive coefficient. Consequently, all of the functions $F_i(z)$ have the same radius of convergence R, and have the same type of singularity at $z = R$. It follows, by equations (2.7) and (2.8), that all of the functions $G_x(z)$ have a common radius of convergence ρ, and that $\rho \leq R$.

2.3. FINITENESS OF G AT THE RADIUS OF CONVERGENCE

Because our primary interest is in the asymptotic behavior of the coefficients of the Green's function $G(z)$, it behooves us to look further into the relation between its radius ρ of convergence and the radius R of convergence of the first-passage generating functions $F_i(z)$. In principle, the relevant information is already embedded in the functional equations (2.8) and (2.9); however, it is easier for us to exploit a structural property of the group Γ to deduce some restrictions on the nature of the singularities. The structural property is this: Any irreducible random walk on the the the free product Γ of three or more copies of $\mathbb{Z}_2$ must be transient. This follows from a stronger theorem of Kesten [15], which asserts that the *spectral radius R* of the transition kernel of an irreducible random walk must be greater than 1.

Proposition 2.1 *Let R be the common radius of convergence of the first-passage generating functions $F_i(z)$ and let ρ be the radius of convergence of the Green's function $G(z)$. Then $R = \rho$, and*

$$G(R) < \infty. \tag{2.10}$$

Proof. The proof uses the fact that any irreducible random walk on Γ must be transient. First note that for any two elements $x, y \in \Gamma$ there is a positive probability path leading from x to y. Let m be the length of such a path and $c_m(x, y) > 0$ its probability; then for every $n \geq 0$,

$$P\{X_{n+m} = y\} \geq c_m(x, y) P\{X_n = x\}.$$

Consequently, for every positive argument s of the Green's functions,

$$G_y(s) \geq c_m(x, y) s^m G_x(s). \tag{2.11}$$

Suppose now that $G(R) = \infty$; then by the "Harnack" inequalities (2.11), the ratios $G_x(s)/G(s)$ remain bounded, and bounded away from zero, as $s \to R-$. Hence, by a diagonal argument, there is a sequence $s_n \to R-$ such that

$$\varphi_x = \lim_{n \to \infty} \frac{G_x(s_n)}{G(s_n)}$$

exists. The Harnack inequalities guarantee that the function φ is everywhere positive. Furthermore, by equations (2.7), the function φ is $\Gamma-$ invariant, that is, the ratios φ_{xy}/φ_x depend only on y. Most important, the function $x \mapsto \varphi_x$ is $R-$ harmonic, that is, for every $x \in \Gamma$,

$$\varphi_x = R \sum_y \varphi_y P\{\xi_1 = y^{-1}x\}.$$

This follows from the hypothesis that $G(s) \to \infty$ as $s \to R-$, since by the Markov property the Green's function satisfies the relations

$$G_x(z) = \delta_{0,x} + z \sum_y G_y(z) P\{\xi_1 = y^{-1}x\}.$$

Since φ is Γ–invariant and R–harmonic, the Doob h–transform of the transition probability kernel of the random walk X_n by the ratios $R\varphi_x/\varphi_y$ is Γ–invariant. The h–transform is the transition kernel defined by

$$q(x,y) := RP\{\xi_1 = y^{-1}x\}\frac{\varphi_x}{\varphi_y} = q(\emptyset, x^{-1}y).$$

The iterates of this transition probability kernel satisfy

$$q^{(n)}(\emptyset, \emptyset) = R^n P\{X_n = \emptyset\}.$$

Since φ_x is Γ–invariant, the kernel $q(x,y)$ is the transition probability kernel of a right random walk on Γ. The hypothesis that $G(R) = \infty$ and the relation between $q(x,y)$ and the transition kernel of the original random walk now implies that the random walk with transition probability kernel q is recurrent. This contradicts the fact that all irreducible random walks on Γ are transient. $\qquad\square$

Corollary 2.1 *For an irreducible random walk on Γ,*

$$p_{\emptyset} + \sum_{x\neq\emptyset} p_x F_{x^{-1}}(R) < 1/R. \tag{2.12}$$

2.4. BEHAVIOR OFF THE REAL AXIS

Proposition 2.2 *For any* aperiodic, *irreducible random walk on a non amenable discrete group, the Green's function $G(z)$ is regular at every point on the circle of convergence $|z| = R$ except $z = R$.*

Proof. *Exercise.* See [1] for the solution. $\qquad\square$

3. Lagrangian Systems of Functional Equations

3.1. A THEOREM OF FLAJOLET AND ODLYZKO

We have already seen in the case of a scalar Lagrange equation(1.2) that the asymptotic behavior of the coefficients of the solution $F(z)$ is controlled by the behavior of F near its smallest positive singularity $z = R$. The same is true for Lagrangian systems (1.1). To extract the asymptotics from the behavior at the singularity, we shall call upon a Tauberian theorem of *Flajolet* and *Odlyzko* [7].

Theorem 3.1 *(Flajolet & Odlyzko) Let $G(z) = \sum_{n=0}^{\infty} a_n z^n$ be a power series with radius of convergence R. Suppose that G has an analytic continuation to the* Pac-Man *domain*

$$\Delta_{\rho,\phi} = \{z : |z| < \rho \quad and \quad |\arg(z - R)| > \phi\}, \tag{3.1}$$

where $\rho > R$ and $\phi < \pi/2$, and suppose that as $z \to R$ in $\Delta_{\rho,\phi}$,

$$G(R) - G(z) \sim K(R - z)^{\alpha} \tag{3.2}$$

for some $K \neq 0$ and $\alpha \notin \{0, 1, 2, \ldots\}$. Then as $n \to \infty$,

$$a_n \sim \frac{K}{\Gamma(-\alpha)R^n n^{\alpha+1}}. \tag{3.3}$$

Thus, to determine the asymptotic behavior of the coefficients it suffices to determine the nature of the lead singularity.

3.2. LAGRANGIAN SYSTEMS

Suppose now that $F(z)$ is a Banach-space valued analytic function that satisfies a functional equation of the form

$$F(z) = zQ(F(z)), \tag{3.4}$$

where Q is a holomorphic mapping of the Banach space to itself such that $Q(0) \neq 0$. (A mapping $Q : B \to B$ is said to be *holomorphic* if it is infinitely differentiable [in the sense of [28], section 4.5], and if for every analytic function $F : \mathbb{C}^m \to B$ the composition $Q(F)$ is analytic.) We shall refer to systems of the form (3.4) as *Lagrangian systems*. Much of the theory extends, with very little change, to the more general class of functional equations

$$F(z) = Q(z, F(z));$$

however, in the interest of simplicity we shall restrict attention to systems of the more special type (3.4). When the system of generating functions $F_i(z)$ satisfying the equations (1.1) is finite, with (say) I members, then the Banach space of interest is finite-dimensional, to wit, $\mathbb{C}^I$, with the usual norm — this is the case, for instance, for the multitype Galton-Watson tree discussed in section 1.4, and also for nearest-neighbor random walks on *finite* free products of $\mathbb{Z}_2$. More examples will be considered in section 4.2.1 below. For random walks on *infinite* free products, random walks on *finite* free products whose step distribution has infinite support, and multitype Galton-Watson trees with denumerably many types, it is necessary to consider infinite systems (1.1). In such cases the choice of an appropriate Banach space will not always be obvious. See section 6 below for an extended example.

3.3. ANALYTIC CONTINUATION OF A BANACH SPACE-VALUED FUNCTION

Let B be a (complex) Banach space, and let $Q : B \to B$ be a holomorphic mapping. Under what conditions will the functional equation (3.4) have an analytic solution $F(z)$ in a neighborhood of the complex plane containing the origin $z = 0$? Clearly, the equation admits the solution $F(0) = 0$ at $z = 0$. By the Implicit Function Theorem ([28], ch. 4) for Banach-valued functions, this solution admits an analytic continuation to a neighborhood of $z = 0$, as the linearized system

$$dF = Q(F)\, dz + z\frac{\partial Q}{\partial F}\, dF \qquad (3.5)$$

is solvable for dF in terms of dz when $z = 0$. (Here and throughout the notes $\partial Q/\partial F$ or $\partial Q/\partial w$ will denote the Jacobian operator of the mapping Q.) Furthermore, analytic continuation of the solution $F(z)$ is possible along any curve in the complex plane starting at $z = 0$ on which the linear operator

$$I - z\frac{\partial Q}{\partial F} \qquad (3.6)$$

remains invertible. This will be the case as long as the spectral radius of the operator

$$z\mathcal{L}(z) := z\left(\frac{\partial Q}{\partial F}\right)_{F(z)} \qquad (3.7)$$

remains less than one. Thus, singular points of the analytic function $F(z)$ can only occur at those points where the spectral radius of $z\mathcal{L}(z)$ attains or exceeds the value 1.

In general, the dependence of the spectrum of the operator $(\partial Q/\partial F)_{F(z)}$ on the parameter z may be quite complicated, even though the mapping Q and the function $F(z)$ are holomorphic. Moreover, in general the spectrum need not be purely discrete. These complications arise only in the infinite-dimensional case. Therefore, we shall first discuss in detail the finite-dimensional case.

3.4. THE FINITE-DIMENSIONAL CASE

We now restrict attention to the the finite-dimensional case, where the Banach space $B = \mathbb{C}^I$ for some integer $I < \infty$. Discussion of the infinite-dimensional case will be resumed in section 5 below. Assume that the components Q_i of the mapping $Q(w) = Q(w_1, w_2, \ldots w_I)$ are given by convergent power series with nonnegative coefficients. Note that this has been the case in all the examples encountered thus far; moreover, all of the component functions $F_i(z)$ have also been defined by power series with nonnegative coefficients. The following proposition shows that this is no accident.

Proposition 3.1 *If the components $Q_i(w)$ of the mapping Q are given by convergent power series with nonnegative coefficients then the system (3.4) has a unique analytic solution in a neighborhood of $z = 0, F(z) = 0$ whose components $F_i(z)$ are given by convergent power series with nonnegative coefficients.*

Proof. We have already seen that there is a unique analytic solution $F(z)$ in a neighborhood of the origin, by the Implicit Function Theorem, so what must be proved is the non negativity of the power series coefficients of (the components of) $F(z)$. For the purposes of this proof, $F \leq G$ will mean that the power series coefficients of $F(z)$ are dominated by those of $G(z)$. Define a series of approximate solutions to (3.4) by setting $F_0(z) = 0$ and for each $n \geq 0$,

$$F_{n+1}(z) = zQ(F_n(z)). \qquad (3.8)$$

Because the mapping $Q(w)$ is holomorphic, its Jacobian $\partial Q/\partial w$ is uniformly bounded in (sup) norm for $\|w\| \leq C$, for any $C < \infty$; consequently, the mapping which sends a function $H(z)$ to the function $zQ(H(z))$ is contractive for small z and H. Therefore, the sequence of functions F_n defined by (3.8) is uniformly norm convergent for $|z| \leq \varepsilon$, provided $\varepsilon > 0$ is sufficiently small, and the limit function $F(z)$ is the solution to (3.4).

Since the mapping Q is nonnegative, $F_1 \geq F_0 = 0$, and hence, by induction, $F_{n+1} \geq F_n$ for every $n \geq 0$. In particular, each $F_n(z)$ has nonnegative power series coefficients. Since the functions $F_n(z)$ converge uniformly in norm to $F(z)$, the Cauchy Integral Formula implies that $F(z)$ has nonnegative power series coefficients. $\qquad \square$

Corollary 3.1 *Under the hypotheses of Proposition 3.1, the entries of the matrix-valued function $\mathcal{L}(z) = (\partial Q/\partial w)_{F(z)}$ are given by convergent power series with nonnegative coefficients. Therefore, for positive arguments s, the entries of $\mathcal{L}(s)$ are nonnegative, nondecreasing, and convex in s. Moreover, the spectral radius $\lambda(s)$ is nondecreasing and continuous in s for nonnegative arguments s. Finally,*

$$|\lambda(z)| \leq \lambda(|z|), \qquad (3.9)$$

and so the function $F(z)$ has an analytic continuation to any disc $|z| \leq r$ such that $r\lambda(r) < 1$.

For positive arguments s, the matrix $\mathcal{L}(s)$ has nonnegative entries, and so its spectral radius coincides with its largest nonnegative eigenvalue. In general, this eigenvalue need not be simple, nor need it be the only eigenvalue on the circle of radius $\lambda(s)$ centered at 0. However, if $\mathcal{L}(s)$ is a *Perron-Frobenius* matrix (a nonnegative matrix M such that some power M^n has all entries positive), then by the Perron-Frobenius theorem ([22]) the eigenvalue $\lambda(s)$ must be simple, and there can be no other eigenvalue of modulus $\lambda(s)$.

Define the *dependency graph* of the system (3.4) to be the directed graph $\mathcal{G}$ with vertex set $[I] := \{1, 2, \ldots, I\}$ such that for each pair (i, j) of vertices there is a directed edge from j to i if and only if the variable w_j occurs as a factor in a term of $Q_i(w)$ with positive coefficient. Let A be the incidence matrix of $\mathcal{G}$, that is, the $I \times I$ matrix whose (i, j)th entry is 1 if there is a directed edge from i to j and 0 otherwise. Say that the system (3.4) is *irreducible aperiodic* if the incidence matrix A of its dependency graph is Perron-Frobenius. Clearly, the (i, j)th entry of $\mathcal{L}(s)$ is positive for $s > 0$ if and only if the (i, j)th entry of A is positive, and so $\mathcal{L}(s)$ is a Perron-Frobenius matrix if and only if A is.

Proposition 3.2 *Assume that the components $Q_i(w)$ of the mapping Q are given by convergent power series with nonnegative coefficients, and assume further that the Lagrangian system (3.4) is irreducible and aperiodic. Then the matrix $\mathcal{L}(s)$ is Perron-Frobenius for all $s > 0$ such that the spectral radius $\lambda(s) \leq 1/s$. Furthermore, the lead eigenvalue $\lambda(z)$ and the corresponding right and left eigenvalues $h(z)$ and $w(z)$, normalized so that*

$$w(z)^T h(z) = w(z)^T \mathbf{1} = 1, \tag{3.10}$$

are analytic in a neighborhood of the segment $[0, R)$ and continuous on $[0, R]$, where

$$R = \min\{s : s\lambda(s) = 1\}. \tag{3.11}$$

Finally, if $Q_i(0) > 0$ for some index i then $R < \infty$ and $F(R)$ is finite.

Proof. That $\mathcal{L}(s)$ is Perron-Frobenius for $0 < s \leq R$ follows from the discussion preceding the statement of the proposition. Because the mapping $z \mapsto \mathcal{L}(z)$ is analytic for $|z| < R$, and because the lead eigenvalue $\lambda(s)$ is simple for $0 < s \leq R$, by the Perron-Frobenius theorem, results of regular perturbation theory (see [21], section XII.2) imply that the mappings

$$
\begin{aligned}
z &\mapsto \lambda(z) \\
z &\mapsto h(z), \quad \text{and} \\
z &\mapsto w(z)
\end{aligned}
$$

have analytic continuations to a neighborhood of $(0, R)$, and are continuous at $z = R$. Now suppose that $Q_i(\mathbf{0}) > 0$ for some i; then by (3.5), the derivative $F_i'(0) > 0$, and so $F_i(s) > 0$ for all $0 < s \leq R$. Since the system (3.4) is irreducible and aperiodic, it follows that $F_j(s) > 0$ for all indices j and $0 < s \leq R$. Consequently, the Jacobian matrix $\mathcal{L}(s) = (\partial Q / \partial w)_{F(s)}$ is not identically zero, and so the spectral radius of $s\mathcal{L}(s)$ cannot remain bounded below 1 for all $s > 0$; thus, $R < \infty$. Finally, $F(s)$ must remain bounded as $s \to R-$ because (a) if $Q(w)$ is linear then the components of $F(s)$ grow at most quadratically in s, by (3.4); and (b) if $Q(w)$ is nonlinear then $(\partial Q / \partial w)_{F(s)}$ has entries that grow at least linearly in $F(s)$, and so it would be impossible for $F(s)$ to grow unboundedly without having the spectral radius $s\lambda(s)$ exceed the value 1. $\qquad\square$

We are now in a position to determine the nature of the singularity at $s = R$:

Theorem 3.2 *Assume that the components $Q_i(w)$ of the mapping Q are given by convergent power series with nonnegative coefficients, and assume further that the Lagrangian system (3.4) is irreducible and aperiodic. If $Q_i(\mathbf{0}) > 0$ for some index i, and if the mapping $Q(w)$ is not linear, then for each index i there exists $C_i > 0$ such that*

$$F_i(R) - F_i(z) \sim C_i\sqrt{R - z} \qquad \text{as } z \to R. \qquad (3.12)$$

Proof. Denote by h and w the right and left Perron-Frobenius eigenvectors of $R\mathcal{L}(R)$; by the Perron-Frobenius theorem, the entries of h and w are strictly positive. Define projection operators P_U, P_V by

$$\begin{aligned} P_U x &= (w^T x)h \\ P_V x &= x - P_U x. \end{aligned}$$

Denote by $\mathcal{V}$ the range of the projection P_V. Observe that because all points of the spectrum of $R\mathcal{L}(R)$ other than the simple eigenvalue 1 are of modulus < 1, the restriction of the operator $I - P_V R\mathcal{L}(R)$ to the subspace $\mathcal{V}$ is invertible. We will show that as $z \to R$,

$$P_U(F(R) - F(z)) = \{C\sqrt{R - z}\}h + o(\sqrt{R - z}) \qquad \text{and} \qquad (3.13)$$

$$P_V(F(R) - F(z)) = O(R - z) \qquad (3.14)$$

where $C > 0$. The desired relations (3.12) will then follow, by the strict positivity of the eigenvectors w and h, since the functions $F_i(s)$ are positive for positive arguments s.

The key to relations (3.13)–(3.14) is that the linearization (3.5) of the Lagrangian equations (3.4) becomes singular at $z = R$: in particular, it cannot be solved for dF in terms of dz. Hence, we shall look a thigher order terms in the expansion of (3.4) around $z = R$. Write

$$
\begin{array}{llll}
V(z) & = & P_V F(z), & \Delta V & = & V(R) - V(z), \\
U(z) & = & P_U F(z) = (w^T F(z))h, & \Delta U & = & U(R) - U(z), \\
u(z) & = & w^T F(z), & \Delta F & = & F(R) - F(z), \\
& \text{and} & & \Delta z & = & R - z.
\end{array}
$$

The functional equation (3.4) and Taylor's theorem (in its multivariate form) imply that

$$
\Delta F = (F(R)/R)\Delta z + R\mathcal{L}(R)\Delta F - \mathcal{H}(\Delta F, \Delta F) + \text{higher order terms},
\tag{3.15}
$$

where each component of $\mathcal{H}$ is a symmetric bilinear quadratic form. Because the power series coefficients of the mapping Q are nonnegative, so are the coefficients of $\mathcal{H}$, and because Q is nonlinear, at least some of these coefficients are positive. Applying the projection operators P_U, P_V to equation (3.15) and using the bilinearity of $\mathcal{H}$ yields

$$
\Delta U = (U(R)/R)\Delta z + P_U R\mathcal{L}(R)\Delta U - \mathcal{H}(\Delta U, \Delta U) + O(|\Delta U||\Delta V| + |\Delta V|^2)
\tag{3.16}
$$

and

$$
\Delta V = (V(R)/R)\Delta z + P_V R\mathcal{L}(R)\Delta V + O((\Delta U)^2) + O(|\Delta U||\Delta V| + |\Delta V|^2).
\tag{3.17}
$$

Since P_U is the projection onto the Perron-Frobenius eigenspace, the second term on the right side of (3.16) cancels the left side, and so the equation may be rewritten as

$$
\kappa(\Delta u)^2 = (u(R)/R)\Delta z + \text{higher order terms}.
\tag{3.18}
$$

Observe that the higher order terms are $O(|\Delta u||\Delta V| + |\Delta V|^2)$ or $o(\Delta z)$; also, because the entries of the eigenvector w are positive, $u(R) > 0$. Because the restriction of $I - P_V R\mathcal{L}(R)$ to the subspace $\mathcal{V}$ is invertible, the linearization of (3.17) can be solved for ΔV in terms of Δz, and so

$$
|\Delta V| = O(|\Delta z| + |\Delta u|^2).
\tag{3.19}
$$

Thus, to prove relations (3.13) and (3.14), it suffices to show that the constant κ in (3.18) is positive. But this follows from the fact, noted above, that the coefficients of the Hession $\mathcal{H}$ are positive, because each entry ΔF_i of ΔF has the form

$$\Delta F_i = (\Delta u)h_i + \text{a linear function of } \Delta V.$$

$\square$

The Tauberian theorem of Flajolet and Odlyzko now yields the following corollary.

Corollary 3.2 *Under the hypotheses of Theorem 3.2, the coefficients of the power series* $F_i(z) = \sum_n a_{n,i} z^n$ *obey the asymptotic relations*

$$a_{n,i} \sim C_i/R^n n^{3/2}. \tag{3.20}$$

3.5. CONSEQUENCE: LOCAL LIMIT THEOREM FOR NNRW ON $\mathcal{T}^d$

As a first example, consider nearest-neighbor random walk on a homogeneous tree $\mathcal{T}^d$, or equivalently, on the free product of d copies of $\mathbb{Z}_2$. Assume that the holding probability $p_\emptyset > 0$, so that the random walk is aperiodic; then the Green's function $G(z)$ is regular at every point on the circle of convergence $|z| = R$ except at $z = R$. Assume also that the nearest-neighbor transition probabilities $p_i > 0$, so that the random walk, and the corresponding Lagrangian system (2.9), are irreducible. Then by Theorem 3.2, the first-passage generating functions $F_i(z)$ have square-root singularities (3.12), and so by Corollary 3.2 their coefficients satisfy

$$P\{\tau_i = n\} \sim C_i/R^n n^{3/2}. \tag{3.21}$$

Furthermore, by Proposition 2.1, the Green's function is finite at $z = R$, and so by (2.8), it too must have a square-root singularity, that is,

$$G(R) - G(z) \sim \kappa\sqrt{R - z}. \tag{3.22}$$

Therefore, by Corollary 3.2,

$$P\{X_n = \emptyset\} \sim C/R^n n^{3/2}. \tag{3.23}$$

This was first proved by Gerl and Woess [8], using somewhat different methods.

4. Random Walks on Regular Languages

4.1. DEFINITION

A random walk on a free product takes values in the set of all finite words whose letters are elements of the set of indices of the groups in the free product. It evolves by successive modifications of the letters at the end of the word (in the nearest-neighbor case, these are just deletions or adjunctions), where the modifications are determined by random inputs (in the nearest-neighbor case, words of length 0 or 1). The rules that determine how a particular input is used to modify the current state of the random walk are just the group multiplication laws. These rules are *local*: for nearest-neighbor random walks, they use the last letter of the current state; and for finite-range random walks, where the step distribution is supported by the set of words of length $\leq K$, the rules use only the last K letters of the current state.

A natural and useful generalization is to Markov chains on the set of finite words over a finite or denumerable alphabet which evolve in the manner just described, *except* that the rules determining how inputs are used to modify the current state are no longer given by a group multiplication law. We shall call such Markov chains *random walks on regular languages*, for reasons that we shall explain shortly. Roughly, a random walk on a regular language is a Markov chain on the set of all finite words from a finite alphabet A whose transition probabilities obey the following rules: (1) Only the last two letters of a word may be modified in one jump, and almost one letter may be adjoined or deleted. (2) Probabilities of modification, deletion, and/or adjunction depend only on the last two letters of the current word. More precisely, a RWRL is a Markov chain whose transition probabilities satisfy:

$$P(X_{n+1} = x_1 x_2 \ldots x_m a'b' \mid X_n = x_1 x_2 \ldots x_m ab) = p(a'b'|ab)$$

$$P(X_{n+1} = x_1 x_2 \ldots x_m a' \mid X_n = x_1 x_2 \ldots x_m ab) = p(a'|ab)$$

$$P(X_{n+1} = x_1 x_2 \ldots x_m a'b'c \mid X_n = x_1 x_2 \ldots x_m ab) = p(a'b'c|ab)$$

$$P(X_{n+1} = a'b' \mid X_n = a) = p(a'b'|a)$$

$$P(X_{n+1} = a' \mid X_n = a) = p(a'|a) \tag{4.1}$$

$$P(X_{n+1} = \emptyset \mid X_n = a) = p(\emptyset|a)$$

$$P(X_{n+1} = a \mid X_n = \emptyset) = p(a|\emptyset) \qquad \text{and}$$

$$P(X_{n+1} = \emptyset \mid X_n = \emptyset) = p(\emptyset|\emptyset).$$

Here $\emptyset$ denotes the empty word. We do *not* assume that the transition probabilities are all positive, nor do we assume that there exist positive-probability paths connecting any two words. Later, however, we shall impose certain further restrictions regarding aperiodicity and irreducibility.

Denote by $\mathcal{L}$ the set of all words that are accessible from the root $\emptyset$ by positive probability paths. It is not difficult to see that $\mathcal{L}$ is a *regular language*, that is, there is a finite-state automaton $\mathcal{M}$ such that $\mathcal{L}$ is the set of all words accepted by $\mathcal{M}$. (See [13] for background on finite-state automata and regular languages.) This explains the term *random walk on regular language*.

4.2. EXAMPLES

4.2.1. *Finite-Range Reflecting Random Walks on $\mathbb{Z}_+$*
Consider a Markov chain on $\mathbb{Z}_+$ with transition probabilities

$$P(X_{n+1} = x + k \mid X_n = x) = p_k \qquad \text{for } |k| \leq K \text{ and } x \geq K;$$
$$P(X_{n+1} = x + k \mid X_n = x) = p_{x,k} \qquad \text{for } 0 \leq x < K \text{ and } -x \leq k \leq K.$$
$$(4.2)$$

Such a process behaves as a (homogeneous) random walk with increments bounded by K outside a finite neighborhood $\{0, 1, \cdot, K-1\}$ of the origin. It is not difficult to see that a Markov chain governed by transition probabilities (4.2) is equivalent to a random walk on the regular language

$$\mathcal{L} = \{a^m b \mid m \in \mathbb{Z}_+ \text{ and } b \in \{0, 1, 2, \cdots, K - 1\}\}.$$

The word $a^m b$ corresponds to the integer $mK + b$. Since the increments in X_n are of magnitude $\leq K$, only the last two letters of the representing word $aa \cdots ab$ need be modified to account for any possible change of state. Local limit theorems for Markov chains with transition probabilities (4.2) were proved, by Wiener-Hopf methods, in [17].

4.2.2. *Finite-Range Random Walks on Homogeneous Trees*
Let X_n be a finite-range random walk on the free product $\Gamma = \mathbb{Z}_2^d$, as defined in section 2 above. Assume that the step distribution has support contained in the set of all words of length $\leq K$ from the alphabet $[d]$. Then the random walk X_n admits a description as a random walk on the regular language consisting of finite words $a_1 a_2 \ldots a_m b$, with $m \geq 0$, such that (i) each a_i is a word of length K in the letters $[d]$; (ii) b is a word of length $0 \leq |b| \leq K - 1$; and (iii) no cancellations between successive a_i, a_{i+1} (or between a_m, b) are possible.

Observe that if the transition probabilities are modified at finitely many elements of Γ, then the resulting Markov chain, a *perturbed random walk*, although no longer a "random walk" (in the usual sense, that is, that the transition probabilities are group-invariant), is nevertheless still a random walk on a regular language. We will see that the transition probabilities of such modified random walk satisfy local limit theorems, but not necessarily with the same $3/2$-power law as in (3.23).

4.2.3. Random Walk on the Modular Group $PSL(2, \mathbb{Z})$

The *modular group* is the quotient group $PSL(2, \mathbb{Z}) = SL(2, \mathbb{Z})/\{\pm I\}$ where $SL(2, \mathbb{Z})$ is the (multiplicative) group of 2×2 matrices with integer entries and determinant 1, and I is the 2×2 identity matrix. (For information about the modular group and its connections with the theory of elliptic modular forms, and also for specific facts used in the subsequent discussion, see [19], especially Chapter XI.) A *(right) random walk* on $PSL(2, \mathbb{Z})$ is a sequence

$$X_n = \xi_1 \xi_2 \ldots \xi_n \tag{4.3}$$

where $\xi_1 \xi_2 \ldots$ are i.i.d. random variables valued in $PSL(2, \mathbb{Z})$ whose distribution has finite support. We will show that any such random walk has a description as a random walk on a regular language: The modular group is finitely generated, with generators

$$U = \begin{pmatrix} 1 & 1 \\ 0 & 1 \end{pmatrix}, \ U^{-1} = \begin{pmatrix} 1 & -1 \\ 0 & 1 \end{pmatrix}, \ \text{and } T = T^{-1} = \begin{pmatrix} 0 & -1 \\ 1 & 0 \end{pmatrix}. \tag{4.4}$$

(More accurately, the group is generated by the *equivalence classes* $\langle T \rangle := \{\pm T\}$ and $\langle U \rangle := \{\pm U\}$. Henceforth, we shall not distinguish between matrices and their equivalence classes in $PSL(2, \mathbb{Z})$.) These generators satisfy the relations

$$T^2 = (TU)^3 = I. \tag{4.5}$$

Thus, every element of $PSL(2, \mathbb{Z})$ may be written as a finite word in the letters T, U, U^{-1} in which no two T's appear consecutively, and no U is adjacent to a U^{-1}. It follows that the modular group is also generated by T, W, W^2, where $W = TU$. The relations (4.5) translate as $T^2 = W^3 = I$. Every element $M \in PSL(2, \mathbb{Z})$ has are presentation as a finite word

$$M = T^a W^{n_1} T W^{n_2} T W^{n_3} \cdots T W^{n_k} T^b \tag{4.6}$$

where $a, b = 0$ or 1 and each n_i is either 1 or 2.

The modular group contains free normal subgroups of finite index, and so it has a representation as a regular language. Following is a sketch of how such a representation may be obtained (see [19], section XI.3E for details). Let Γ' be the *commutator subgroup* of $PSL(2, \mathbb{Z})$, that is, the subgroup consisting of all finite products of *commutators*. (The *commutators* of any group are the elements of the form $aba^{-1}b^{-1}$.) The commutator subgroup of any finite or countable group is a normal subgroup, and so Γ' is a normal subgroup of $PSL(2,\mathbb{Z})$. The group Γ' is free (see [19], Theorem XI.3E), with generators

$$X = TWTW^2, \qquad\qquad X^{-1} = WTW^2T,$$
$$Y = TW^2TW, \qquad \text{and} \qquad Y^{-1} = W^2TWT. \qquad (4.7)$$

The index of Γ' in $PSL(2,\mathbb{Z})$ (that is, the number of distinct cosets of Γ') is 6. Thus, since Γ' is a normal subgroup, one may choose 6 distinct elements $A_1, A_2, \ldots, A_6$ of $PSL(2,\mathbb{Z})$, one from each coset, such that every element of $PSL(2,\mathbb{Z})$ has a representation

$$M = WA_i, \qquad (4.8)$$

where W is a reduced word in the (free) generators X, X^{-1}, Y, Y^{-1}. Observe that there may be some cancellation of letters at the end of W; but since there are only 6 possibilities for A_i, there are only finitely many possible types of cancellation, involving at most a bounded number of letters at the end of W. Thus, there is a finite subset $\mathcal{B} \subset PSL(2,\mathbb{Z})$ such that every element of $PSL(2,\mathbb{Z})$ has a representation as a *reduced* word

$$M = W'B_j, \qquad (4.9)$$

where W' is a reduced word in X, X^{-1}, Y, Y^{-1} and $B_j \in \mathcal{B}$. This exhibits $PSL(2,\mathbb{Z})$ as a regular language $\mathcal{L}$, with alphabet $\mathcal{A} = \{X, X^{-1}, Y, Y^{-1}\} \cup \mathcal{B}$.

Now consider a right random walk (4.3) whose jumps ξ_i are i.i.d. from a distribution on $PSL(2,\mathbb{Z})$ with finite support. Because there are only finitely many possible values of ξ_{n+1}, there are only finitely many different ways that multiplication by ξ_{n+1} can modify the end of the word X_n. In particular, there is a finite integer K such that all positive-probability transitions $X_n \to X_{n+1}$ involve only the last K letters of X_n, and such that the transition probabilities depend only on the last K letters of the current state X_n. By replacing the regular language $\mathcal{L}$ by a regular language $\mathcal{L}'$ whose alphabet consists of the words of length $\leq K$ in $\mathcal{L}$, one obtains a representation of the random walk X_n as a random walk on a regular language.

A local limit theorem for *nearest-neighbor* random walk on the modular group was proved by Woess [26], using different and somewhat more difficult arguments than those employed here. Later [27], Woess also proved that, for any nearest-neighbor random walk on a discrete group containing a free normal subgroup of finite index, the Green's function is algebraic; however, his arguments do not allow one to deduce the type of the leading singularity.

4.3. THE GREEN'S FUNCTION OF A RWRL

Assume now that X_n is a random walk on a regular language $\mathcal{L}$. For $x, y \in \mathcal{L}$, define the *Green's function(s)* $G_{xy}(z)$ as follows:

$$G_{xy}(z) := \sum_{n=0}^{\infty} P^x\{X_n = y\} z^n = E^x \left(\sum_{n=0}^{\infty} z^n \mathbf{1}\{X_n = y\} \right). \qquad (4.10)$$

Thus, $G_{xy}(z)$ is the expected discounted number of visits to y given that the initial state is x, where z is the discount factor. Since the coefficients in the power series are all probabilities, the series converge absolutely and uniformly in the closed unit disc $|z| \leq 1$, and so the radius of convergence is at least 1. Later we shall prove that, under suitable irreducibility hypotheses, the power series of all of the Green's functions $G_{xy}(z)$ have the same radius R of convergence.

The Green's functions $G_{xy}(z)$ are determined by a finite system of generating functions $H_{ab,c}(z)$, the *reentry* generating functions, indexed by words ab and c of lengths 1 and 2, respectively. These are defined as follows:

$$H_{ab,c}(z) := E^{ab} z^T \mathbf{1}\{X_T = c\} = \sum_{n=1}^{\infty} z^n P^{ab}\{T = n \text{ and } X_n = c\} \qquad (4.11)$$

where

$$T := \min\{n : |X_n| < |X_0|\}. \qquad (4.12)$$

Observe that some — or even all — of the generating functions $H_{ab,c}(z)$ may be identically zero, as there may be no positive-probability paths from ab to c through words of length ≥ 2. Those that are not identically zero we shall refer to as the *nondegenerate* reentry generating functions.

The relationship between the Green's functions and the reentry generating functions $H_{ab,c}(z)$ is a routine consequence of the Markov property. Let x and y be words of lengths $|x|, |y| \leq 1$. If $X_0 = x$, then X_1 must be a word of length $0, 1$, or 2; if X_1 is a word of length 2, then there can be no more visits to y until after the Markov chain X_n returns from the set of words of length ≥ 2. Hence, by the Markov property, if $|x| \leq 1$ and $|y| \leq 1$, then

$$G_{xy}(z) = \delta_x(y) + z \sum_{w : |w| \leq 1} p(w|x) G_{wy}(z) + z \sum_{ab} \sum_c p(ab|x) H_{ab,c}(z) G_{cy}(z),$$

$$(4.13)$$

where $\sum_{ab}$ is over all two-letter words, and $\sum_c$ is overall one-letter words.

The equations (4.13) may be written compactly in matrix form as follows. Define $\mathbf{G}(z)$ to be the matrix-valued function of z with entries $G_{xy}(z)$, indexed by words x, y of lengths ≤ 1, and define $\mathbf{P}$ to be the matrix of transition probabilities $p(y|x)$, where again x and y are words of length ≤ 1. Define $\mathbf{K}(z)$ to be the matrix-valued function of z with entries $K_{xy}(z)$, indexed by words x, y of length ≤ 1, defined by

$$
\begin{aligned}
K_{xy}(z) &= \textstyle\sum_{ab} p(ab|x) H_{ab,y}(z) && \text{if } |y| = 1; \\
&= 0 && \text{if } |y| = 0.
\end{aligned}
\tag{4.14}
$$

Then equations (4.13) are equivalent to the matrix equations

$$
\begin{aligned}
\mathbf{G}(z) &= \mathbf{I} + z\mathbf{P}\mathbf{G}(z) + z\mathbf{K}(z)\mathbf{G}(z) \qquad \Longleftrightarrow \\
\mathbf{I} &= (\mathbf{I} - z\mathbf{P} - z\mathbf{K}(z))\,\mathbf{G}(z)
\end{aligned}
\tag{4.15}
$$

where $\mathbf{I}$ is the identity matrix. Notice that for all z of sufficiently small modulus $|z|$, the matrix $(z\mathbf{P} + z\mathbf{K}(z))$ has norm less than one, and so the matrix $(\mathbf{I} - z\mathbf{P} - z\mathbf{K}(z))$ is invertible; thus,

$$
\mathbf{G}(z) = (\mathbf{I} - z\mathbf{P} - z\mathbf{K}(z))^{-1}
\tag{4.16}
$$

for all z in a disc centered at $z = 0$.

4.4. THE LAGRANGIAN SYSTEM OF A RWRL

Recall (see equations (4.1)) that, for a random walk on a regular language, only the last two letters of the current word play a role in the transition to the next word. Consequently, for any initial state $\xi = x_1 x_2 \ldots x_m ab \in \mathcal{L}$, the letters $x_1 x_2 \ldots x_m$ are ignored up to time T. It follows that for any word $\xi = x_1 x_2 \ldots x_m ab \in \mathcal{L}$, and any $c \in A$,

$$
H_{ab,c}(z) = E^{\xi} z^{T} \mathbf{1}\{X_T = x_1 x_2 \ldots x_m c\}.
\tag{4.17}
$$

The homogeneity relations (4.17) lead directly to a system of quadratic equations relating the functions $H_{ab,c}$. These equations derive from the Markov property: Starting from any word w, the random walk must make an initial jump to another state. If this first jump is to a word whose length is less than that of the initial word w, then $T = 1$; otherwise, the initial jump must be either to a word w' of the same length, or to a word w'' with one additional letter. In the latter case, T is the first time after the initial jump that *two* letters are removed; in the former, T is the first time that a single letter is removed. Summing over all possible initial jumps and applying the identity (4.17) yields the following system of equations:

$$
\begin{aligned}
H_{ab,c}(z) = {}& zp(c|ab) + zp(ab|ab)H_{ab,c}(z) + z \sum_{d,e \in A} p(de|ab)H_{de,c}(z) \\
& + z \sum_{d,e,f \in A} p(def|ab) \sum_{g \in A} H_{ef,g}(z)H_{dg,c}(z),
\end{aligned}
\tag{4.18}
$$

We shall abbreviate this system of equations as

$$\mathbf{H}(z) = z\mathbf{Q}(\mathbf{H}(z)) \tag{4.19}$$

where $\mathbf{H}(z)$ is the vector of *nondegenerate* first-passage generating functions $H_{ab,c}(z)$ (that is, those that are *not* identically zero), and $\mathbf{Q}$ is the vector of quadratic polynomials $Q_{ab,c}$ in the variables $H_{ab,c}$ occurring on the right sides of the equations (4.18) above. Observe that $\mathbf{Q}$ is quadratic (that is, each of its components is quadratic) and its coefficients are nonnegative. Consequently, the solution set of (4.19) is an *algebraic curve*, and so the functions $H_{ab,c}(z)$ are *algebraic functions*. Most important, the system (4.19) is of the generic form (3.4), and so the results of section 3 apply.

The hypotheses of Theorem 3.2 require that the Lagrangian system be irreducible and aperiodic, that is, that the incidence matrix of the dependency graph be a Perron-Frobenius matrix. Following is a relatively simple set of conditions on the transition probabilities of the RWRL that will guarantee that the Lagrangian system (4.18) is irreducible and aperiodic. These are not in any sense minimal, nor do they cover all cases of the examples discussed in section 4.2 above.

Assumption 4.1 (Aperiodicity) *For all words* $x_1 x_2 \ldots x_m$,

$$P(X_{n+1} = x_1 x_2 \ldots x_m \mid X_n = x_1 x_2 \ldots x_m) > 0 \tag{4.20}$$

This is equivalent to assuming that all of the transition probabilities $p(ab \mid ab), p(a \mid a)$, and $p(\emptyset \mid \emptyset)$ are positive.

Assumption 4.2 (Irreducibility-A) *Let* $\mathcal{L}$ *be the set of all words that may be reached from* $\emptyset$ *via positive-probability paths. Then* $\forall w \in \mathcal{L}$ *there is a positive-probability path from* w *to* $\emptyset$.

Assumption 4.3 (Irreducibility-B) *For every word* $x = x_1 x_2 \ldots x_m \in \mathcal{L}$ *of length 2 (or longer) and every triple abc of letters there exist words*

$$w_1 = x_1 x_2 \ldots x_m y_1 y_2 \ldots y_n ab \in \mathcal{L} \quad \text{and}$$
$$w_2 = x_1 x_2 \ldots x_m y_1 y_2 \ldots y_n c \in \mathcal{L}$$

and positive probability paths

$$x \to w_1 \to w_2 \to x$$

through words prefaced by x, w_1, *and* x, *respectively.*

Proposition 4.1 *If the transition probabilities of a RWRL satisfy Assumptions 4.1, 4.2, and 4.3 above, then the Lagrangian system (4.18) is irreducible and aperiodic. Furthermore, the mapping* $\mathbf{Q}$ *will be nonlinear and will satisfy the restriction* $\mathbf{Q}(\mathbf{0}) \neq \mathbf{0}$.

Proof. Assumption 4.1 ensures that the incidence matrix of the dependency graph will have positive entries on the diagonal (because the second term on the right side of (4.18) will be positive). Hence, the system will be aperiodic. Assumption 4.3 guarantees that, for each pair of triples $a'b'c'$ and abc it is possible to make a succession of substitutions in (4.18) (in the last sum) so as to obtain an equation for $H_{a'b',c'}(z)$ in which $H_{ab,c}(z)$ appears as a factor in a term with positive coefficient. Thus, the system will be irreducible. Moreover, the right side of (4.18) will include quadratic terms with positive terms, so $\mathbf{Q}$ will be nonlinear. Finally, Assumption 4.2 implies that at least some of the transition probabilities $p(c|ab)$ will be positive, so $\mathbf{Q}(0) \neq 0$ (because the first term on the rightside of at least one of the equations (4.18) will be positive). □

Corollary 4.1 *Under Assumptions 4.1, 4.2, and 4.3, for every triple of letters abc, there are constants $C_{ab,c} > 0$ so that*

$$P^{ab}\{T = n\,;\, X_T = c\} \sim \frac{C_{ab,c}}{\underline{\rho}^n n^{3/2}}, \qquad (4.21)$$

where $\underline{\rho} \geq 1$ is the smallest positive singularity of the Lagrangian system (4.18).

4.5. LOCAL LIMIT THEOREMS FOR RWRL

For nearest-neighbor random walks on homogeneous trees, the spectral radius R (the radius of convergence of the Green's function) coincides with the smallest positive singularity ρ of the associated Lagrangian system. For random walks on regular languages, this need not be the case. Furthermore, the lead singularity of the Green's function need not be a square-root singularity, as it is for the Lagrangian system, and consequently the return probabilities need not obey a 3/2−power law, as they must for random walks on homogeneous trees. In fact, there are five distinct possibilities. These are enumerated in the following theorem.

Theorem 4.1 *Assume that the transition probabilities of the RWRL X_n satisfy the assumptions 4.1, 4.2, and 4.3. Then the spectral radius R of the random walk is bounded above by the lead singularity ρ of the associated Lagrangian system. The random walk may be positive recurrent, null recurrent, or transient. In the null recurrent case, $R = 1$ and for any two words $x, y \in \mathcal{L}$ there exists a positive constants C_{xy} such that*

$$P^x\{X_n = y\} \sim C_{xy} n^{-1/2}. \qquad (4.22)$$

In the transient case, $R > 1$, and for suitable positive constants C_{xy} one of the following laws holds for all pairs $x, y \in \mathcal{L}$:

$$P^x\{X_n = y\} \sim C_{xy} R^{-n}; \quad or \tag{4.23}$$

$$P^x\{X_n = y\} \sim C_{xy} R^{-n} n^{-1/2}; \quad or \tag{4.24}$$

$$P^x\{X_n = y\} \sim C_{xy} R^{-n} n^{-3/2}. \tag{4.25}$$

All five cases are possible. This may be verified by considering nearest-neighbor reflecting random walks on the nonnegative integers $\mathbb{Z}_+$, whose Green's functions may be obtained in closed form by solving quadratic equations (*Exercise!*).

Proof of Theorem 4.1 (Sketch) We shall consider only words $x, y \in \mathcal{L}$ of length 1, and leave as an exercise for the reader to complete the argument. First, observe that the Green's functions $G_{xy}(z)$ all have the same radius of convergence. This follows from the irreducibility assumptions 4.2–4.3, which guarantee that there exist positive-probability paths from x to y and back, because this implies that there is a positive constant c and integers k, l such that for all $n \geq 0$,

$$\begin{aligned}
P^x\{X_{n+k} = y\} &\geq cP^x\{X_n = x\}, \\
P^x\{X_{n+k} = y\} &\geq cP^y\{X_n = y\}, \\
P^x\{X_{n+l} = x\} &\geq cP^x\{X_n = y\}, \qquad \text{and} \\
P^x\{X_{n+k} = x\} &\geq cP^y\{X_n = x\}.
\end{aligned}$$

Recall that the matrix $\mathbf{G}(z)$ of Green's functions $G_{xy}(z)$ indexed by words of length 1 is determined by the reentry generating functions by equations (4.16). The matrix $\mathbf{P}$ in (4.16) is constant, and the entries of $\mathbf{K}(z)$ are positive linear combinations of the reentry generating functions $H_{ab,c}(z)$; consequently, the radius R of convergence of $\mathbf{G}(z)$ cannot exceed the radius ρ of convergence of $\mathbf{H}(z)$. In particular, if the spectral radius of $\rho\mathbf{P} + \rho\mathbf{K}(\rho)$ is less than one, then (4.16) implies that $\mathbf{G}(z)$ must have the same type of singularity — a square-root singularity – at $z = \rho$ as does $\mathbf{H}(z)$. This implies that $R = \rho \geq 1$ and that the asymptotic relations (4.25) must hold, by the Tauberian theorem of Flajolet and Odlyzko.

It is also possible that the spectral radius of $s\mathbf{P} + s\mathbf{K}(s)$ attains the value 1 at some $s = R < \rho$. If $R < \rho$, then since $\mathbf{K}(z)$ is analytic at $z = \rho$, the function $\mathbf{G}(z)$ will have a pole at $z = R$. This pole must be simple, because $\mathbf{P} + \mathbf{K}(z)$ is a Perron-Frobenius matrix at $z = R$. Thus, in this case, the asymptotic relations (4.23) hold, possibly with $R = 1$. (If $R = 1$ then the RWRL is positive recurrent.)

The last possibility is that the spectral radius of $s\mathbf{P} + s\mathbf{K}(s)$ attains the value 1 at $s = R = \rho$. In this case the singularity of $\mathbf{G}(z)$ at $z = R$ is such that for each entry $G_{xy}(z)$,

$$G_{xy}(R) - G_{xy}(z) \sim \frac{c_{xy}}{\sqrt{R - z}}. \tag{4.26}$$

The proof of this is somehat delicate, and uses the fact that the matrix $R\mathbf{P} + R\mathbf{K}(R)$ is a Perron-Frobenius matrix. See [18] for details. In brief, the reasoning is as follows: by (4.16), the function $\mathbf{G}(z)^{-1}$ behaves near $z = R$ like

$$\mathbf{M}\sqrt{R - z} + \mathbf{N}(z)$$

where $\mathbf{M}$ is a rank-one constant matrix and $\mathbf{N}(z)$ is analytic near $z = R$, and so the inverse of $\mathbf{G}(z)$ blows up like $1/\sqrt{R - z}$. Theorem 3.1 implies that if the Green's function $G_{xy}(z)$ behaves like (4.26) at the singularity $z = R$, then the coefficients must satisfy (4.24).

It remains to show that if the transition probabilities obey the asymptotic law (4.25) then $R > 1$. Note that (4.25) implies that the Markov chain X_n is transient. Consequently, there is at least one two-letter word ab such that $\sum_y H_{ab,y}(1) < 1$ (that is, so that the probability of return to the set of one-letter words from ab is less than one). The irreducibility assumptions ensure that the number of steps until X_n ends with the letters ab has an exponentially decaying tail (uniformly in all starting states, since the alphabet is finite). At each such time, the RWRL may choose to never the set of words of length $< |X_n|$: this occurs with positive (conditional) probability independent of the letters of X_n that precede the closing ab. Thus, the probability that $|X_n| \leq 1$ must decay at least exponentially, that is $R > 1$. □

5. Infinite-Dimensional Lagrangian systems

In section 3 we showed that for a large class of *finite-dimensional* Lagrangian systems the lead singularity is of the square-root type (3.12). This, together with the Tauberian theorem of Flajolet and Odlyzko, implies that the power series coefficients of the component functions satisfy a $3/2$−power law. In this section, we consider the simplest infinite-dimensional case, where the spectrum of the Jacobian operator $\partial Q/\partial F$ near the spectral radius consists of eigenvalues of finite multiplicity. In section 6 below we will show how this analysis may be used to obtain local limit theorems for random walks on *infinite* free products.

5.1. LYAPUNOV-SCHMIDT REDUCTION

Recall that by the Implicit Function Theorem, the Lagrangian system

$$F(z) = Q(F(z)) \tag{5.1}$$

has a solution $z = 0$, $F(0) = 0$ that admits an analytic continuation along any curve on which the spectral radius of the operator

$$z\mathcal{L}(z) = z\left(\frac{\partial Q}{\partial w}\right)_{w=F(z)} \tag{5.2}$$

remains less than 1. Unfortunately, the spectral radius of a continuous, operator-valued function need not be itself continuous (although it must be at least lower semi-continuous), and so the singular points of the solution $F(z)$ of (3.4) may not in general be located by searching for the points z where the spectrum of $z\mathcal{L}(z)$ includes the value 1. However, in certain problems the spectral radius of $\mathcal{L}(s)$ will for positive s be an isolated eigenvalue of finite multiplicity, and in such cases the spectral radius of $\mathcal{L}(z)$ will vary continuously with z near the positive axis $s > 0$. When this happens, the singular behavior of $F(z)$ at $z = R$ will essentially be determined by a finite-dimensional section, and classical methods of algebraic geometry (the Weierstrass preparation theorem, Newton diagrams, Puiseux expansions, etc.) may be used. This program is known in nonlinear analysis as *Lyapunov-Schmidt reduction*; it is commonly used in bifurcation theory (see, for instance, [28], chapter 8).

Hypothesis 5.1 There exists $R \in (0, \infty)$ such that
 (a) the spectral radius of $s\mathcal{L}(s)$ is less than 1 for all $s \in [0, R)$;
 (b) $\lim_{z \to R} F(z) = F(R)$ exists and is finite;
 (c) 1 is an isolated eigenvalue of $R\mathcal{L}(R)$ with finite multiplicity.

Condition (a) guarantees that $F(z)$ has a unique analytic continuation along the line segment $[0, R)$, and condition (b) implies that the Jacobian operator $\mathcal{L}(z)$ has a limit $\mathcal{L}(R)$ as $z \to R$ (from inside the disk). Condition (c) implies that the Banach space B may be decomposed as a direct sum

$$B = V \oplus W, \tag{5.3}$$

where V is the space of eigenvectors of $R\mathcal{L}(R)$ with eigenvalue 1, and that there is a projection operator $P_V : B \to V$ with range V that commutes with $\mathcal{L}(R)$, given by

$$P_V = \frac{1}{2\pi i} \oint (\zeta - R\mathcal{L}(R))^{-1} \, d\zeta, \tag{5.4}$$

where the contour integral extends over a circle in the complex plane surrounding the point $\zeta = 1$ that contains no other points of the spectrum of $R\mathcal{L}(R)$ (see [14], Theorem 6.17). It then follows that

$$P_W = I - P_V \tag{5.5}$$

is a projection operator with range W that also commutes with $R\mathcal{L}(R)$. That 1 is an isolated point of the spectrum of $R\mathcal{L}(R)$ implies that 1 is *not* in the spectrum of the restriction to W of $P_W R\mathcal{L}(R)$, equivalently, $I - P_W R\mathcal{L}(R)$ is invertible on W.

Proposition 5.1 *Assume that Hypothesis 5.1 holds, and let P_V be the projection operator defined by (5.4). Then in some neighborhood of $z = R$, the V-valued function $P_V F(z)$ satisfies a functional equation of the form*

$$P_V F(z) = zQ_V(z, P_V F(z)), \tag{5.6}$$

where $Q_V(z, v)$ is a function of dim $V + 1$ complex variables that is holomorphic in a neighborhood of $z = R$, $v = P_V F(R)$.

Proof. Consider the mapping $K : \mathbb{C} \times V \times W \to W$ defined by $K(z, v, w) = w - zP_W Q(v + w)$. This is holomorphic, and takes the value 0 at the point $\omega := (R, P_V F(R), P_W F(R))$. I claim that there is a holomorphic mapping $W(z, v)$, valued in W and defined in a neighborhood of $z = R$, $v = P_V F(R)$, such that for all z, v in this neighborhood,

$$W(z, v) = zP_W Q(v + W(z, v)),$$

and such that all zeros of $K(z, v, w)$ near ω are of the form $(z, v, W(z, v))$. This follows from the Implicit Function Theorem for the mapping K, because the Jacobian of the mapping $w \mapsto K(z, v, w)$ is

$$I - zP_W(\partial Q / \partial F)_{v+w}$$

and this is nonsingular at $z = R$, $v = P_V F(R)$, since $P_W R\mathcal{L}(R) = R\mathcal{L}(R)P_W$ and $I - P_W R\mathcal{L}(R)$ is invertible on W. Now consider the functional equation (3.4). Applying the projections P_V and P_W to both sides, and using the relation $I = P_V + P_W$, one obtains

$$\begin{aligned} P_V F(z) &= zP_V Q(P_V F(z) + P_W F(z)) \quad \text{and} \\ P_W F(z) &= zP_W Q(P_V F(z) + P_W F(z)). \end{aligned}$$

The solution to the second of these equations must be at $P_W F(z) = W(z, P_V F(z))$, and so the first may be rewritten in the form

$$P_V F(z) = zP_V Q(P_V F(z) + W(z, P_V F(z))).$$

Since the function $W(z, v)$ is holomorphic in its arguments, this is the desired functional equation. $\qquad\qquad\square$

5.2. SIMPLE EIGENVALUES AND SQUARE-ROOT SINGULARITIES

Hypothesis 5.1 requires only that the spectrum of the Jacobian operator $\mathcal{L}(R)$ have an isolated lead eigenvalue of finite multiplicity. If the lead eigenvalue is simple, as will often be the case, the conclusions of Proposition 5.1 can be substantially strengthened.

Hypothesis 5.2 There exists $R \in (0, \infty)$ such that

(a) the spectral radius of $s\mathcal{L}(s)$ is less than 1for all $s \in [0, R)$;
(b) $\lim_{z \to R} F(z) = F(R) \neq 0$ exists and is nonzero and finite;
(c) 1 is an isolated eigenvalue of $R\mathcal{L}(R)$ with multiplicity 1.
(d) The projection $P_V F(R)$ of $F(R)$ on the 1$-$eigenspace V is *nonzero*.

When Hypothesis 5.2 holds, the projection P_V defined by (5.4) will have a one-dimensional range, and so there will exist nonzero elements $h \in B$ and $v \in B^*$ (here B^* denotes the Banach space dual to B) such that for all $\varphi \in B$,

$$P_V \varphi = \langle v, \varphi \rangle h. \qquad (5.7)$$

Note that since P_V is a projection and h is in the range of P_V,

$$\langle v, h \rangle = 1. \qquad (5.8)$$

Proposition 5.2 *Assume that Hypothesis 5.2 holds, and let h and v be such that (5.7) holds. Define*

$$v(z) = \langle v, F(z) \rangle. \qquad (5.9)$$

Then there exist an integer $m \geq 1$ and a function $A(\zeta)$ holomorphic in a neighborhood of $\zeta = 0$, satisfying $A(0) = 0$, such that in some neighborhood of $z = R$,

$$v(R) - v(z) = A((R - z)^{1/m}). \qquad (5.10)$$

Remark. The representation (5.10) implies that the function $v(z)$ has an analytic continuation to a slit disk centered at $z = R$ of the form

$$\{z : |z - R| < \varepsilon \quad \text{and} \quad |\arg(z - R)| > 0\} \qquad (5.11)$$

and that its asymptotic behavior near $z = R$ is of the form (3.2) for some rational α.

Proof. Set $u(\xi) = v(R) - v(R - \xi)$. By Proposition 5.1, the function $P_V F(z) = v(z)h$ satisfies a functional equation (5.6) in a neighborhood of $z = R$. This may be rewritten in ξ and u as a functional equation for $u(\xi)$ of the form

$$K(\xi, u(\xi)) = 0 \qquad (5.12)$$

where $K(\xi, u)$ is a function of two variables (ξ, u) that is holomorphic in a neighborhood of the origin in $\mathbb{C}^2$, and $K(0,0) = 0$. We may assume that the power series expansion of the function $K(\xi, u)$ contains a term au^d, where $a \neq 0$, that is not divisible by ξ, for if this were not the case then K could be replaced by K/ξ^m for some m. (The power series for K must include terms divisible by u, because the functional equation (5.12) derives from (5.6), which holds for $z < R$ near R.) Thus, by the Weierstrass Preparation Theorem ([11], Chapter 1) there exists a Weierstrass polynomial

$$\Phi(\xi, u) = u^d + \sum_{j=1}^{d} a_j(\xi) u^{d-j} \qquad (5.13)$$

such that in some neighborhood of $(0,0) \in \mathbb{C}^2$ the zero sets of K and Φ coincide. (Note: The definition of a Weierstrass polynomial requires that each of the coefficients $a_j(\xi)$ be holomorphic in ξ.)

Since the ring of holomorphic functions near $(0,0)$ is a unique factorization domain ([11], Chapter 1), the polynomial Φ may be factored into irreducible Weierstrass polynomials:

$$\Phi(\xi, u) = \prod_{i=1}^{r} \Phi_i(\xi, u). \qquad (5.14)$$

Here each Φ_i is irreducible in the ring of locally holomorphic functions; this means that in some neighborhood of the origin, $\Phi_i(\xi, u)$ is, for each fixed $\xi \neq 0$, irreducible in the ring $\mathbb{C}[u]$ of polynomials in the variable u. Consequently, the derivative $\partial\Phi_i/\partial u$ cannot have a zero in common with Φ_i (except when $\xi = 0$), and so by the Implicit Function Theorem, for each $\xi \neq 0$ and each u such that $\Phi_i(\xi, u) = 0$, the equation $\Phi_i(\xi', u') = 0$ defines a branch of an analytic function $u(\xi')$ near $\xi' = \xi$ and $u' = u$. For at least one of the indices i (say $i = 1$), one of the branches of the analytic function $u(\xi)$ defined by $\Phi_i = 0$ coincides with the function $u(\xi) := v(R) - v(R - \xi)$ for $\xi > 0$ small.

The result (5.10) now follows by a standard argument in the theory of algebraic functions of a single variable (see, for instance, [12], vol. 2, section 12.2). If the branch $u(\xi)$ is followed around a small contour $\xi \in \gamma$ surrounding 0, then after a finite number m of circuits $u(\xi)$ will return to the original branch (because for each ξ the polynomial $\Phi_1(\xi, u)$ has only m roots, where m is the degree of Φ_1 in the variable u). Consequently, u is an analytic function of ξ^m in a neighborhood of $\xi = 0$ (observe that $\xi = 0$ is are movable singularity because all roots of $\Phi_1(\xi, u) = 0$ approach zero as $\xi \to 0$), and the representation (5.10) follows. □

Let $\mathcal{H}(v, w)$ be the Hessian (second differential) form of the mapping Q at the point $F(R)$, that is, the symmetric bilinear form $\mathcal{H} : B \times B \to B$ defined by

$$\mathcal{H}(v, v) = \left(\frac{d^2}{d\varepsilon^2} Q(F(R) + \varepsilon v) \right)_{\varepsilon = 0}. \tag{5.15}$$

Theorem 5.1 *Let $F(z)$ be the solution of the Lagrangian equation (3.4), and assume that Hypothesis 5.2 holds. Assume that the Hessian $\mathcal{H}(h, h)$ has a nonzero projection on the subspace V, that is,*

$$\langle v, \mathcal{H}(h, h) \rangle \neq 0. \tag{5.16}$$

Then the function $v(z) := \langle v, F(z) \rangle$ has a square-root singularity at $z = R$, that is, for some nonzero constant C,

$$v(R) - v(z) \sim C\sqrt{R - z} \tag{5.17}$$

as $z \to R$ in a slit domain (5.11). Furthermore, the function $P_W F(z)$ satisfies

$$P_W F(R) - P_W F(z) = O(|R - z|) \tag{5.18}$$

near $z = R$.

Remark. Theorem 5.1 shows that under Hypothesis 5.2 and (5.16), the function $v(z)$ satisfies most of the hypotheses of the Flajolet-Odlyzko Transfer Theorem, with $\alpha = 1/2$: the only hypothesis that requires separate verification is that v has no singularity on the circle $|z| = R$ except that at $z = R$. Puiseux expansions for other linear functionals of $F(z)$ can be deduced from (5.17) and (5.18).

Proof of Theorem 5.1 The proof is modelled on that of Theorem 3.2 above. Proposition 5.2 implies that $v(z)$ has a Puiseux expansion in powers of $(R - z)^\alpha$ for some positive rational number α. Since $P_W F(z) = W(z, v(z)h)$ where $W(z, v)$ is the holomorphic function constructed in the proof of Proposition 5.1, it follows that $P_W F(z)$ also has a Puiseux expansion in powers of $(R - z)^\alpha$. We must show that α is a multiple of $1/2^k$ for some $k \geq 1$, and that the first nonzero terms in the expansions are as advertised.

Write

$$\begin{aligned} W(z) &= P_W F(z), & \Delta z &= R - z, \\ V(z) &= P_V F(z) = v(z)h & \Delta F &= F(R) - F(z), \end{aligned}$$

etc. Recall that $v(R) \neq 0$, by (d) of Hypothesis 5.2. The functional equation (3.4) and Taylor's theorem ([28], section 4.6) imply that

$$\Delta F = (F(R)/R)\Delta z + R\mathcal{L}(R)\Delta F - \frac{1}{2}\mathcal{H}(\Delta F, \Delta F) + \mathcal{R}_3 \qquad (5.19)$$

where the remainder $\mathcal{R}_3$ is $o(\Delta z + \|\Delta F\|^2)$ as $\Delta z \to 0$. Apply the projections P_V, P_W to both sides and use the bilinearity of the Hessianform to obtain

$$\Delta W = (W(R)/R)\Delta z + P_W R\mathcal{L}(R)\Delta W - P_W \mathcal{H}(\Delta V, \Delta V)/2 + \mathcal{R}_W \quad (5.20)$$

and

$$\Delta V = (V(R)/R)\Delta z + P_V R\mathcal{L}(R)\Delta V - P_V \mathcal{H}(\Delta V, \Delta V)/2 + \mathcal{R}_V \quad (5.21)$$

where $\mathcal{R}_W, \mathcal{R}_V = O(\|\Delta V\| \|\Delta W\| + \|\Delta W\|^2) + o(\Delta z)$. Since P_V is the projection onto the $1-$eigenspace of $R\mathcal{L}(R)$, the second term on the right side of (5.21) cancels the left side, and so the equation reduces to

$$(\Delta v)^2 \langle \nu, \mathcal{H}(h,h) \rangle = 2(v(R)/R)\Delta z + \mathcal{R}_V' \qquad (5.22)$$

where $\mathcal{R}_V = \mathcal{R}_V' h$. Recall that the operator $P_W - P_W R\mathcal{L}(R)$ is invertible on the Banach space W, with inverse (say) $\mathcal{M}$, so equation (5.20) may be rewritten as

$$\Delta W = \mathcal{M}(W(R)/R)\Delta z - (\Delta v)^2 \mathcal{M}(\mathcal{H}(h,h)) + \mathcal{R}_W. \qquad (5.23)$$

There are now two possibilities: (a) $\|\Delta W\| = o(|\Delta v|)$, or (b) there is a sequence of points $z_n \to R$ along which $\|\Delta w\| \geq \varepsilon|\Delta v|$ for some $\varepsilon > 0$. I claim that possibility (b) cannot occur: If so, equation (5.23) would imply that $\|\Delta W\| = O(\Delta z)$; this would imply that the remainder term in equation (5.22) is $o(\Delta z)$; and this would make (5.22) impossible, since $v(R) \neq 0$ and $\langle \nu, \mathcal{H}(h,h) \rangle \neq 0$. Thus, (a) must hold. But (a) implies that the remainder term $\mathcal{R}_V'$ is of smaller order of magnitude than the other terms in equation (5.22). It follows that $(\Delta v)^2 \sim C\Delta z$ for some nonzero constant C. This implies that the leading non constant term in the Puiseux expansion of $v(z)$ is $C\sqrt{R-z}$ for some nonzero C. That $P_W F(z)$ has Puiseux expansion (5.18) with first nonzero term proportional to $(R-z)$ now follows from (5.5) together with (5.23). $\qquad \square$

6. Application: Local Limit Theorem for RW on Infinite Free Products

We now turn our attention to random walks on *infinite* free products. For simplicity, we shall consider only the nearest-neighbor case. In the nearest-neighbor case, the methods of [8] provide another approach to the study of transition probabilities; however, this approach breaks down in the non-nearest-neighbor case, whereas the method we shall develop may be adapted to finite-range random walks, and even to some infinite-range random walks.

Nearest-neighbor random walk on a countable free product Γ of copies of $\mathbb{Z}_2$ was defined in section 2 above. Recall that the Green's function of a nearest-neighbor random walk (defined by equation (2.2)) is determined by the *first-passage* generating functions — see equation (2.8). The first-passage generating functions are themselves inter related by the Lagrangian system (2.9) of quadratic equations. This system involves infinitely unknowns $F_i(z)$, and so analysis of the lead singularity requires the use of the machinery of section 5.

6.1. HOLOMORPHIC CHARACTER OF THE LINK FUNCTION

Theorem 5.1 requires that the function $F(z)$ take values in a Banach space B, that the link function $Q : B \to B$ be holomorphic, and that the conditions of Hypothesis 5.2 be satisfied. Of these, the most critical is that 1 should be an isolated, simple eigenvalue of the Jacobian operator $R\mathcal{L}(R)$, and the proof of this will occupy most of the argument.

Recall that, by Corollary 2.1, the function $F(z) = (F_i(z))$ takes values in the Banach space B of bounded sequences (with sup norm), for all $|z| \leq R$. The Lagrangian system (2.9) has the special form

$$F(z) = z(M(F(z)) + N(F(z)) \times F(z)) \qquad (6.1)$$

where $M, N : B \to B$ are bounded linear operators and the operation $\times$ is coordinate wise multiplication. Now any linear operator $L : B \to B$ is clearly holomorphic, and by the product rule ([28], sec. 4.3) the coordinate wise product of holomorphic mappings is holomorphic, so the implied link function Q for the system (2.9) is holomorphic on B.

6.2. SPECTRUM OF THE JACOBIAN OPERATOR $\mathcal{L}(s)$

Recall from equation (3.7) that the operator $\mathcal{L}(s)$ is the Jacobian $\partial Q/\partial w$ evaluated at $w = F(s)$. In our case the Lagrangian system specializes to (2.9); the function $F(z)$ has coordinates $F_i(z)$, indexed by $i \in \mathbb{N}$. Thus, by (2.9), the Jacobian in matrix form has entries

$$\begin{aligned} \mathcal{L}(s)_{ij} &= p_j F_i(s) & \text{for} \quad j \neq i; \text{ and} \\ &= p_\emptyset + \sum_{j \neq i} p_j F_j(s) & \text{for} \quad j = i. \end{aligned} \qquad (6.2)$$

This is clearly a positive operator for $s \geq 0$, as all entries in the matrix are nonnegative. Moreover, since the functions $F_i(s)$ are nondecreasing in s, the norms and spectral radii of the operators $\mathcal{L}(s)$ are monotone in s.

The key to the spectral analysis of the Jacobian operator is that it decomposes as the sum of a compact operator and a scalar multiple of the identity. To see this, observe that

$$\mathcal{L}(s) = \mathcal{K}(s) + \mathcal{D}(s) + H(s)I \tag{6.3}$$

where

$$
\begin{aligned}
\mathcal{K}(s)_{ij} &= & p_j F_i(s), \\
\mathcal{D}(s)_{ij} &= & -2p_i F_i(s)\delta_{ij}, \quad \text{and} \\
H(s) &= & p_\emptyset + \sum_j p_j F_j(s).
\end{aligned}
\tag{6.4}
$$

Observe that by equation (2.8),

$$H(z) = (G(z) - 1)/zG(z) \tag{6.5}$$

where $G(z)$ is the Green's function; note for future reference that $(G(s) - 1)/sG(s)$ is nondecreasing in s for $0 \le s \le R$, and that

$$H(R) < 1/R \tag{6.6}$$

by Corollary 2.1.

Lemma 6.1 Let R be the common radius of convergence of the Green's function $G(z)$ and the first-passage generating functions $F_i(z)$. For each $s \in [0, R]$, the operators $\mathcal{K}(s)$ and $\mathcal{D}(s)$ are compact.

Proof. The operator $\mathcal{K}(s)$ has rank one and so is trivially compact. The operator $\mathcal{D}(s)$ is diagonal, with diagonal entries $-2p_i F_i(s)$, and so it maps the unit ball of B (the sequences with entries bounded in absolute value 1) into the set of sequences with entries bounded by $2p_i F_i(R)$. This set is compact, since the values $F_i(R)$ are bounded and $\sum_i p_i < \infty$. □

Corollary 6.1 *For each $s \in [0, R]$, the operator $\mathcal{L}(s)$ and its adjoint $\mathcal{L}(s)^*$ have purely discrete spectrum. Each element of the spectrum not equal to $H(s)$ is an eigenvalue of finite multiplicity (the same for both $\mathcal{L}(s)$ and $\mathcal{L}(s)^*$), and $H(s)$ is the only possible accumulation point of the spectrum. The spectral radius of $\mathcal{L}(s)$ varies continuously with s for $0 \le s \le R$. There is a unique $\sigma \in (0, R]$ such that the spectral radius of $\sigma\mathcal{L}(\sigma)$ is 1, and at least for $s > \sigma - \varepsilon$, for some $\varepsilon > 0$, the spectral radius of $\mathcal{L}(s)$ is an eigenvalue of finite multiplicity.*

Note: Later it will be shown that $\sigma = R$.

Proof. The Riesz-Schauder theory of compact operators (see, for example, [14], Ch. 4) asserts that the spectrum of a compact operator consists of at most countably many isolated eigenvalues of finite multiplicity (the same multiplicity for both the operator and its adjoint) that accumulate only at

0. (For a compact operator on an infinite-dimensional Banach space, 0 must also be an element of the spectrum). Since addition of a scalar multiple of the identity to an operator merely shifts its spectrum, the statements about the spectrum of a particular $\mathcal{L}(s)$ follow.

To see that the spectral radius $\rho(s)$ of $\mathcal{L}(s)$ must vary continuously with s, observe that either $\rho(s) = H(s)$ or $\rho(s)$ is an isolated eigenvalue of finite multiplicity. Isolated eigenvalues of finite multiplicity must vary continuously ([14], Chapter 6), but so does the Green's function; hence, the spectral radius is continuous in s. The spectral radius of $s\mathcal{L}(s)$ cannot remain bounded away from 1 for $s \in [0, R]$, because if so then by the Implicit Function Theorem, the function $F(z)$ would have an analytic continuation to a neighborhood of $z = R$. Thus, by the Intermediate Value Theorem of calculus, for some $\sigma \in [0, R]$ the spectral radius of $\sigma\mathcal{L}(\sigma)$ attains the value 1.

Recall now that the function $H(s)$ is nondecreasing and is bounded above by $1/R$, by (6.6). Consequently, when the spectral radius of $s\mathcal{L}(s)$ attains (or exceeds) the value 1 it must coincide with an eigenvalue of finite multiplicity. Since the spectral radius of $s\mathcal{L}(s)$ is nondecreasing in s, the last assertion of the corollary follows.   □

6.3. SIMPLICITY OF THE LEAD EIGENVALUE

Proposition 6.1 *The eigenvalue 1 of the operator $T = \sigma\mathcal{L}(\sigma)$ has multiplicity one. Furthermore, there are* strictly *positive right and left eigenvectors h and ν.*

The proof will consist of several lemmas. By Corollary 6.1, 1 is an eigenvalue of T with finite multiplicity, and no elements of the spectrum have modulus greater than one. *A priori* it is possible that 1 is not the only element of the spectrum on the unit circle; however, Corollary 6.1 implies that the spectrum of T has only finitely many points on the unit circle, and all are eigenvalues of finite multiplicity.

Lemma 6.2 There exist positive right and left eigenvectors h and ν for T with eigenvalue 1.

Proof. Because the operator T is positive, if $T\varphi = \varphi$ then $T|\varphi| \geq |\varphi|$. Since V contains nonzero elements, it follows that there is a nonnegative vector g such that $Tg \geq g$. By the positivity of T, the sequence $g_n := T^n g$ is nondecreasing, and so in particular each g_n is positive. If it could be shown that the sequence g_n converges in norm, then the limit h would be a positive eigenvector of T with eigenvalue 1.

Claim: The sequence g_n converges in $B-$norm.

Proof. Let $\lambda_1, \lambda_2, \ldots, \lambda_r$ be the eigenvalues of modulus 1, with $\lambda_1 = 1$, and let $V_1, V_2, \ldots V_r$ be the corresponding eigenspaces. Define projections $P_1, P_2, \ldots P_r$ by

$$P_j = \frac{1}{2\pi i} \oint_{\gamma_j} (\zeta I - T)^{-1} \, d\zeta,$$

where γ_j is a circle surrounding λ_j that encircles no other points of the spectrum. Each P_j is a projection whose range is the eigenspace V_j (see [14], Theorem 6.17). The projections P_j commute with each other and with T (since they are linear combinations of powers of T). The range U of complementary projection

$$P_U = I - P_1 - P_2 - \cdots - P_r$$

is an invariant subspace of T, and the spectral radius of TP_U is strictly less than one. Now

$$\begin{aligned} T^n g &= \sum_{j=1}^{r} T^n P_j g + T^n P_U g \\ &= \sum_{j=1}^{r} \lambda^n P_j g + T^n P_U g. \end{aligned}$$

Since each λ_j has absolute value 1, any increasing subsequence of the integers has a subsequence along which λ_j^n converges; consequently, along any such subsequence, $T^n g$ converges in norm. The limit is of necessity a positive element of B. Finally, there can be only one possible limit, because since $g_n \leq g_{n+1}$ for all n, if there were two subsequential limits ψ_A and ψ_B then $\psi_A \leq \psi_B$ and $\psi_B \leq \psi_A$.   □

A similar argument shows that there is a positive left eigenvector.   □

The following lemma will imply that positive eigenvectors of T and T^* are *strictly* positive, and this in turn will imply that the lead eigenspaces are one-dimensional.

Lemma 6.3 Let $\mathcal{M}_m$ be the finite section of $\mathcal{L}(s)$ indexed by pairs i, j such that $1 \leq i, j \leq m$. Then for each $m = 3, 4, \ldots$ and every $s \in (0, R]$, the matrix $\mathcal{M} = \mathcal{M}_m$ is a Perron-Frobenius matrix, that is, some positive power of $\mathcal{M}$ has strictly positive entries.

Proof. By our standing hypotheses on the transition probabilities of the random walk, $p_i > 0$ for each index i and $p_\emptyset > 0$, and so all of the entries of $\mathcal{M}_m$ are positive.   □

Corollary 6.2 *Positive eigenvectors of T and T^* are strictly positive.*

Proof. Suppose that $Th = h$ is a positive eigenvector. Since h is not identically 0, some entry, say the $(j; y)$ entry, is strictly positive. By Lemma 6.3, for each $m \geq 1$ some positive power $\mathcal{M}_m^n$ of the finite section $\mathcal{M}_m$ of T has all entries positive; consequently, if $m \geq j$ then all entries of $T^n h$ indexed by $(i; x)$, where $i \leq m$, are positive. Since $m \geq j$ is arbitrary, it follows that h is strictly positive. A similar argument shows that positive eigenvectors of T^* must be strictly positive.   □

Proof of Proposition 6.1 Denote by V and V^* the $1-$eigenspaces of the operators T and T^*. By Corollary 6.1, V and V^* are finite-dimensional subspaces of B and B^* of the same dimension. By Corollary 6.2, there are strictly positive right and left eigenvectors $h \in V$ and $\nu \in V^*$.

Claim 1. $\varphi, \psi \in V \implies |\varphi \vee \psi| \in V$.

Proof. Since the operator T is positive,

$$T|\varphi \vee \psi| \geq |T\varphi| \vee |T\psi| = |\varphi| \vee |\psi|. \tag{6.7}$$

To see that the inequality is actually an equality, apply the positive linear functional ν to both sides:

$$\langle \nu, |\varphi \vee \psi| \rangle = \langle T^* \nu, |\varphi \vee \psi| \rangle = \langle \nu, T|\varphi \vee \psi| \rangle \geq \langle \nu, |\varphi \vee \psi| \rangle.$$

Since ν is *strictly* positive, it must be that equality holds in (6.7).   □

Claim 2. Let $\varphi \in V$ be real-valued. Then φ is strictly positive, or strictly negative, or identically 0.

Proof. Suppose first that there are indices $\alpha = (i; x)$ and $\beta = (j; y)$ such that $\varphi(\alpha) > 0 > \varphi(\beta)$. By their reducibility of T (Lemma 6.3), there exists and integer $n \geq 1$ such that the (α, β) entry of T^n is positive. Now since $\varphi \in V$ it is an eigenvector of T^n with eigenvalue 1; by Claim 1, the positive part of φ is also an eigenvector with eigenvalue 1. Thus,

$$
\begin{aligned}
\varphi(\alpha) &= \sum_{\gamma \in J} T_{\alpha,\gamma}^n \varphi(\gamma) \qquad \text{and} \\
\varphi(\alpha) &= \sum_{\gamma \in J_+} T_{\alpha,\gamma}^n \varphi(\gamma)
\end{aligned}
$$

where J_+ denotes the set of indices γ such that $\varphi(\gamma) \geq 0$ and J the set of *all* indices γ. The two sums can be equal only if $\sum_J$ contains no negative terms. But n was chosen so that there would be at least one negative term, namely, the term $\gamma = \beta$. This proves that either $\varphi \geq 0$ or $\varphi \leq 0$.

    A similar argument shows that if $\varphi \in V$ is nonnegative, then either $\varphi \equiv 0$ or φ is strictly positive.   □

    Let $\varphi \in V$ be any right eigenvector of T with eigenvalue 1; I will show that φ is a scalar multiple of h. Without loss of generality, assume that φ is real-valued (because if $\varphi \in V$ then its real and imaginary parts are both in V, as T is positive) and positive (by Claim 2). Define $a \geq 0$ to be the supremum of all nonnegative real numbers b such that $b\varphi \leq h$. If the vector $a\varphi$ is not identically equal to h then there is an index $\alpha = (i; x)$ such that $a\varphi(\alpha) < h(\alpha)$, and so for $b > a$ sufficiently near a it must be the case that $b\varphi(\alpha) < h(\alpha)$. But then $h - b\varphi$ is a real-valued element of V with both negative and positive entries, contradicting Claim 2.   □

6.4. SINGULARITY OF F AT $z = \sigma$

Lemma 6.4 Let $\mathcal{H}$ be the Hessian of the link mapping Q associated with the Lagrangian system (2.9), and let h and ν be the positive eigenvectors of $T = \sigma\mathcal{L}(\sigma)$ and T^* with eigenvalue 1. Then

$$\langle \nu, \mathcal{H}(h, h) \rangle > 0. \tag{6.8}$$

Proof. The Hessian operator is the matrix of second partial derivatives of the link mapping Q. The first partial derivatives are given inequations (6.2); from these equations it is apparent that the nonzero second partials are

$$\frac{\partial^2 Q_i}{\partial F_i \partial F_j} = p_j \qquad \text{for} \quad j \neq i \tag{6.9}$$

Hence,

$$\mathcal{H}(h, h)_{i;x} = \sum_{j \neq i} h_i h_j p_j. \tag{6.10}$$

Since the eigenvector h has strictly positive entries, this sum is strictly positive for all i. Since ν is positive, (6.8) follows.   □

Corollary 6.3 *For every positive $\mu \in B^*$, the function $\langle \mu, F(z) \rangle$ has a square-root singularity at $z = \sigma$. Consequently,*

$$\sigma = R. \tag{6.11}$$

Proof. Proposition 6.1 and Lemma 6.4 imply that the hypotheses of Theorem 5.1 are satisfied. Consequently, the function $\langle \nu, F(z) \rangle$ has a square-root singularity at $s = \sigma$. But the common radius of convergence of the power series $F_i(z)$ is R; therefore, it must be that $\sigma = R$. Consequently, For any positive $\mu \in B^*$,

$$\langle \mu, F(z) \rangle = \langle \mu, h \rangle \langle \nu, F(z) \rangle + \langle \mu, P_W F(z) \rangle.$$

By Theorem 5.1, the projection $P_W F(z)$ has a Puiseux series around $z = R$ whose first term is linear in $R - z$, and by Proposition 6.1, the lead eigenvectors ν and h are *strictly* positive. Thus, $\langle \mu, h \rangle > 0$, and so the

Puiseux expansion of $\langle \mu, F(z) \rangle$ has as its first nonconstant term a nonzero multiple of $\sqrt{R-z}$.   □

6.5. SINGULARITY OF THE GREEN'S FUNCTION AT $z = R$

Proposition 6.2 *Let $G(z)$ be the Green's function of an aperiodic, irreducible, quasi-nearest-neighbor random walk on a countable free product of finite groups, and let R be its radius of convergence. Then $G(z)$ has a square-root singularity at $z = R$, that is, it has a Puiseux expansion in a neighborhood of $z = R$ of the form*

$$G(z) - G(R) = \sum_{n=2^k}^{\infty} g_n (R - z)^{n/2^{k+1}} \qquad (6.12)$$

whose first nonconstant term is $g_{2^k} \sqrt{R - z}$, with $g_{2^k} < 0$.

Proof. The relation (2.8) exhibits G as a meromorphic function of a positive linear combination $\mathcal{F}$ of the first-passage generating functions F_i. By Proposition 2.1, $G(z)$ is finite at $z = R$, and so near $z = R$ the function $G(z)$ is in fact an *analytic* function of $\mathcal{F}(z)$. By Corollary 6.3, every positive linear functional of $F(z)$ has a square-root singularity at $z = R$; consequently, the same is true of G.   □

6.6. LOCAL LIMIT THEOREM

By Proposition 2.2 the Green's function $G(z)$ has nosingularity on the circle of convergence except at $z = R$, and by the preceding proposition at $z = R$ the function $G(z)$ admits the expansion (6.12). Therefore, the Flajolet-Odlyzko Transfer Theorem implies the following Local Limit Theorem.

Theorem 6.1 *For nearest-neighbor random walk on a countable free product of copies of $\mathbb{Z}_2$ with positive holding probability $p_\emptyset$, there turn probabilities satisfy*

$$P\{X_n = \emptyset\} \sim \frac{C}{R^n n^{3/2}}. \qquad (6.13)$$

6.7. REMARKS

The crux of the preceding argument is the analysis of section 6.2, which establishes that the Jacobian operator $\mathcal{L}(s)$ has discrete spectrum. Only in this part of the argument does the detailed structure of the Lagrangian system (2.9) play a significant role, and only the decomposition (6.3) of the Jacobian as a sum of a compact operator and a multiple of the identity is needed. It is not difficult to show that a similar decomposition holds for the corresponding Lagrangian systems of non-nearest-neighbor random walks, and of random walks on free products of arbitrary finite groups. Other applications of the results of section 5 will be published elsewhere.

Acknowledgement. Supported by NSF grant DMS-0071970.

References

1. Cartwright, D. (1991) Singularities of the Green's Function of a Random Walk on a Discrete Group.
2. Cartwright, D. and Soardi, P.M. (1986) Random Walks on Free Products, Quotients and Amalgams. *Nagoya Math. J.* **Vol.102**, pp. 163–180.
3. de la Harpe, P. (2000) *Topics in Geometric Group Theory*, University of Chicago Press, Chicago, IL.
4. Drmota, M. (1997) Systems of Functional Equations, *Random Structures and Algorithms*, **Vol.10**, pp. 103–124.
5. Erdelyi, A. (1956) *Asymptotic Expansions*, Dover Publications, NY.
6. Feller, W. (1965) *An Introduction to Probability Theory and its Applications*, **Vol.2**, Wiley & Sons, NY.
7. Flajolet, P. and Odlyzko, A. (1990) Singularity Analysis of Generating Functions, *SIAM J. Discrete Math.*, **Vol.3**, pp. 216–240.
8. Gerl, P. and Woess, W. (1986) Local Limits and Harmonic Functions for Non-isotropic Random Walks on Free Groups, *Probability Theory & Reated Fields*, **Vol.71**, pp. 341–355.
9. Good, I.J. (1960) Generalizations to Several Variables of Lagrange's Expansion, with Applications to Stochastic Processes, *Proc. Cambridge Philos. Soc.*, **Vol.56**, pp. 367–380.
10. Goulden, I.P. and Jackson, D.M. (1983) *Combinatorial Enumeration*, Wiley & Sons, NY.
11. Griffiths, P. and Harris, J. (1978) *Principles of Algebraic Geometry*, Wiley & Sons, NY.
12. Hille, E. (1959) *Analytic Functions*, **Vol.1–2**, Ginn, Boston.
13. Hopcraft, J. and Ullman, J. (1979) *Introduction to Automata Theory, Languages, and Computation*, Addison-Wesley, Reading, MA.
14. Kato, T. (1966) *Perturbation Theory for Linear Operators*, Springer Verlag, NY.
15. Kesten, H. (1959) Full Banach Mean Values on Countable Groups, *Math. Scand.*, **Vol.7**, pp. 146–156.
16. Lalley, S. (1993) Finite Range Random Walk on Free Groups and Homogeneous Trees, *Annals of Probability*, **Vol.21**, pp. 2087–2130.
17. Lalley, S. (1995) Return Probabilities For Random Walk on a Half-Line, *J. Theoretical Probability*, **Vol.8**, pp. 571–599.
18. Lalley, S. (2001) Random Walks on Regular Languages and Algebraic Systems of Generating Functions, *Algebraic Methods in Statistics and Probability (Notre Dame, IN, 2000)*, Amer. Math. Soc., Providence, RI.
19. Lehner, J. (1964) *Discontinuous Groups and Automorphic Forms*, American Math. Soc., Providence, RI.
20. Lothaire, M. *et al* (1983) *Combinatorics on Words*, Addison-Wesley, Reading, MA.
21. Reed, M. and Simon, B. (1978) *Methods of Mathematical Physics IV: Analysis of Operators*, Academic Press, NY.
22. Seneta, E. (1981) *Nonnegative Matrices and Markov Chains*, 2nd ed., Springer Verlag, NY.

23. Stanley, R. (1999) *Enumerative Combinatorics*, **Vol.2**, Cambridge University Press, Cambridge.
24. Wilf, H. (1990) *Generatingfunctionology*, Academic Press, Boston, MA.
25. Woess, W. (2000) *Random Walks on Infinite Graphs and Groups*, Cambridge University Press, Cambridge.
26. Woess, W. (1982) A Local Limit Theorem for Random Walks on Certain Discrete Groups, *Probability Measures on Groups*, Oberwolfach, 1981), pp. 467–477, Lecture Notes in Math., **Vol.928**, Springer Verlag, Berlin-New York.
27. Woess, W. (1987) Context-Free Languages and Random Walks on Groups, *Discrete Math.*, **Vol.67** pp. 81–87.
28. Zeidler, E. (1986) *Nonlinear Functional Analysis and its Applications. I: Fixed Point Theorems*, Springer Verlag, NY.

RECURRENT MEASURES AND MEASURE RIGIDITY

ELON LINDENSTRAUSS

Courant Institute of Mathematical Sciences
251 Mercer St., New York, NY 10012, USA
`elonbl@member.ams.org`

Abstract. We study maps which preserve a foliation and a metric on this foliation. Such maps arise when studying multiparameter abelian actions, and also in the study of arithmetic quantum unique ergodicity. We also discuss measurable dynamics in which neither the measure nor the measure class is preserved, but nonetheless the system has complicated orbit structure.

1. Introduction

The purpose of this paper is to introduce, in what I hope is a friendly way, some new ideas and techniques in the study of certain algebraic actions. These ideas have had implications in other fields, namely in the study of automorphic forms, where they have been used to prove arithmetic cases of the quantum unique ergodicity conjecture.

This paper is not intended to be a survey, but more of a tutorial, and can serve as a companion to the paper [19] where a proof of an arithmetic case of the quantum unique ergodicity conjecture is given. However, the scope of this note is wider, and I hope will stimulate further research by other people in various directions.

Much of the research exposed was motivated directly or indirectly by Furstenberg's study of the simultaneous action of the maps $x \mapsto px$ mod 1 and $x \mapsto qx$ mod 1 on $\mathbb{R}/\mathbb{Z}$ for p, q integers which are multiplicatively independent, for example p, q relatively prime, and in particular by his conjecture that the only nonatomic $\times p, \times q$-invariant measure on $\mathbb{R}/\mathbb{Z}$ is the Lebesgue measure. However, I focus on one particular element which arrises naturally when considering these problems: the study of maps on

A. Maass et al. (eds.), Dynamics and Randomness II, 123–145.
© 2004 *Kluwer Academic Publishers.*

spaces which preserve a generalized foliation, and what is most important on this foliation an intrinsic metric is preserved.

Most treatments of ergodic theory, which is essentially a synonym for the study of actions on measure spaces, preferably with complicated orbits, deal with actions which at the very least preserve the measure class of the given measure. However, it seems that much can be said even without this assumption. One can get interesting dynamics even for singular actions. The only thing which is really needed is some form of recurrence which gives the complicated orbits which are the life and blood of ergodic theory.

Such general situations appear naturally in the study of concrete maps preserving the metric structure of a foliation, and this plays a crucial role in the proof of arithmetic quantum unique ergodicity. The study of singular actions on measure spaces (perhaps an appropriate name would be recurrent dynamics) seems to be both interesting and useful, and I give a more comprehensive discussion of this topic than the minimum that is needed for the applications.

Acknowledgement. It is a pleasure to thank the Departamento de Ingeniería Matemática in the Universidad de Chile for their hospitality during the II Workshop on Dynamics and Randomness where I had the pleasure of giving a series of talks on this subject. In [19] I give a long list of people whose help was indispensable to me. Since this list is quite long, I would merely like to say that I am at least as grateful now as I was then to them all. This work has been written while supported by a Long Term Prize Fellowship of the Clay Mathematics Institute, and for this support I am also very grateful. Finally, I would also like to acknowledge the support of NSF grant DMS-0140497.

2. T-Spaces

Let X be a locally compact, σ-compact metric space. If we are given a continuous action of a locally compact group T the orbits of T give us a partition of X into equivalence classes. Let $T_x = \{g \in T : g.x = x\}$ be the **stabilizer** of $x \in X$. These equivalence classes have an additional structure: the equivalence class of a point $x \in X$ can be identified with T/T_x. The nicest case is when $T_x = \{e\}$, i.e. T acts freely at x, where the orbit of x is naturally identified with T.

Let us try to be a bit more explicit: given a x with $T_x = \{e\}$, the map $t_{x,T} : g \mapsto g.x$ is a bijection between T and the orbit of x. If $y = g_0.x$ is a different point on the same orbit, we get a different map $t_{y,T}$ from T to the same orbit. These two maps differ by a right translation, i.e. $t_{x,T}(gg_0) = t_{y,T}(g)$. Thus it would be more precise to say that the orbit of

x can be naturally identified with T up to a right translation. For example, if we start with some metric d_T on T which is right invariant under T, we can define using it a metric on the orbit of x: $D(g.x, h.x) = d_T(g, h)$ which does not depend on the particular choice of x but only on the orbit.

While we shall stick with group actions in this paper, it is better to view the results we present in a more general context. Indeed, for most of what we present below, we do not actually need an action. What we need is an equivalence relation where the equivalence classes are modeled on a fixed homogeneous space T (or, for exceptional points, modeled on a quotient of T by some discrete group). In order to get the same description of the equivalence class starting from different base points, we need T be homogeneous in the sense that there is a locally compact second countable group G which acts transitively and continuously on T. Roughly, what we want is an identification of the equivalence class of a point x with T which is well-defined up to the action of an element from G; a precise definition is given in the Appendix A. For this reason, even though we will consider only the case of group actions, we will think of the equivalence class generated by the action as a foliation of X and in particular we use leaf and orbit interchangeably for the orbit of a point under the group T. A space X together with a continuous action of a group T is called a T-space, and the more general construction which we will define in the appendix is called a (T, G)-space. We make the following regularity assumption on our T spaces: for every compact $X_0 \subset X$, there is an open neighborhood of the identity $U \subset T$ so that $G_x \cap U = \{e\}$ for all $x \in X_0$.

Now let us introduce another player to the game: an arbitrary Radon measure μ (i.e. a measure defined on Borel sets which is finite on compact sets) on X. To make sure we actually have a T action we make the mild assumption that for μ almost every $x \in X$, the stabilizer T_x is trivial.

3. Recurrent Measures

The most basic and fundamental dynamical property is recurrence, and this turns out to be precisely what is needed in order to have interesting dynamics for μ. We stress that we do not assume μ is T-invariant, or even that the measure class of μ is invariant (the latter condition is usually called quasi invariance of μ).

Definition 3.1 μ is T-recurrent if for every set $B \subset X$ with $\mu(B) > 0$, for almost every $x \in B$,

$$\forall \text{ compact } C \subset T, \quad \exists g \in T \setminus C \quad \text{satisfying } g.x \in B. \qquad (3.1)$$

For example, by Poincare recurrence, if μ is any T-invariant probability measure then μ is automatically T-recurrent. If μ is quasi invariant, or if it is an invariant infinite measure, μ may or may not be T-recurrent. For μ quasi invariant and $T = \mathbb{R}$ or $\mathbb{Z}$, what we call recurrence is often called conservativity of μ. Note that if a measure μ is T-recurrent, so is every measure in its measure class. In §8 we show how the notion of recurrence arises naturally in the study of the quantum unique ergodicity question.

A stronger condition than recurrence, which is of course fundamental in the theory of measure preserving actions, is ergodicity:

Definition 3.2 μ is T-ergodic if for every set $B \subset X$ with $\mu(B) > 0$, (3.1) holds for almost every $x \in X$ (not just for $x \in B$).

We mention that the ergodic decomposition holds in this general context: in particular, every T-recurrent measure μ can be presented as $\mu = \int \mu_\xi d\nu(\xi)$ with each μ_ξ T-ergodic, and there is some additional information about how μ_ξ is related to μ in terms of the conditional measures described in §4. Precise statements and proofs will appear elsewhere; we mention also [8] where the ergodic decomposition is proved in the context of quasi-invariant measures and T countable (the MR review of this paper by Scot Adam is itself a short research paper).

The definition of recurrence given above its usually quite convenient when one tries to use the recurrence of a measure. However, it is usually easier to verify the recurrence condition in an equivalent but seemingly very different form involving the conditional measures on T-leaves (a.k.a. T-orbits).

4. Conditional Measures on Leaves

When studying the relation between a measure μ on a T-space and the T action, an important tool is a system of locally finite measures induced by μ: the conditional measures on T-leaves.

These measures can be thought alternatively as either measures on the T-leaves or, since for almost every x we have assume that T_x is trivial, a measure on T. We follow [19] in choosing the latter point of view, so for us this system of conditional measures is a Borel measurable map $x \mapsto \mu_{x,T}$ from X to the space $\mathcal{M}_\infty(T)$ of locally finite possibly infinite measures on T. Note that $\mathcal{M}_\infty(T)$, equipped with the weak star topology as the dual to compactly supported continuous functions, is a metrizable locally compact space.

In order to define these measures, we first record some basic fact from measure theory. If $\mathcal{A}$ is a countably generated sigma ring of Borel subsets of X, we define the atom of a point $x \in X$ by

$$[x]_\mathcal{A} = \bigcap_{A \in \mathcal{A}: x \in A} A.$$

Since we are only considering countably generated sigma rings, $\mathcal{A}$ has a maximal element, namely $X' = \cup \mathcal{A}$, and $\mathcal{A}$ is a sigma algebra of subsets of X'. It is easy to see that if $\mathcal{A}$ is generated by the sets $A_1, A_2, \ldots$ then

$$[x]_{\mathcal{A}} = \bigcap_{i : x \in A_i} A_i \cap \bigcap_{i : x \notin A_i} (X' \setminus A_i).$$

Definition 4.1 We say that two countably generated sigma rings $\mathcal{A}$ and $\mathcal{B}$ are **compatible** if for every $x \in \cup \mathcal{A}$ we have that $[x]_{\mathcal{A}} \cap (\cup \mathcal{B})$ is a subset of a countable union of atoms of $\mathcal{B}$ and vice versa.

Suppose now that μ is a Borel probability measure on X. Then μ induces a system of probability measures $\mu_x^{\mathcal{A}}$ on X with the following properties:

(1) for every x, $\mu_x^{\mathcal{A}}([x]_{\mathcal{A}}) = 1$.
(2) for every $x, x' \in X$ with $[x]_{\mathcal{A}} = [x']_{\mathcal{A}}$ we have that $\mu_x^{\mathcal{A}} = \mu_{x'}^{\mathcal{A}}$.
(3) the map $x \mapsto \mu_x^{\mathcal{A}}$ is Borel measurable (hence, by (2), $\mathcal{A}$ measurable).
(4) for every Borel set $B \subset X$ and $A \in \mathcal{A}$,

$$\mu(A \cap B) = \int_A \mu_x^{\mathcal{A}}(B) dx.$$

up to a set of measure zero, these conditions uniquely determined this system of conditional measures $\mu_x^{\mathcal{A}}$.

An important property of this construction is that it behaves nicely when one switches between compatible sigma rings:

Proposition 4.1 *If $\mathcal{A}$ and $\mathcal{B}$ are compatible countably generated sigma rings, then for μ almost every $x \in (\cup \mathcal{A}) \cap (\cup \mathcal{B})$*

$$\frac{\mu_x^{\mathcal{A}}|_{[x]_{\mathcal{A}} \cap [x]_{\mathcal{B}}}}{\mu_x^{\mathcal{A}}([x]_{\mathcal{A}} \cap [x]_{\mathcal{B}})} = \frac{\mu_x^{\mathcal{B}}|_{[x]_{\mathcal{A}} \cap [x]_{\mathcal{B}}}}{\mu_x^{\mathcal{B}}([x]_{\mathcal{A}} \cap [x]_{\mathcal{B}})}$$

(in particular, the denominators in both sides of this equation are nonzero almost surely).

We would like to find a similar construction that will give us well-defined measures on orbits of T. Unfortunately, in all interesting situations, there is no countably generated sigma ring whose atoms are full T-orbits. What is easy, however, is to construct countably generated sigma rings whose atoms are arbitrarily large chunks of T-orbits and this will be sufficient for our purposes.

First we introduce some notations and definitions: for every $x \in X$, let $t_{x,T} : T \mapsto X$ be the map

$$t_{x,T} : g \mapsto g.x.$$

In order to avoid having to write explicit proportionality constants we shall use the notation $\mu \propto \nu$ to denote the fact that $\mu = c\nu$ for some $c > 0$.

Definition 4.2 A set $B \subset X$ is an **open T-plaque around** x if it is a subset of the T orbit of x and $t_x^{-1}(B)$ is an open neighborhood of the identity in T with compact closure. B is an **open T-plaque** if there is some $x \in X$ for which B is an open T-plaque around x. A sigma ring $\mathcal{A}$ is an **open T-stack** if it is a countably generated sigma ring of Borel subsets, and for every $x \in \cup\mathcal{A}$ the atom $[x]_{\mathcal{A}}$ is an open T-plaque.

Exercise 4.1 Show that every two open T-stacks are compatible in the sense of Definition 4.1. (Hint: use the fact that T is locally compact.)

We are finally in a position to define the conditional measures $\mu_{x,T}$. These measures are typically infinite, and there is no canonical way to normalize them. In order to get a definition which uniquely determines $\mu_{x,T}$ (up to on a set of $x \in X$ of measure zero) we will fix some arbitrary open neighborhood U_0 of the identity in T and demand that $\mu_{x,T}(U_0) = 1$. It is not immediately clear that the conditions defining these measures are compatible; we will explain below at least why they determine the system of measures. For a full proof of the theorem, see [19][Sect. 3], though this is by no means the first, and quite likely not the last place where such a theorem is proved.

Theorem 4.1 *There is a measurable map $x \mapsto \mu_{x,T}$ from X to $\mathcal{M}_\infty(T)$ satisfying the following conditions:*

$\Diamond$-1. for every $x \in X$, $\mu_{x,T}(U_0) = 1$.
$\Diamond$-2. for every open T-stack $\mathcal{A}$, for a.e. $x \in \cup\mathcal{A}$,

$$(t_{x,T})_*(\mu_{x,T})|_{[x]_{\mathcal{A}}} \propto \mu_x^{\mathcal{A}}.$$

$\Diamond$-3. there is a set of full measure $X' \subset X$ so that for every $x \in X$ and $g \in T$ satisfying that both $x, g.x \in X'$ it holds that

$$\mu_{x,T} \propto (\mu_{gx,T}).g.$$

Furthermore, the map $x \mapsto \mu_{x,T}$ is uniquely determined by conditions (1) and (2) up to a set of $x \in X$ of measure zero.

In order to verify the uniqueness of $\mu_{x,T}$ we need to construct good open T stacks. In fact, we will exhibit a countable collection of such stacks,

$$\mathcal{A}_1, \mathcal{A}_2, \mathcal{A}_3, \ldots \tag{4.1}$$

so that for a.e. $x \in X$ and open relatively compact neighborhood of the identity $U \subset T$ there is an open T-stack $\mathcal{A}$ from this collection of stacks for which $[x]_{\mathcal{A}} \supset U.x$. In view of $(\Diamond - 2)$, this will be sufficient to determine $\mu_{x,T}$ a.e. up to normalization, which will be fixed by $(\Diamond - 1)$. Incidentally, this construction, together with Proposition 4.1 is almost sufficient to prove existence of $\mu_{x,T}$ satisfying $(\Diamond-1) - (\Diamond-2)$; the main remaining ingredient is the following lemma:

Lemma 4.1 *For any open T-stack $\mathcal{A}$ and every open neighborhood of the identity $U \subset T$, for a.e. $x \in \cup \mathcal{A}$,*

$$\mu_x^{\mathcal{A}}(U.x) > 0.$$

The proof of this lemma is yet another variant on the liar's paradox (a person who claims he always lies cannot be saying the truth) and is omitted. Deducing $(\lozenge - 3)$ from $(\lozenge - 2)$. is also reasonably straightforward.

What we do not omit (at least not entirely) is the construction of this collection of open T stacks. These stacks have some additional nice properties, and are called **flowers** in [19]. They have two parameters: an arbitrarily large open relatively compact symmetric neighborhood of the identity U, and a small ball $B \subset X$ and have the following three properties:

♣-1. $A = \cup \mathcal{A}$ is open, relatively compact, and $B \subset \cup \mathcal{A}$.

♣-2. for every $y \in A$, the atom $[y]_{\mathcal{A}} = A \cap U^4.y$. In particular, $[y]_{\mathcal{A}}$ is an open T-plaque.

♣-3. if $y \in B$ then $[y]_{\mathcal{A}} \supset U.y$.

Given a $x_0 \in X$ for which $t_{x,T}$ is injective (equivalently, T_x is trivial), it is fairly straightforward to construct an U, B-flower around it, i.e. with $x_0 \in B$. Indeed, it is clear that if B is a sufficiently small open neighborhood of x_0 then for every $y \in B$ we have that $t_{y,T}$ is injective on U^{10}, and in addition $t_{y,T}^{-1}(B) \cap U^{10} \subset U$. Now take $A = U \cdot B$, which is indeed open and relatively compact, and $\mathcal{A}$ to be the sigma ring of subsets of A generated by those open subsets $O \subset A$ so that for every $y \in O$ it holds that $U^2.ya \subset O$ (in particular $A \in \mathcal{A}$). With these definitions, $\mathcal{A}$ is evidently countably generated, and verifying (♣-1)-(♣-3) is straightforward.

To construct the sequence of stacks (4.1), take an increasing sequence of symmetric, relatively compact, open neighborhoods of the identity in T, say $U_1, U_2, \ldots \nearrow T$, and then for every U_i a countable collection of balls $B_{i1}, B_{i2}, \ldots$ which cover all points $x \in X$ with a trivial T-stabilizer so that each for each B_{ik}, the construction above gives a (U_i, B_{ik})-flower, say $\mathcal{A}_{ik}$. Taken together these flowers form the desired sequence of stacks.

We end this section with two (related) examples.

First, suppose that μ is T-invariant. Then $\mu_{x,T}$ is a.e. the right invariant Haar measure on T normalized to give measure one to U_0. This is so important for us we will formally state this as a proposition:

Proposition 4.2 *μ is T-invariant if, and only if, for μ-almost every x the conditional measure $\mu_{x,T}$ is a left invariant Haar measure on T.*

The proof of this proposition is fairly straightforward (see, e.g., [19][Prop. 4.3]), for instance by considering an arbitrarily (U, B)-flower and comparing the measure of a subset $S \subset B$ with the measure of $g.S$ for an arbitrarily $g \in U$.

Now relax our assumptions to assume that μ is only quasi-invariant. Then the Radon-Nikodym derivative

$$c(x, g) = \frac{dg^{-1}.\mu}{d\mu}(x)$$

gives us a much studied multiplicative cocycle (i.e. a function satisfying $c(x, hg) = c(gx, h)c(x, g)$). The following exercise gives the conditional measures $\mu_{x,T}$ in terms of this cocycle:

Exercise 4.2 Show that if μ is quasi invariant then for almost every x and $A \subset G$, the conditional measure $\mu_{x,T}$ is proportional to the measure $A \mapsto \int_A c(x, g)d\eta_T(g)$ with η_T a left invariant Haar measure on T.

5. A Definition of Recurrence Via Conditional Measures

Now that we have developed the theory of conditional measures on T-orbits, we are ready to give an alternative characterization of recurrence. This alternative characterization works only for finite measures (see the example below); however this is a fairly mild restriction since recurrence depends only on measure class and every Radon measure on a locally compact sigma compact space as a probability measure in its measure class.

Proposition 5.1 ([19][Prop 4.1]) *A probability measure μ is T-recurrent if, and only if, for μ-almost every x we have that*

$$\mu_{x,T}(T) = \infty.$$

For the case of a $\mathbb{Z}$-action which preserves the measure class of μ this is the Halmos Recurrence Theorem (see §1.1 in [1]).

Example 5.1 Consider the following very simple example where $X = T$ is a noncompact locally compact metric group, with the T acting on itself by left multiplication. Let $\mu = \eta_T$ be a left Haar measure on T. This measure is clearly not recurrent. However in this case $\mu_{x,T} \propto \eta_T$, in particular infinite.

Remark 5.1 There is a connection between entropy and recurrence. For example, suppose $X = \Gamma\backslash G$ with G a linear algebraic group over $\mathbb{R}$ and Γ a discrete subgroup of G. The group G as well as every subgroup of it act on X by right translations $g.(\Gamma h) = \Gamma hg^{-1}$. Let $A = \{a_t\}$ be a one parameter subgroup of G diagonalizable over $\mathbb{R}$, and

$$U = \{g \in G : a_{-t}ga_t \to e \quad \text{as } t \to \infty\}.$$

Then U is a unipotent subgroup of G (i.e. a group every element of which has only 1 as an eigenvalue), and a A-invariant and ergodic probability measure μ on X is U-recurrent if and only if the entropy of μ with respect to the A-action is positive. If μ is not U-recurrent then necessarily $\mu_{x,U}$ is a.e. the delta measure at $e \in U$. This is a consequence of much more general results regarding diffeomorphisms due to Ledrappier and Young [24]. See [19][Thm. 7.6] for more details.

6. Ergodic Theory for Recurrent Measures

Many of the results traditionally proved regarding dynamics of quasi-invariant measures should probably carry over to general recurrent measures, with $\mu_{x,T}$ serving as a substitute for the Radon-Nikodym cocycle.

   Ergodic theorems have been a recurrent theme since the pioneering work of von Neumann and Birkhoff for $\mathbb{Z}$-actions. Even for measure preserving actions, it is still an active area of research today. Given a sequence of relatively compact sets $F_n \subset T$, one considers the **ergodic averages**

$$S_n[f](x) = \frac{1}{\eta_T(F_n)} \int_{F_n} f(g.x) d\eta_T(g) \qquad (6.1)$$

with η_T denoting as usual left Haar measure on T.

   For actions of amenable groups, proving the mean ergodic theorem, i.e. that $S_n[f](x)$ converges to a T-invariant function in L^p for every $f \in L^p$ is both elegant and simple ($p \geq 1$, with $p = 1$ the most natural). Of course, one needs some assumption on the averaging sets F_n, but in this case the very natural assumption that F_n is a Folner sequence suffice (see [30] for a wealth of information regarding amenable groups actions).

   The pointwise ergodic theorem, i.e. that $S_n[f](x)$ converges to a T-invariant function pointwise almost everywhere for every $f \in L^p$, particularly $p = 1$, turns out to be substantially more delicate. One difficulty, even for $T = \mathbb{Z}$, is that not every Folner sequence works. A general pointwise ergodic theorem for amenable groups (and more background material) is given in [18].

   There has also been substantially work in proving ergodic theorems for actions of nonamenable groups, such as free groups, lattices in semisimple Lie groups, and the semisimple Lie groups themselves; one difference from the amenable case is that these results seem to be limited to $f \in L^p$ for $p > 1$; they also require working with very special (but natural) averaging sets F_n. For more information, see e.g. the survey [28] (see also [3] for a more probabilistic approach via Markov processes).

   This comes as a background to what I actually want to discuss in this section, and that is what can be done in the non-measure preserving case. Very little seems to be known about this subject.

First, one needs to reformulate the ergodic averages to take into account how the action and the measure are related, and depend on μ as follows:

$$S_{n,\mu}[f](x) = \frac{1}{\mu_{x,T}(F_n)} \int_{F_n} f(g.x) d\mu_{x,T}(g);$$

if μ is quasi-invariant then by Exercise 4.2 we have that

$$S_{n,\mu}[f](x) = \left(\int_{F_n} f(g.x) \frac{d(g^{-1}.\mu)}{d\mu}(x) d\eta_T(g) \right) \bigg/ \left(\int_{F_n} \frac{d(g^{-1}.\mu)}{d\mu}(x) d\eta_T(g) \right).$$

The classical literature seems to treat only the case $T = \mathbb{Z}, \mathbb{R}$ (and μ quasi invariant). In this case there is a very nice pointwise ergodic theorem due to Hurewitz [1][Thm. 2.2.1]:

Theorem 6.1 (Hurewicz's Ergodic Theorem) *Let μ be a T-quasi invariant T-recurrent measure for $T = \mathbb{Z}, \mathbb{R}$. Then for any $f, h \in L^1(\mu)$ there is a T-invariant function $F_{f,h}$*

$$\frac{S_{n,\mu}[f](x)}{S_{n,\mu}[h](x)} \to F_{f,h}(x) a.e. \qquad as\ n \to \infty$$

where we take the averaging sets F_n to be either the one-sided or two sided intervals $[-n, n]$ or $[0, n]$.

The pointwise ergodic theorem is usually proved via a maximal inequality. For any function f on X we define the maximal function $M_{f,\mu}$ (M_f in the non-measure preserving case) by

$$M_{f,\mu}(x) = \sup_n S_{n,\mu}[f](x).$$

For any measure preserving action of an amenable group, for suitable Folner sequences F_n (for example, for $T = \mathbb{Z}^n$ or $\mathbb{R}^n$, balls of radius n centered at the origin) one has, for every $\lambda > 0$ (see [18])

$$\mu\{x : M_f(x) > \lambda\} \le C\lambda^{-1} \|f\|_1.$$

It is important that the constant C in (6) is absolute and depends only on the Folner sequence F_n and the group T but not on the action. In the measure preserving case, the L^1 pointwise ergodic theorem follows directly from the maximal inequality (6) using a standard argument. However, the maximal inequality itself is often just as useful, and has a certain intrinsic appeal in that it is very effective.

The arguments of [18] are loosely related to the Vitali Covering Theorem (indeed, for some amenable groups such as groups of polynomial growth one can prove (6) by an almost literal application of this covering argument [40, 29]).

This kind of arguments do not seem to work in the non-measure preserving case. However, there is a more powerful, though less general, covering theorem available to analysts: the Besicovitch Covering Theorem (see [25][Thm. 2.7]). For groups and averaging sequences F_n for which the Besicovitch covering theorem holds, the maximal inequality can be proved for general μ. This plays an important technical role in the proof of measure rigidity for partially rigid maps in [19]. In particular, taking $T = \mathbb{R}^n$ we have the following

Theorem 6.2 ([21]) Let $T = \mathbb{R}^n$ and F_r to be Euclidean balls of radius around 0. Let X be T-space and μ any measure on X. Let

$$M_{f,\mu}(x) = \sup_{r \in \mathbb{R}^+} S_{r,\mu}[f](x)$$

(note that we allow arbitrary $r > 0$). Then there is some absolute constant C depending only on n so that for any $f \in L^1(\mu)$

$$\mu\left\{x : M_{f,\mu}(x) > \lambda\right\} \le C\lambda^{-1} \|f\|_1.$$

A special case of this theorem with some extra assumptions can be found in the appendix of [19] (which is also joint with D. Rudolph).

7. Maps Preserving a (T, G)-Structure

The most rigid dynamics imaginable is that of an isometry ϕ acting on a compact metric space X. In this section we investigate partially rigid maps, and the limitations imposed by this rigidity on the invariant measures.

Suppose X is a locally compact T-space for a locally compact group T as above. Let G be a group that acts transitively on T, and $G^0 = \{\alpha \in G : \alpha(e) = e\}$ where $e = e_T$ is the identity in T. More generally, one can consider a (T, G)-space as in the appendix.

Definition 7.1 Let X be as above and $\phi : X \to X$ a homeomorphism preserving T-leaves. Let G^0 be a closed subgroup of $\mathrm{Aut}T$, the group of automorphisms of T. T acts on itself by right translation, and let G be the group of transformations generated by G_0 and this action of T. We say that ϕ **preserves the** (T, G)-**structure** of X if for every $x \in X$ there is an $\alpha_x \in G^0$ so that

$$\phi(h.x) = \alpha_x(h).\phi(x) \qquad \text{for every } t \in T. \tag{7.1}$$

The most interesting case is when G^0 is compact, in which case we can find a G invariant metric on T. In this situation we will say that ϕ is **partially rigid** or **rigid on T-leaves**. The reason we use G and not G^0 in the above definition is so that our notion of (T,G)-structure for a group T will be compatible with that introduced for more general T in the appendix.

We give three examples of partially rigid maps:

Example 7.1 Let $X = \mathbb{R}/\mathbb{Z}$, $T = \mathbb{Z}[p^{-1}]/\mathbb{Z}$ (equipped with the discrete topology) and $\phi : x \mapsto qx \bmod 1$ for p, q relatively prime. Define a metric on T by setting $d_T(a,b) = p^{l_{a-b}}$ where l_{a-b} is the smallest integers such that $p^l(a-b) = 0 \bmod 1$, and let G be the group of isometries of T with respect to this metric. Then ϕ satisfies (7.1). Strictly speaking, however, this is not an example of a partially rigid map since ϕ is not a homeomorphism, a slight defect which is easily corrected by replacing X by its two-sided completion $\tilde{X}$.

Example 7.2 Let $A \in \mathrm{SL}(n,\mathbb{Z})$ be irreducible (i.e. its characteristic polynomial is irreducible over $\mathbb{Q}$). Let $X = \mathbb{R}^n/\mathbb{Z}^n$ and $\phi : X \to X$ the map $x \mapsto Ax \bmod \mathbb{Z}^n$. Let T be a A-invariant linear subspace of $\mathbb{R}^n$, which acts on X by translations. Then ϕ preserves the $(T, \mathrm{Aff}T)$-structure of X.

ϕ is rigid on T-leaves iff all of the eigenvalues of A corresponding to the invariant subspace T are complex with absolute value 1 (in particular, T has to be even dimensional).

Both Example 7.1 and Example 7.2 can be grouped together as automorphisms of solenoids; see e.g. [23]. Another important class of examples arises from the following construction:

Example 7.3 Let G be a linear algebraic group, $\Gamma < G$ a discrete subgroup, $X = \Gamma \backslash G$. Let A and T be two closed subgroups of G which centralize each other, i.e. $at = ta$ for every $t \in T, a \in A$. Both A and T act on X by translations, and every $a \in A$ acts in a rigid manner on T-leaves.

We will pay particular attention to the following special case: take $G = \mathrm{PGL}(2,\mathbb{R}) \times T$, and A a Cartan subgroup of $\mathrm{PGL}(2,\mathbb{R})$, say the group of diagonal matrices.

We are interested in understanding the invariant measures for a partially rigid map. The prototype to the kind of the result we are after is the following theorem of B. Host:

Theorem 7.1 ([10]) *Let $X = \mathbb{R}/\mathbb{Z}$, $\phi : x \to qx \bmod 1$, and $T = \mathbb{Z}[p^{-1}]/\mathbb{Z}$ as in Example 7.1. The only T-recurrent ϕ-invariant probability measure on X is the Lebesgue measure.*

We know that T-recurrence is equivalent to having infinite conditional measures on T-leaves. Host has also shown that for any not necessarily ϕ-invariant measure μ, under a more quantitative condition regarding these conditional measures, for μ almost every $x \in X$ it holds that $\{q^n x \bmod 1\}$ is equidistributed in X.

Host's theorem can be used to give a simple proof to the following theorem of Rudolph (itself a partial result towards Furstenberg's conjecture):

Theorem 7.2 ([35]) *The only measure μ on $X = \mathbb{R}/\mathbb{Z}$ invariant and ergodic under the $\mathbb{Z}_+^2$ action generated by the maps $\phi_1 : x \mapsto qx \bmod 1$ and $\phi_2 : x \to px \bmod 1$ for p, q relatively prime and which has positive entropy with respect to the map ϕ_2 is Lebesgue measure.*

Indeed, it is straightforward to verify that if μ has positive entropy with respect to $x \mapsto px \bmod 1$ then μ is $\mathbb{Z}[p^{-1}]/\mathbb{Z}$-recurrent (see also Remark 5.1), and Theorem 7.1 applies.

This connection between questions about measure rigidity of multiparameter actions and measure rigidity of partially rigid maps is very typical. There have been many extensions of Host's theorem [26, 11] including to the case of p, q not relatively prime (when ϕ is **not** T-rigid) [17]. See also the related [20].

Using a related technique, Klaus Schmidt and myself have been able to prove a similar theorem in the context of solenoidal automorphisms. In particular, we show the following:

Theorem 7.3 ([23]) *Let ϕ be an automorphism of the torus $\mathbb{R}^n/\mathbb{Z}^n$ associated to a matrix $A \in \mathrm{SL}(n, \mathbb{Z})$ as in Example 7.2. Assume A is totally irreducible (i.e. A^l is irreducible for every l), and T an A-invariant linear subspace of $\mathbb{R}^n$ with all the corresponding eigenvalues of absolute value 1. Then again the only T-recurrent ϕ-invariant probability measure on X is Lebesgue measure.*

We remark that non-hyperbolic irreducible toral automorphism have long served as a basic example for a partially hyperbolic map. Despite many efforts, including the recent advances [9], this seemingly innocent class of dynamical systems is far from understood; Theorem 7.3 is yet another surprise arising from it.

It is interesting that even in the completely hyperbolic case there are nontrivial restrictions on conditional measures for (e.g.) irreducible toral automorphisms, and these play a major role in recent work with Manfred Einsiedler [7] regarding the classification of measures on a torus invariant under several independent toral automorphisms.

Anatole Katok and Ralf Spatzier generalized Rudolph's theorem mentioned above [14, 15] to a rather general context which includes multidimensional actions on both tori and locally homogeneous spaces. The uniform treatment of these seemingly diverse examples substantially changed the flavor of the subject, and influenced many, including myself (for a careful account of the methods, see [13]). It should be noted, however, that in most cases the theorem of Katok and Spatzier is substantially weaker than the analogue of Rudolph's theorem. At the core of the method of [14] is a lemma ([14, Lem. 5.4] or [13, Lem. 3.2]). The argument in the lemma is quite general, and can be used to deduce the following:

Theorem 7.4 (Generalized Katok-Spatzier Rigidity Lemma) *Let $\phi : X \to X$ be a map preserving a (T, G)-structure with G^0 compact, and let μ be a ϕ-invariant and* **ergodic** *probability measure on X. Then for almost every $x \in X$ there is a closed subgroup $H < G$ so that H acts transitively on the support of $\mu_{x,T}$ and so that*

$$h_* \mu_{x,T} \propto \mu_{x,T} \qquad \text{for every } h \in H. \tag{7.2}$$

To give a flavor of the kind of arguments used in fairly simple (but non the less interesting) case, we deviate from our practice by giving a full proof of this theorem.

Proof. To simplify matters, it would help to choose the arbitrary normalizing set $U_0 \subset T$ from ($\Diamond$-1) to be invariant under the compact group G^0. Even better, instead of normalizing $\mu_{x,T}$ so that $\mu_{x,T}(U_0) = 1$, choose some G^0-invariant nonnegative compactly supported continuous function f_0 on G which is positive on an open neighborhood of the identity and normalize $\mu_{x,T}$ by demanding that $\int f_0 d\mu_{x,T} = 1$.

Since $x \mapsto \mu_{x,T}$ is a Borel measurable map between locally compact spaces by Luzin's theorem there is for every $\epsilon > 0$ a compact set $K_1 = K_1(\epsilon)$ of measure $\geq 1 - \epsilon$ on which this map is continuous.

Let $K_2 = K_1 \cap \text{supp}(\mu|_{K_1})$. Then $\mu(K_2) = \mu(K_1)$.

For every $y \in K_2$ and $r > 0$ the measure of $K_1 \cap B_r(y)$ ($B_r(y)$ denoting the ball of radius r around y) is positive. Since ϕ is ergodic, this means that for a.e. $x \in X$ there is an arbitrarily large n so that $\phi^n(x) \in K_1 \cap B_r(y)$.

Let $K_3 \subset K$ be a set of measure $\geq 1 - 2\epsilon$ so that for every $x \in K_3$ and $y \in K_2$ there is a sequence n_i so that $\phi^{n_i}(x) \to y$ and $\phi^{n_i}(x) \in K_1$ for all i. We may also assume that K is a subset of the set X'_ϵ of Theorem 4.1 ($\Diamond - 3$).

Since ϕ preserves (T, G)-structure, it follows easily from the definitions that for a.e. x

$$\mu_{\phi^n(x),T} = \alpha_*(\mu_{x,T}) \qquad \text{for some } \alpha \in G^0 \text{ depending on } n, \tag{7.3}$$

and we may as well assume (7.3) holds for every $x \in K_3$ and n.

Take two points $x, y \in K_3$ with y in the T-leaf of x (say $y = g.x$), and appropriate n_i as above. Then there are $\alpha_i \in G^0$ so that

$$\mu_{\phi^{n_i}(x),T} = (\alpha_i)_*(\mu_{x,T}).$$

Since G^0 is compact, without loss of generality $\alpha_i \to \alpha \in G^0$, and because $z \mapsto \mu_{z,T}$ is continuous on K_1 we have that

$$\mu_{y,T} = \lim_{i \to \infty} \mu_{\phi^n(x),T} = \alpha_*(\mu_{x,T}).$$

Now we use ($\Diamond$-1) to see that $\mu_{y,T} \propto \mu_{x,T}.g$ hence there is a $\beta \in G$ (namely the composition of right translate by g with α^{-1}) satisfying $\beta(e) = g$ and $\beta_*\mu_{x,T} \propto \mu_{x,T}$.

A moment's reflection shows that we can take $K_3(\epsilon)$ to be monotonically increasing as $\epsilon \to 0$. This means that for every $x \in K_3(\epsilon)$, for every $g \in t_{x,T}^{-1}(K_3(\epsilon') \cap T.x)$ for $\epsilon' < \epsilon$ we can find a $\beta \in G$ preserving $\mu_{x,T}$ (up to a multiplicative scalar) such that $\beta(e) = g$.

One can verify (see Exercise 7.1 below) that a.s.

$$\mathrm{supp}\mu_{x,T} \subset \bigcup_\epsilon t_{x,T}^{-1}(K_3(\epsilon') \cap T.x) \tag{7.4}$$

and the proof is complete once one shows that the subgroup of G which preserves $\mu_{x,T}$ up to a multiplicative scalar is closed. $\qquad\square$

Exercise 7.1   1. where did we use in the proof of Theorem 7.4 our choice of G^0-invariant normalization (whether by function or set)?
   2. prove (7.4).
   3. show that indeed the subgroup of G which preserves $\mu_{x,T}$ up to a multiplicative scalar is closed.

It is possible to get by with slightly less than ergodicity in the proof of Theorem 7.4 — it is not hard to isolate exactly what is needed for the argument to work — and this is useful for example for studying totally non-symplectic $\mathbb{Z}^d$ actions (see [13]).

It is an interesting question whether, for T unimodular, the proportionality constant in (7.2) can be anything but one. This would be a strange situation indeed, but except for the case $T = \mathbb{R}$ (for which the techniques of [21] are applicable) I do not know of any argument which precludes it.

Example 7.4 Let ϕ_t be a flow (i.e. an $\mathbb{R}$-action) on X, and take $\phi = \phi_1$ to be the time one map for this flow. Let μ be an ϕ-invariant and ergodic probability measure on X; we assume that a.e. x is not a periodic point of the flow ϕ_t. Then X is a $\mathbb{R}$-space, with ϕ_1 preserving the $\mathbb{R}$-leaves and acting rigidly on them.

One can easily see that in this case $\mu_{x,\mathbb{R}}$ is either Lebesgue measure or counting measure on the integers, which of course is compatible with Theorem 7.4. In particular, $\mu_{x,\mathbb{R}}$ is never a probability measure. Contrast this with Theorem 7.1 and Theorem 7.3.

The main drawback of Theorem 7.4, and it is a serious one, is the assumption of ergodicity. This is one of the rare instances where a theorem about invariant measures cannot be reduced to studying ergodic components, and in many cases, particularly for locally homogeneous spaces such as in Example 7.3, the non-ergodic case is much harder.

We end this section with the main result of [19], which fits nicely in this framework. Let G and Γ be as in Example 7.3. A measure μ on $\Gamma\backslash G$ is said to be algebraic if it is the L-invariant measure on a closed L orbit in $\Gamma\backslash G$ for some closed $L < G$.

Theorem 7.5 ([19][Thm. 1.1]) *Let $G = \mathrm{PGL}(2,\mathbb{R}) \times L$, where L is an S-algebraic group, $H < G$ be the $\mathrm{PGL}(2,\mathbb{R})$ factor of G. Take Γ to be a discrete subgroup of G (not necessarily a lattice) such that $\Gamma \cap L$ is finite. Suppose μ is a probability measure on $X = \Gamma\backslash G$, invariant under elements of the diagonal group $A = \left\{ \begin{pmatrix} * & 0 \\ 0 & * \end{pmatrix} \right\}$. Assume that*

1. *All ergodic components of μ with respect to the A-action have positive entropy.*
2. *μ is L-recurrent.*

then μ is a linear combination of algebraic probability measures invariant under H.

Note the assumption (1) about entropy which is absolutely essential for the proof. In view Remark 5.1, this condition is equivalent to μ being U-recurrent for $U = \left\{ \begin{pmatrix} 1 & * \\ 0 & 1 \end{pmatrix} \right\}$.

The proof of this theorem is beyond the scope of these notes. Despite the conceptual similarity, the techniques of [10] and [23] do not apply. It is somewhat similar in spirit to the proof of Theorem 7.4, but is substantially more involved, and draws heavily from ideas developed by Ratner in the study of unipotent flows [32, 31]; to learn more about this deep and fascinating subject, see e.g. [33] and the recent and comprehensive [27].

The proof of Theorem 7.5 also uses a technical but important lemma from [5], which implies that there is a subset $X' \subset X$ of full measure so that for every two $x, x' \in X'$ with $x' \in L.x$ the conditional measures $\mu_{x,U} = \mu_{x',U}$. This allows us to use the fact that μ is L recurrent almost as if μ was actually invariant under a subgroup of L.

Theorem 7.5 has an interesting application in a completely different subject. Namely, in conjunction with a result with Jean Bourgain [2], it proves a special but important case of the quantum unique ergodicity conjecture. This conjecture deals with properties of eigenfunctions of the Laplacian, and the arithmetic case to which Theorem 7.5 is applicable deals with the special case of Hecke-Maass and more general automorphic forms.

We also remark that combining the main result of [5] with the methods of [19] we have been able to prove a theorem along the line of Rudolph's theorem for the action of the diagonal group on $SL(n, \mathbb{R})/SL(n, \mathbb{Z})$, which in particular implies that the set of exceptions to Littlewood's conjecture, i.e. those $(\alpha, \beta) \in \mathbb{R}^2$ satisfying $\varliminf n \, \|n\alpha\| \, \|n\beta\| > 0$, which presumably is empty, has at most Hausdorff dimension zero [6].

8. Automorphic Forms, Quantum Unique Ergodicity, and Hecke Recurrence

Take $X = \Gamma \backslash PGL(2, \mathbb{R})$, where Γ is either a congruence subgroup of $PGL(2, \mathbb{Z})$ or of certain lattices that arise from quaternionic division algebras over $\mathbb{Q}$ that are unramified over $\mathbb{R}$. The later lattices are slightly harder to define but have the advantage that X is compact.

In both cases, for all but finitely many prime p which we exclude from now on, there is a map T_p from X to $(p+1)$-tuples of points of X called the **Hecke correspondence**. For example, if $\Gamma = PGL(2, \mathbb{Z})$, one can define T_p of a point $x \in X$ as follows: choose some $g \in PGL(2, \mathbb{R})$ so that $x = \Gamma g$. Then

$$T_p(x) = \left\{ \Gamma \begin{pmatrix} p & 0 \\ 0 & 1 \end{pmatrix} g, \Gamma \begin{pmatrix} p & 0 \\ 1 & 1 \end{pmatrix} g, \dots, \Gamma \begin{pmatrix} p & 0 \\ p-1 & 1 \end{pmatrix} g, \Gamma \begin{pmatrix} 1 & 0 \\ 0 & p \end{pmatrix} g \right\}. \tag{8.1}$$

Notice that while each individual element of the right hand side depend on the choice of g, the set of $p + 1$ points depends only on x.

Using the Hecke correspondence we can define operators called the **Hecke operators** on $L^2(X)$ by

$$T_p(f)[x] = \sum_{y \in T_p(x)} f(y) \tag{8.2}$$

which play a very important role in the spectral theory of X. They are self adjoint, and for distinct primes p, q we have that T_p and T_q commute.

Here is an alternative way to think of these correspondences. Again, consider for simplicity $\Gamma = PGL(2, \mathbb{Z})$. Consider $G = PGL(2, \mathbb{R}) \times PGL(2, \mathbb{Q}_p)$. Then $\tilde{\Gamma} = GL(2, \mathbb{Z}[p^{-1}])$ can be embedded as a dense subgroup of both $PGL(2, \mathbb{R})$ and $PGL(2, \mathbb{Q}_p)$. The diagonal embedding of $\tilde{\Gamma}$ in G is discrete (in fact is a lattice, i.e. has finite covolume), and it is well-known that

$$X = \Gamma \backslash PGL(2, \mathbb{Z}) \cong \tilde{\Gamma} \backslash PGL(2, \mathbb{R}) \times PGL(2, \mathbb{Q}_p)/PGL(2, \mathbb{Z}_p) \tag{8.3}$$

(with $\mathrm{PGL}(2, \mathbb{Z}_p)$ embedded into G by $g \mapsto (e, g)$). A similar construction works for more general congruence lattices, which is precisely why we are interested in such lattices.

It is well known that $\mathrm{GL}(2, \mathbb{Q}_p)/\mathrm{PGL}(2, \mathbb{Z}_p)$ can be naturally identified with a $p + 1$-regular tree [37]. To every $x \in \tilde{\Gamma} \backslash G/\mathrm{PGL}(2, \mathbb{Z}_p)$ there is attached the "coset" $x\mathrm{PGL}(2, \mathbb{Q}_p)/\mathrm{PGL}(2, \mathbb{Z}_p)$ i.e. a $p+1$ regular combinatorial tree embedded in $\Gamma \backslash G/\mathrm{PGL}(2, \mathbb{Z}_p)$. In view of the isomorphism (8.3), we see that for every $x \in X$ there corresponds $p + 1$ regular combinatorial tree, which we will call the **Hecke tree**, embedded in X. The $p+1$ nearest neighbors to x on this tree are precisely the $p + 1$ points given by (8.1).

In the notations of Appendix A, this construction gives X the structure of a $(T, \mathrm{PGL}(2, \mathbb{Q}_p))$-space with $T = \mathrm{GL}(2, \mathbb{Q}_p)/\mathrm{PGL}(2, \mathbb{Z}_p)$ — a $p + 1$ regular tree.

We now describe the quantum unique ergodicity conjecture of Rudnick and Sarnak, and explain how it is related to Theorem 7.5.

Conjecture 8.1 (Rudnick and Sarnak [34]) Let M be a compact hyperbolic surface, and let ϕ_i a sequence of linearly independent eigenfunctions of the Laplacian Δ on M, normalized to have L^2-norm one. Then the probability measures $\tilde{\mu}_i$ defined by $\tilde{\mu}_i(A) = \int_A |\phi_i(x)|^2 \, d\eta_M(x)$ tend in the weak star topology to the uniform measure η_M.

A. I. Šnirel'man, Y. Colin de Verdière and S. Zelditch have shown in great generality (specifically, for any manifold on which the geodesic flow is ergodic) that if one omits a subsequence of density 0 the remaining $\tilde{\mu}_i$ do indeed converge to η_M [38, 4, 43]. An important component of their proof is the **microlocal lift** of any weak star limit $\tilde{\mu}$ of a subsequence of the $\tilde{\mu}_i$. The microlocal lift of $\tilde{\mu}$ is a measure μ on the unit tangent bundle SM of M whose projection to M is $\tilde{\mu}$, and most importantly it is always invariant under the geodesic flow on SM. We shall call any measure μ on SM arising as a microlocal lift of a weak star limit of $\tilde{\mu}_i$ a **quantum limit**. Thus a slightly stronger form of Conjecture 8.1 is the following conjecture, also due to Rudnick and Sarnak:

Conjecture 8.2 (Quantum Unique Ergodicity Conjecture) For any compact hyperbolic surface M the only quantum limit is the uniform measure η_{SM} on SM.

If $M = \Gamma \backslash \mathbb{H}$, then SM can be identified with $\Gamma \backslash \mathrm{PGL}(2, \mathbb{R})$; under this identification the geodesic flow becomes the flow associated to the group $A < \mathrm{PGL}(2, \mathbb{R})$ of diagonal matrices. We further restrict ourselves now to the case $M = \Gamma \backslash \mathbb{H}$, for Γ one of the arithmetic lattices considered above. If Γ is a lattice arising from a quaternionic division algebra over $\mathbb{Q}$, then M is a compact hyperbolic surface, precisely the kind of surface considered in Conjectures 8.1 and 8.2. If Γ is a congruence sublattice of $\mathrm{PGL}(2, \mathbb{Z})$

then M has finite volume, but is not compact. A special property of these very special surfaces of finite volume is that the asymptotics of the number of eigenfunctions of the Laplacian in $L^2(M)$ with eigenvalues less than some bound behaves according to the Weyl law governing the behavior of this quantity for compact surfaces of the same area. We remark that the continuous spectrum, which in this case is given by Eisenstein series, was shown by Luo, Sarnak and Jakobson to obey the natural analogue of the quantum unique ergodicity conjecture in this case [22, 12].

When looking at eigenfunctions of the Laplacian on arithmetic surfaces, it is natural to consider joint eigenfunctions of both the Laplacian and of all Hecke operators. Since the Hecke operators are self adjoint operators that commute with each other and with the Laplacian one can always find an orthonormal basis of the subspace of $L^2(M)$ which corresponds to the discrete part of the spectrum of the Laplacian consisting of such joint eigenfunctions. Presumably, for e.g. $\Gamma = \mathrm{PGL}(2, \mathbb{R})$ the spectrum of the Laplacian on M is simple (more generally, for congruence lattices, the spectrum is expected to have bounded multiplicities which makes little difference) and if this holds then the eigenfunctions of the Laplacian are automatically eigenfunctions of all Hecke operators.

Joint eigenfunctions of the Laplacian are called Hecke-Maass cusp forms, and play an extremely important role in modern analytic number theory. It is natural to consider the following special case of Rudnick and Sarnak's quantum unique ergodicity conjecture:

Question 8.1 (Arithmetic Quantum Unique Ergodicity) Let M be $\Gamma \backslash \mathbb{H}$ for Γ a congruence that this as above, and let ϕ_i a sequence of linearly independent joint eigenfunctions of the Laplacian Δ and all Hecke operators on M, normalized to have L^2-norm one. What are the possible microlocal lifts of weak star limits of the probability measures $\tilde{\mu}_i$ defined as above by $\tilde{\mu}_i(A) = \int_A |\phi_i(x)| \, d\eta_M(x)$ (we shall refer to these lifts as arithmetic quantum limits)?

Let us consider what we know about these arithmetic quantum limits. They are measures on X, which, if we want we can think of as $\tilde{\Gamma} \backslash \mathrm{PGL}(2, \mathbb{R}) \times T$ for $T = \mathrm{PGL}(2, \mathbb{Q}_p)/\mathrm{PGL}(2, \mathbb{Z}_p)$. They are invariant under the action of $A < \mathrm{PGL}(2, \mathbb{R})$. What can we say about how arithmetic quantum limits interact with the T-structure of X?

It turns out that in the arithmetic case, the quantum limits are T recurrent. Strictly speaking, we have not properly defined what this means since T is not a group. The discussion in sections §3– §6 generalizes to the framework of (T, G)-spaces (as defined in Appendix A) as is done in [19].

Set $\hat{X} = \tilde{\Gamma} \backslash G$, $K_p = \mathrm{PGL}(2, \mathbb{Z}_p)$ and $G_p = \mathrm{PGL}(2, \mathbb{Q}_p)$. A measure μ on X can be thought of as a K_p invariant measure $\tilde{\mu}$ on $\hat{X}$. Not surprisingly, μ is T-recurrent iff $\tilde{\mu}$ is G_p-recurrent.

Suppose μ is a quantum limit that comes from lifting a weak star limit of $|\phi_n(x)|^2 d\eta_M$ with all $\phi_n(x)$ eigenfunctions of the Hecke operator T_p. It turns out that μ can also be represented as a weak star limit of $|\Phi_n(x)|^2 d\eta_X$ with $\Phi_n(x) \in L^2(X)$ eigenfunctions of T_p with $\|\Phi_n\| = 1$ (see [42, 16]). Purely for formal reasons consider Φ_n as K_p-invariant functions $\tilde{\Phi}_n$ on $\tilde{X}$. A function $\tilde{\Phi}$ on $\tilde{X}$ can be convolved with a function (say continuous, compactly supported) on G_p. The bi-K_p invariant functions on G_p form a commuting algebra with respect to convolutions, the Hecke algebra, and Φ_n is an eigenfunction of the Hecke operator T_p iff $\tilde{\Phi}_n$ is an eigenfunction of the Hecke algebra.

Let $\tilde{\mu}_n$ be the measure on $\tilde{X}$ defined by $|\tilde{\Phi}_n|^2 d\eta_{\tilde{X}}$. Since $\eta_{\tilde{X}}$ is G_p invariant, It follows e.g. from Exercise 4.2 that $(\tilde{\mu}_n)_{x,G_p}$ (which, we recall is a measure on G_p is given (up to a normalizing constant) by the measure with density $|\tilde{\Phi}_n(g.x)|^2 d\eta_{G_p}$.

The function $f(g) = \tilde{\Phi}_n(g.x)$ is a K_p invariant eigenfunction of the Hecke algebra (since it is K_p-invariant, we can think of f as a function on the $p+1$-regular tree T, and the fact is it is an eigenfunction of the Hecke algebra translates simply to f being an eigenfunction of the graph theoretic Laplacian on the tree T.

There are no K_p-invariant eigenfunctions of the Hecke algebra in $L^2(G_p)$ (this is well known, and can be shown elementarily by studying eigenfunctions of the tree Laplacian), so $(\tilde{\mu}_n)_{x,G_p}$ is infinite. Indeed, it can be shown that this holds uniformly in n and x, i.e. for every M, there is a compact $C \subset G_p$ so that $(\tilde{\mu}_n)_{x,G_p}(C) > M$. Because of this uniformity, it can be deduced that $\tilde{\mu}_{x,G_p}$ is infinite, hence $\tilde{\mu}$ is G_p recurrent (equivalently, μ is T-recurrent). See [19][Sect. 8] for more details; the basic idea in a less elaborate form appeared in [34].

At this stage we know that any quantum limit μ is A-invariant and T-recurrent. At the current state of technology, this is not enough to identify μ. But so far, we have used only one Hecke operator. By using the combinatorics of Hecke correspondences for many primes simultaneously, jointly with J. Bourgain we have proved

Theorem 8.1 ([2]) *Every ergodic component of an arithmetic quantum limit has positive entropy (actually greater than some explicit lower bound).*

We now have all the ingredients to apply Theorem 7.5 and we get

Theorem 8.2 (Arithmetic Quantum Unique Ergodicity) *Let M be $\Gamma\backslash\mathbb{H}$ for Γ a congruence lattice as above. Then any arithmetic quantum limit for M is $c\eta_M$ for some $c \in [0,1]$; in the compact case, $c = 1$.*

This (almost) proves the arithmetic case of the quantum unique ergodicity conjecture of Rudnick and Sarnak (the only missing piece is showing $c = 1$ also in the non compact case).

Lior Silberman and Akshay Venkatesh have generalized the methods developed in [2] to more general groups and using [5, 6, 19] obtained generalizations to Theorem 8.2 above [39]. They have also pointed out to me a gap in the passage from the cocompact case to the finite volume case (i.e. Γ congruence subgroup of $SL(2, \mathbb{Z})$) in [2]. Conveniently, they also came up with a way of fixing this gap.

We mention that ergodic theory is not the only approach to arithmetic quantum unique ergodicity. In his thesis Thomas Watson [41] linked the arithmetic quantum unique ergodicity question with subconvexity estimates for automorphic L-functions at the critical line, and there are many other related results including [36] where it is shown using analytical techniques that a certain subsequence of the Hecke Maass forms, the CM-forms, satisfy arithmetic quantum unique ergodicity. However, at the moment, ergodic theory does indeed seem to give the best results in this particular direction.

Appendix A. (T,G)-Spaces

In this appendix, we give a precise definition to what we mean by a (T, G)-space, with T a homogeneous space for the locally compact group G.

Definition A.1. A locally compact metric space X is said to be a (T, G)-space if there is some open cover $\mathfrak{T}$ of X by relatively compact sets, and for every $U \in \mathfrak{T}$ a continuous map $t_U : U \times T \to X$ with the following properties:

A-1 For every $x \in U \in \mathfrak{T}$, we have that $t_U(x, e) = x$.

A-2 For any $x \in U \in \mathfrak{T}$, for any $y \in t_U(x, T)$ and $V \in \mathfrak{T}$ containing y, there is a $\theta \in G$ so that

$$t_V(y, \cdot) \circ \theta = t_U(x, \cdot). \qquad (A.1)$$

In particular, for any $x \in U \in \mathfrak{T}$, and any $y \in t_U(x, T), V \in \mathfrak{T}(y)$ we have that $t_U(x, T) = t_V(y, T)$, and $t_U(x, \cdot)$ is injective if and only if $t_V(y, T)$ is.

A-3 There is some $r_U > 0$ so that for any $x \in U$ the map $t_U(x, \cdot)$ is injective on $\overline{B_{r_U}^T(e)}$.

A-4 For a dense set of $x \in X$, for some (equivalently for every) $U \in \mathfrak{T}(x)$ the map $t_U(x, \cdot)$ is injective.

Note that if X is a (T, G)-space and $G < H$ then X is automatically also a (T, H)-space. The most interesting case is when G acts on T by isometries. If the stabilizer in G of the point $e \in T$ is compact than it is always possible to find a metric on T so that G acts by isometries.

Example A.2. Suppose that G is a locally compact metric group, acting continuously on a locally compact metric space X. Suppose that this action is locally free (i.e. G_x is a discrete subgroup of G for every x) and free on a dense set of $x \in X$.

Then X is a (G, G)-space with $t_U(x, g) = g.x$ (note that this has trivial dependence on U) for some countable atlas of open sets $\mathfrak{T}$ which satisfy (A-3) above. If X is compact we can simply take $\mathfrak{T} = \{X\}$.

Note that we implicitly identify G with the subgroup of left translations in the group of homeomorphism of G.

References

1. Aaronson, J. (1997) *An Introduction to Infinite Ergodic Theory*, **Vol.50** of *Mathematical Surveys and Monographs*, American Mathematical Society, Providence, RI.
2. Bourgain, J. and Lindenstrauss, E (2003) Entropy of Quantum Limits, *Comm. Math. Phys.*, **Vol.233(1)**, pp. 153–171.
3. Bufetov, A.I. (2002) Convergence of Spherical Averages for Actions of Free Groups, *Ann. of Math.*, **Vol.(2), 155(3)**, pp. 929–944.
4. Colin de Verdière, Y. (1985) Ergodicité et Fonctions Propres du Laplacien, *Comm. Math. Phys.*, **Vol.102(3)**, pp. 497–502.
5. Einsiedler, M. and Katok, A. (2003) Invariant Measures on G/Γ for Split Simple Lie Groups G, *Comm. Pure Appl. Math.*, **Vol.56(8)**, pp. 1184–1221, dedicated to the memory of Jürgen K. Moser.
6. Einsiedler, M., Katok, A. and Lindenstrauss, E. (2003) Invariant Measures and the Set of Exceptions to Littlewoods Conjecture, submitted (44 pages).
7. Einsiedler, M. and Lindenstrauss, E. (2003) Rigidity Properties of Z^d-Actions on Tori and Solenoids, *Electron. Res. Announc. Amer. Math. Soc.*, **Vol.9**, pp. 99–110.
8. Greschonig, G. and Schmidt, K. (2000) Ergodic Decomposition of Quasi-Invariant Probability Measures, *Colloq. Math.*, 84/85, part 2 pp. 495–514, dedicated to the memory of Anzelm Iwanik.
9. Rodriguez Hertz, F. (2002) Stable Ergodicity of Certain Linear Automorphisms of the Torus, arXiv:math.DS/0212345.
10. Host, B. (1995) Nombres Normaux, Entropie, Translations, *Israel J. Math.*, **Vol.91(1-3)**, pp. 419–428.
11. Host, B. (2000) Some Results of Uniform Distribution in the Multidimensional Torus, *Ergodic Theory Dynam. Systems*, **Vol.20(2)**, pp. 439–452.
12. Jakobson, D. (1997) Equidistribution of Cusp Forms on $PSL_2(\mathbb{Z}) \backslash PSL_2(\mathbb{R})$, *Ann. Inst. Fourier (Grenoble)*, **Vol.47(3)**, pp. 967–984.
13. Kalinin, B. and Katok, A. (2001) Invariant Measures for Actions of Higher Rank Abelian Groups, in *Smooth Ergodic Theory and its Applications (Seattle, WA, 1999)*, **Vol.69** of *Proc. Sympos. Pure Math.*, pp. 593–637. Amer. Math. Soc., Providence, RI.
14. Katok, A. and Spatzier, R. J. (1996) Invariant Measures for Higher-Rank Hyperbolic Abelian Actions, *Ergodic Theory Dynam. Systems*, **Vol.16(4)**, pp. 751–778.
15. Katok, A. and Spatzier, R. J. (1996) Corrections to: "Invariant Measures for Higher-Rank Hyperbolic Abelian Actions" [Ergodic Theory Dynam. Systems, **Vol.16(4)**, pp. 751–778; MR 97d:58116]. *Ergodic Theory Dynam. Systems*, **Vol.18(2)**, pp. 503–507, (1998).
16. Lindenstrauss, E. (2001) On Quantum Unique Ergodicity for $\Gamma \backslash \mathbb{H} \times \mathbb{H}$, *Internat. Math. Res. Notices*, **Vol.17**, pp. 913–933.
17. Lindenstrauss, E. (2001) p-Adic Foliation and Equidistribution, *Israel J. Math.*, **Vol.122**, pp. 29–42.
18. Lindenstrauss, E. (2001) Pointwise Theorems for Amenable Groups, *Invent. Math.*, **Vol.146(2)**, pp. 259–295.

19. Lindenstraussi, E. (2003) Invariant Measures and Arithmetic Quantum Unique Ergodicity, submitted (54 pages).

20. Lindenstrauss, E., Meiri, D. and Peres, Y. (1999) Entropy of Convolutions on the Circle, *Ann. of Math. (2)*, **Vol.149(3)**, pp. 871–904.

21. Lindenstrauss, E. and Rudolph, D. (2003) in preparation.

22. Luo, Wen Zhi and Sarnak, P. (1995) Quantum Ergodicity of Eigenfunctions on $PSL_2(\mathbb{Z})\backslash\mathbb{H}^2$, *Inst. Hautes Études Sci. Publ. Math.*, **Vol.81**, pp. 207–237.

23. Lindenstrauss, E. and Schmidt, K. (2003) Invariant Measures of Nonexpansive Group Automorphisms, *Israel J. Math.* (28 pages), to appear.

24. Ledrappier, F. and Young, L.-S. (1985) The Metric Entropy of Diffeomorphisms. I. Characterization of Measures Satisfying Pesin's Entropy Formula, *Ann. of Math. (2)*, **Vol.122(3)**, pp. 509–539.

25. Mattila, P. (1995) *Geometry of Sets and Measures in Euclidean Spaces*, **Vol.44** of *Cambridge Studies in Advanced Mathematics*, Cambridge University Press, Cambridge, Fractals and Rectifiability.

26. Meiri, D. (1998) Entropy and Uniform Distribution of Orbits in $\mathbf{T}^d$, *Israel J. Math.*, **Vol.105**, pp. 155–183.

27. Morris, D.W. (2003) Ratner's Theorem on Unipotent Flows, arXiv:math.DS/0310402. 120 pages, 14 figures, *Chicago Lectures in Mathematics Series*, University of Chicago Press, submitted.

28. Nevo, A. (1999) On Discrete Groups and Pointwise Ergodic Theory, in *Random Walks and Discrete Potential Theory (Cortona, 1997)*, Sympos. Math., **Vol.XXXIX**, pp. 279–305, Cambridge Univ. Press, Cambridge.

29. Ornstein, D. and Weiss, B. (1983) The Shannon-McMillan-Breiman Theorem for a Class of Amenable Groups, *Israel J. Math.*, **Vol.44(1)**, pp. 53–60.

30. Ornstein, D.S. and Weiss, B. (1987) Entropy and Isomorphism Theorems for Actions of Amenable Groups, *J. Analyse Math.*, **Vol.48**, pp. 1–141.

31. Ratner, M. (1983) Horocycle Flows, Joinings and Rigidity of Products, *Ann. of Math. (2)*, **Vol.118(2)**, pp. 277–313.

32. Ratner, M. (1991) On Raghunathan's Measure Conjecture, *Ann. of Math. (2)*, **Vol.134(3)**, pp. 545–607.

33. Ratner, M. (1992) Raghunathan's Conjectures for $SL(2,\mathbb{R})$, *Israel J. Math.*, **Vol.80(1-2)**, pp. 1–31.

34. Rudnick, Z. and Sarnak, P. (1994) The Behaviour of Eigenstates of Arithmetic Hyperbolic Manifolds, *Comm. Math. Phys.*, **Vol.161(1)**, pp. 195–213.

35. Rudolph, D.J. (1990) ×2 and ×3 Invariant Measures and Entropy, *Ergodic Theory Dynam. Systems*, **Vol.10(2)**, pp. 395–406.

36. Sarnak, P. (2001) Estimates for Rankin-Selberg L-Functions and Quantum Unique Ergodicity, *J. Funct. Anal.*, **Vol.184(2)**, pp. 419–453.

37. Serre, J.-P. (1980) *Trees*, Springer-Verlag, Berlin, translated from the French by John Stillwell.

38. Šnirel'man, A.I. (1974) Ergodic Properties of Eigenfunctions, *Uspehi Mat. Nauk*, **Vol.29(6(180))**, pp. 181–182.

39. Silberman, L. and Venkatesh, A. (2003) Quantum Unique Ergodicity on Locally Symmetric Spaces, in preparation.

40. Tempel'man, A.A. (1967) Ergodic Theorems for General Dynamical Systems, *Dokl. Akad. Nauk SSSR*, **Vol.176**, pp. 790–793.

41. Watson, T. (2001) Rankin Triple Products and Quantum Chaos, Ph.D. thesis, Princeton University.

42. Wolpert, S.A. (2001) The Modulus of Continuity for $\Gamma_0(m)\backslash\mathbb{H}$ Semi-Classical Limits, *Comm. Math. Phys.*, **Vol.216(2)**, pp. 313–323.

43. Zelditch, S. (1987) Uniform Distribution of Eigenfunctions on Compact Hyperbolic Surfaces, *Duke Math. J.*, **Vol.55(4)**, pp. 919–941.

STOCHASTIC PARTICLE APPROXIMATIONS FOR TWO-DIMENSIONAL NAVIER-STOKES EQUATIONS

SYLVIE MÉLÉARD
Université Paris 10
MODALX, 200 Av. de la République
92000 Nanterre

and

Laboratoire de Probabilités et Modèles Aléatoires
UMR 7599
4 place Jussieu, 75252 Paris cedex, France
sylm@ccr.jussieu.fr

Abstract. We present a probabilistic interpretation of some Navier-Stokes equations which describe the behaviour of the velocity field in a viscous incompressible fluid. We deduce from this approach stochastic particle approximations, which justify the vortex numerical schemes introduced by Chorin to simulate the solutions of the Navier-Stokes equations.

After some recalls on the McKean-Vlasov model, we firstly study a Navier-Stokes equation defined on the whole plane. The probabilistic approach is based on the vortex equation, satisfied by the curl of the velocity field. The equation is then related to a nonlinear stochastic differential equation, and this allows us to construct stochastic interacting particle systems with a "propagation of chaos" property: the laws of their empirical measures converge, as the number of particles tends to infinity to a deterministic law with time-marginals solving the vortex equation. Our approach is inspired by Marchioro and Pulvirenti [26] and we improve their results in a pathwise sense.

Next we study the case of a Navier-Stokes equation defined on a bounded domain, with a no-slip condition at the boundary. In this case, the vortex equation satisfies a Neumann condition at the boundary, which badly depends on the solution. We simplify the model by studying in details the case of a fixed Neumann condition and we finally explain how the results should be adapted in the Navier-Stokes case.

A. Maass et al. (eds.), *Dynamics and Randomness II*, 147–197.

1. Introduction

We present in this course a probabilistic interpretation of some Navier-Stokes equations, from which we will deduce stochastic particle approximations and numerical schemes for the solutions of the equations.

The Navier-Stokes equations we consider describe the evolution of the velocity of a viscous and incompressible fluid in dimension two. About twenty years ago, Chorin [9] proposed a vortex method to simulate the solutions of these equations, based on the vorticity and involving cutoff kernels. His approach was not mathematically justified and many authors gave partial proofs of convergence of the algorithm.

The main fact for explaining this approach is that in dimension two, the Navier-Stokes equation can be expressed as a simpler equation for the curl of the velocity, called the vortex equation. In the stochastic framework, this equation appears as a McKean-Vlasov equation, in which the coefficient of the drift term can explode. This remark is the basis of the probabilistic interpretation.

In 1982, Marchioro and Pulvirenti [26] have given a probabilistic interpretation of the Navier-Stokes equation thanks to a nonlinear diffusion, for bounded integrable initial data. They have rigorously introduced a cut-off model and some particle systems, and proved for each fixed time the convergence of the expectations of the empirical measures of the particle systems to the solution of the N.S. equation.

Then an open question was the pathwise convergence of these empirical measures to the law of the nonlinear diffusion, or equivalently, the propagation of chaos for the interacting particle systems.

In 1987, Osada [32] proved a propagation of chaos result for an interacting particle system without cut-off by an analytical method based on generators of generalized divergence form, but only for large viscosities and bounded density initial data. The tightness of the laws of the particle systems was always satisfied and the constraint on the viscosity appeared in the identification of the limiting laws. This result is not satisfying, since the numerical stochastic particle methods are most efficient in the case of small viscosities, case which was not considered by the author. (See for example the comparison between finite volume deterministic methods and stochastic particle methods in [4]).

In this course, our aim is to obtain some pathwise particle approximations of the solution of the Navier-Stokes equation by easily simulable systems. We consider two situations. In the first one, the equation is considered in the whole plane with an integrable and bounded initial condition (cf. [29]). We will interpret the vortex equation in a probabilistic point of view and will deduce some pathwise approximations with a precise rate of convergence. The second case will be devoted to equations in a bounded domain of the space, with a Dirichlet condition at the boundary (a no-slip condition). This will correspond to a vortex equation with a Neumann condition at the boundary, as it has been heuristically proven in [10]. We will study the simplified case in which the Neumann condition is fixed, as developed in [21]. We will explain how we should modify the approach to approximate the solution of the Navier-Stokes equation with no-slip condition on the boundary of the domain.

We will essentially consider the framework introduced by Marchioro and Pulvirenti and will define particle systems with cut-off drift coefficients. Then we will study the convergence of these particle systems when the number of particles tends to infinity and the cut-off parameter tends to zero. In both cases described above, one defines a nonlinear diffusion process associated with the vortex (nonlinear) equation. In the bounded domain case, it is a reflected process with space-time random births at the boundary. At the level of processes, the nonlinearity means that the drift coefficient depends on the law of the diffusion process. We define a coupling between independent copies of this nonlinear process and some interacting particle systems with cut-off drift kernels. We work in the path space and consider initial data which are not necessarily probability densities. So we associate with any sample path a signed weight depending on the initial condition. We makes the size of the system tend to infinity and the size of the cut-off tend to 0 in related asymptotics, and obtain in such a way a propagation of chaos result for the particle system. Then the weighted empirical measures converge, as probability measures on the path space, to a deterministic probability measure of which time marginals are measure solutions of the vortex equation. This result leads to a natural algorithm to simulate the solution of the Navier-Stokes equation.

We will in the second section recall the main results concerning the probabilistic interpretation of the McKean-Vlasov equation and the associated interacting particle systems. We will explain what "propagation of chaos" means. Next, we will describe the two-dimensional Navier-Stokes equation in the whole plane and its relation with the vortex equation. We will then introduce the probabilistic framework and construct the interacting particle approximations. In the fourth section, we will develop the case of a bounded domain. For a Navier-Stokes equation with a no-slip condition at the boundary (which is a physically natural hypothesis), the associated vortex equation induces a nontrivial Neumann boundary condition and leads to a more complicated probabilistic interpretation. We will rigorously study a simpler case and explain what we should do in the initial one.

Notation

- For any integer $1 \leq p \leq +\infty$, we denote by L^p the space $L^p(\mathbb{R}^2)$. We will denote by $|.|$ the euclidian norm in $\mathbb{R}^2$, by $\|.\|_\infty$ the L^∞-norm and by $\|.\|_1$ the L^1-norm in $\mathbb{R}^2$.

- For any polish space E, the space $\mathcal{P}(E)$ will be the space of probability measures on E.

- For $p \in \mathcal{P}(E)$ and for any bounded measurable function f defined on E, we denote by $\langle p, f \rangle$ the integral $\int_E f(x)p(dx)$.

- The letter C will denote a positive real constant which can change from line to line.

2. Recalls on the McKean-Vlasov Model

Let us now recall the classical McKean-Vlasov model. McKean [27] following ideas of Kac, was the first one to rigorously study these equations, and Gärtner [14] introduced the terminology, relying the previous McKean works concerning stochastic systems in weak interaction with the problem of the Vlasov equation as limit of particle systems evolving following the law of the Newtonian mechanics.

2.1. THE MCKEAN-VLASOV EQUATION AND THE ASSOCIATED NONLINEAR MARTINGALE PROBLEM

The nonlinear partial differential equation, called McKean-Vlasov equation, is a nonlinear Fokker-Planck equation given in dimension d by

$$\frac{\partial p_t}{\partial t} = \frac{1}{2} \sum_{i,j=1}^{d} \frac{\partial^2}{\partial x_i \partial x_j}(a_{ij}[x, p_t]p_t) - \sum_{i=1}^{d} \frac{\partial}{\partial x_i}(b_i[x, p_t]p_t), \quad p_0 \in \mathcal{P}(\mathbb{R}^d),$$

$$(2.1)$$

where p_t is for any time t a probability measure on $\mathbb{R}^d$, and for $m \in \mathcal{P}(\mathbb{R}^d)$,

$$b[x, m] = \int_{\mathbb{R}^d} b(x, y)m(dy), \quad b(x, y) \text{ being a vector of } \mathbb{R}^d$$

$$a[x, m] = \sigma[x, m]\sigma[x, m]^*, \text{ and}$$

$$\sigma[x, m] = \int_{\mathbb{R}^d} \sigma(x, y)m(dy), \quad \sigma(x, y) \text{ being a matrix of size (d, k).}$$

The equation is understood in a weak sense. For nice test functions φ, we have

$$\frac{\partial}{\partial t} < p_t, \varphi > \;=\; < p_t, \frac{1}{2} \sum_{i,j=1}^{d} a_{ij}[., p_t] \frac{\partial^2 \varphi}{\partial x_i \partial x_j}(.) + \sum_{i=1}^{d} b_i[., p_t] \frac{\partial \varphi}{\partial x_i}(.) >$$

$$=\; < p_t, \mathcal{L}(p_t)\varphi > \qquad (2.2)$$

where the second order differential generator $\mathcal{L}(m)\varphi$ is defined for φ in $C_b^2(\mathbb{R}^d)$ by

$$\mathcal{L}(m)\varphi(x) = \frac{1}{2} \sum_{i,j=1}^{d} a_{ij}[x, m] \frac{\partial^2 \varphi}{\partial x_i \partial x_j}(x) + \sum_{i=1}^{d} b_i[x, m] \frac{\partial \varphi}{\partial x_i}(x). \qquad (2.3)$$

This equation has been studied from a probabilistic point of view by several authors, in particular McKean [27], Tanaka [38], Léonard [24], Gärtner [14], Sznitman [37] and Méléard [28]. The probabilistic approach consists in looking for underlying stochastic processes of which time marginals are solutions of the nonlinear equation. More precisely, one assumes some Markovian behaviour and one defines these processes as solutions of nonlinear martingale problems, as follows.

Definition 2.1 Let $\{X_t, t \in [0, T]\}$ be the canonical process on $C([0, T], \mathbb{R}^d)$ and let us consider P_0 belonging to $\mathcal{P}(\mathbb{R}^d)$. The probability measure P on $C([0, T], \mathbb{R}^d)$ is a solution of the nonlinear martingale problem $(\mathcal{M}_{MV})$ issued from P_0 if for every $\varphi \in C_b^2(\mathbb{R}^d)$,

$$\varphi(X_t) - \varphi(X_0) - \int_0^t \mathcal{L}(P_s)\varphi(X_s)ds \qquad (2.4)$$

is a P-$(\mathcal{F}_t)$ martingale where $P_s = P \circ X_s^{-1}$, $P_0 = P \circ X_0^{-1}$ and $\mathcal{F}_t = \sigma(X_s, s \le t)$.

Remark 2.2   1) If we take expectations in (2.4), the family $(P_t)_{t \ge 0}$ is a solution of the evolution equation (2.1). The martingale problem gives more information than the evolution equation. It enables to consider multidimensional time-marginals as $P[X_s \in A, X_t \in B]$ or functionals depending on the whole process, as for example hitting times. So we consider the whole Markov process corresponding to the underlying physical model.

   2) This martingale problem defines a class of generalized Markov processes. Given an initial condition $\mu \in \mathcal{P}(\mathbb{R}^d)$, we are looking for a law

P^μ on the canonical space satisfying $P^\mu(X(0) \in A) = \mu(A)$, but we do not demand that $P^\mu(B) = \int P^x(B)\mu(dx)$ (with $P^x = P^{\delta_x}$).

If we denote $p_t(A) = P^\mu(X_t \in A)$, the process $(X_\cdot, P^\mu)$ is Markovian in the sense that for any t, the quantity $P^\mu(X_{t+\cdot} \in B | X_s, s \le t)$ is a function of X_t and p_t and $\forall x \in \mathbb{R}^d$,

$$P^\mu(X_{t+\cdot} \in B | X_t = x) = P^{p_t}(X_\cdot \in B | X(0) = x).$$

Theorem 2.3 *If the coefficients σ and b are Lipschitz continuous on $\mathbb{R}^{2d}$, and if P_0 has a second order moment, the nonlinear martingale problem $(\mathcal{M}_{MV})$ has a unique solution.*

Proof. We prove here a stronger result, namely the existence and uniqueness, pathwise (given X_0 and B) and in law, of the solution of the following nonlinear stochastic differential equation:

$$X_t = X_0 + \int_0^t \sigma[X_s, P_s] dB_s + \int_0^t b[X_s, P_s] ds, \qquad (2.5)$$

where B is a k-dimensional Brownian motion and X_0 independent of B is distributed according to P_0. The nonlinearity appears through P_s, which is the marginal at time s of the law P of X,

This proof is completely detailed in Sznitman [37] Theorem 1.1 and is based on a fixed point theorem. Let $\mathcal{P}(C_T)$ be the space of probability measures on $C_T = C([0, T], \mathbb{R}^d)$ endowed with the weak convergence. This topology is metrisable with the Vaserstein complete metric ρ_T (cf. Rachev [34]), defined for m_1, m_2 by

$$\rho_T(m_1, m_2) = \inf\{\int_{C^T \times C^T} (d_T(x, y) \wedge 1) \, m(dx, dy);$$
$$m \in \mathcal{P}(C_T \times C_T) \text{ with marginals } m_1 \text{ and } m_2\}$$

Here, d_T denotes the uniform metric on C_T. One considers the mapping $\psi : \mathcal{P}(C_T) \to \mathcal{P}(C_T)$ which associates with every $m \in \mathcal{P}(C_T)$ the law of X^m defined by:

$$X_t^m = X_0 + \int_0^t \sigma[X_s^m, m_s] dB_s + \int_0^t b[X_s^m, m_s] ds, \quad t \le T.$$

Observe that if (X) is a solution of (2.5) then the law of (X) is a fixed point of the mapping ψ and conversely.

By pathwise considerations one proves that for $t \le T$,

$$\rho_t^2(\psi(m^1), \psi(m^2)) \le K \int_0^t \rho_u^2(m^1, m^2) du,$$

for $m^1, m^2 \in \mathcal{P}(C_T)$. Then one deduces from the fixed point theorem the existence and uniqueness of the solution P of the martingale problem (2.4) defined on $[0, T]$. Pathwise uniqueness follows immediately due to the Lipschitz continuity of the coefficients. $\square$

2.2. THE STOCHASTIC INTERACTING PARTICLE SYSTEM

Let us now describe a way to approximate the solution P of $(\mathcal{M}_{MV})$ previously obtained by the empirical measures of interacting particle systems. Inspired by the large numbers law, we "replace" the nonlinearity in $(\mathcal{M}_{MV})$ by the empirical measure of particles, which leads to the following definition of interacting (triangular) systems.

Definition 2.4 Let us consider independent $\mathbb{R}^k$-valued Brownian motions $(B^i)_{i \in \mathbb{N}^\star}$ and independent random variables $(X_0^i)_{i \in \mathbb{N}^\star}$ with law P_0 and independent of (B^i). For each $n \in \mathbb{N}^\star$, we define the n-particle system $(X^{1n}, X^{2n}, ..., X^{nn})$ as solution of the stochastic differential system

$$\forall i \in \{1, ..., n\}, \quad X_t^{in} = X_0^i + \int_0^t \sigma[X_s^{in}, \mu_s^n] dB_s^i + \int_0^t b[X_s^{in}, \mu_s^n] ds \quad (2.6)$$

where μ^n is the empirical measure of the system, i.e. the (random) probability measure on the path space C_T defined by:

$$\mu^n = \frac{1}{n} \sum_{i=1}^n \delta_{X^{in}}, \quad (\delta \text{ denoting the Dirac measure}).$$

The stochastic differential system (2.6) means that for every $1 \leq j \leq d$, $1 \leq i \leq n$,

$$X_{j,t}^{in} = X_{j,0}^i + \int_0^t \sum_{h=1}^k \sigma_{jh}[X_s^{in}, \mu_s^n] dB_{h,s}^i + \int_0^t b_j[X_s^{in}, \mu_s^n] ds.$$

It is easy to prove that under the assumptions of Theorem 2.3, (2.6) has for each n a unique pathwise solution. To compare the behaviour of these particles with the nonlinear problem, we use coupling techniques, consisting in comparing the n-particle system with n independent copies of the limiting equation (2.5) constructed on the same probability space. More precisely, we define for $(B^i)_i$ and $(X_0^i)_i$ previously given, the system $(X^i)_{i \in \mathbb{N}^\star}$ of independent processes with distribution P by

$$\forall i \in \mathbb{N}^\star, \quad X_t^i = X_0^i + \int_0^t \sigma[X_s^i, P_s] dB_s^i + \int_0^t b[X_s^i, P_s] ds.$$

Let us prove pathwise estimates comparing $X^{i,n}$ and X^i.

Theorem 2.5 *Let us assume that the functions σ and b are bounded by a real constant M and Lipschitz continuous with Lipschitz constant L.*

Then for all $i \in \{1, ..., n\}$ and for any $T > 0$, one gets

$$\sup_n E(\sup_{t \leq T} |X_t^{i,n}|^2) < +\infty \; ; \; E(\sup_{t \leq T} |X_t^i|^2) < +\infty$$

$$E(\sup_{t \leq T} |X_t^{i,n} - X_t^i|^2) \leq \frac{C_T M^2}{nL^2} \exp(CL^2 T). \tag{2.7}$$

Inequality (2.7) obviously implies that the law of every subsystem of size k (k is a fixed integer), issued from the system $(X^{i,n})$, converges when n tends to infinity to the law $P^{\otimes k}$. This property is called propagation of chaos, which means a propagation of the independence. Although the particles interact together, the initial independence assumption propagates in time for finite subsystems, when the size of the system tends to infinity. That is mainly due to the exchangeability of the laws and to the nature of the interaction, as function of the empirical measure. Such an interaction is called weak interaction or mean field interaction.

Proof. The two first assertions are standard. The form $\frac{K}{n}$ of the convergence rate in the third one has been proved in Sznitman [37], but for what follows in the next sections, we need to know the explicit dependence of K on M and L. Let us detail the computation.

We denote by μ, the empirical measure of the system $(X^1, ..., X^n)$. Using Burkholder-Davis-Gundy's and Holder's inequalities and the exchangeability of the system $((X^{1,n}, X^1), ..., (X^{n,n}, X^n))$, we get for any $t \leq T$,

$$E(\sup_{s \leq t} |X_s^{in} - X_s^i|^2)$$

$$\leq C_1 \left(\sum_{j=1}^d \sum_{h=1}^k \int_0^t E((\sigma_{jh}[X_s^{in}, \mu_s^n] - \sigma_{jh}[X_s^i, P_s])^2) ds \right.$$

$$\left. + \sum_{j=1}^d \int_0^t E((b_j[X_s^{in}, \mu_s^n] - b_j[X_s^i, P_s])^2) ds \right)$$

$$\leq C_2 \left(L^2 \int_0^t E(\sup_{u \leq s} |X_u^{in} - X_u^i|^2) ds \right.$$

$$+ \sum_{j=1}^d \sum_{h=1}^k \int_0^t E((\sigma_{jh}[X_s^i, \mu_s] - \sigma_{jh}[X_s^i, P_s])^2) ds$$

$$\left. + \sum_{j=1}^d \int_0^t E((b_j[X_s^i, \mu_s] - b_j[X_s^i, P_s])^2) ds \right)$$

$$\leq C_3 \left(L^2 \int_0^t E(\sup_{u\leq s} |X_u^{in} - X_u^i|^2) ds \right.$$

$$+ \sum_{j=1}^d \sum_{h=1}^k \int_0^t E((\frac{1}{n}\sum_{J=1}^n \sigma_{jh}(X_s^i, X_s^J) - \sigma_{jh}[X_s^i, P_s])^2) ds$$

$$\left. + \sum_{j=1}^d \int_0^t E((\frac{1}{n}\sum_{J=1}^n b_j(X_s^i, X_s^J) - b_j[X_s^i, P_s])^2) ds \right)$$

A convexity inequality is not sufficient to obtain good estimates in the second and the third terms of the last expression. One remarks that since the variables $(X^i)_{1\leq i\leq n}$ are independent with law P, then

$$E(\sigma_{jh}(X_s^i, X_s^J)|X_s^r, \quad \text{for all } r \neq J) = \sigma_{jh}[X_s^i, P_s], \quad \text{for } i \neq J,$$

and then, for $J \neq k$ and $i \neq J$,

$$E((\sigma_{jh}(X_s^i, X_s^J) - \sigma_{jh}[X_s^i, P_s]) \cdot (\sigma_{jh}(X_s^i, X_s^k)) - \sigma_{jh}[X_s^i, P_s]) = 0$$

Thus, it suffices to consider the n terms of the form $(\sigma_{jh}(X_s^i, X_s^J) - \sigma_{jh}[X_s^i, P_s])^2$. Hence,

$$E(\sup_{s\leq t} |X_s^{i,n} - X_s^i|^2) \leq CL^2 \int_0^t E(|X_r^{i,n} - X_r^i|^2) dr + \frac{C'M^2 t}{n}. \qquad (2.8)$$

If we take $\phi(t) = E(\sup_{s\leq t} |X_s^{i,n} - X_s^i|^2) + \frac{C'M^2}{nCL^2}$, we have

$$\forall t \leq T , \quad \phi(t) \leq \frac{C'M^2}{nCL^2} + CL^2 \int_0^t \phi(r) dr \qquad (2.9)$$

By Gronwall's lemma, we conclude that

$$\phi(t) \leq \frac{C'M^2}{nCL^2} \exp(CL^2 T) \qquad (2.10)$$

where C and C' are two real constants depending on T. $\qquad \square$

2.3. SOME WORDS ABOUT THE NUMERICAL APPROACH

Let us describe the numerical algorithm we deduce from this study to simulate the solution of the McKean-Vlasov equation. We refer to Bossy [2] and Bossy-Talay [5].

We only consider the case of dimension one, all what follows being easily generalized to every finite dimension. We follow [5] and explain how to simulate the cumulative distribution function of the solution P_t of the McKean-Vlasov equation at time t. This algorithm can easily be adapted to simulate other functionals of P_t, as for example moments of different orders. Let us define $V(t,x) = \int_{\mathbb{R}} H(x-y)P_t(dy)$, the cumulative distribution function of P_t where H is the Heavyside function defined by $H(x) = 1_{x \geq 0}$.

From now on, the number n of particles is fixed and the initial law P_0 is assumed to satisfy one of both following assumptions:

(1) P_0 is a Dirac measure at x_0.
(2) P_0 has a continuous density function u_0 such that there exist constants $M > 0$, $\eta \geq 0$ and $\alpha > 0$ with $u_0(x) \leq \eta \exp(-\alpha \frac{x^2}{2})$ for $|x| > M$. (If $\eta = 0$, the measure P_0 has a compact support).

The algorithm starts with an approximation of the initial condition $V(0, .) = V_0(.)$ which is the cumulative distribution function of P_0. One chooses n points $(y_0^1, ..., y_0^n)$ in $\mathbb{R}$ such that the piecewise constant function

$$\bar{V}_0(x) = \frac{1}{n} \sum_{i=1}^{n} H(x - y_0^i)$$

approximates V_0 in $L^1(\mathbb{R})$. In the case (1), one takes $y_0^i = x_0$ and in the case (2), a possible choice is to set $y_0^i = \inf\{y; V_0(y) = \frac{i}{n}\}$ if $i = 1, ..., n-1$, and $y_0^n = \inf\{y; V_0(y) = 1 - \frac{1}{2n}\}$.

Then, we discretize in time the interacting particle system following an Euler scheme. We take $\Delta t > 0$ and K is chosen such that $T = K \Delta t$. The discrete times are denoted $t_k = k \Delta t$ with $1 \leq k \leq [K]$. The Euler scheme leads to the following discrete-time system:

$$Y_{t_k+1}^i = Y_{t_k}^i + \frac{1}{n} \sum_{j=1}^{n} b(Y_{t_k}^i, Y_{t_k}^j) \Delta t + \frac{1}{n} \sum_{j=1}^{n} \sigma(Y_{t_k}^i, Y_{t_k}^j)(B_{t_k+1}^i - B_{t_k}^i)$$

$$Y_0^i = y_0^i \quad , \quad i = 1, ..., n. \tag{2.11}$$

Thus we approximate the empirical measure μ_{t_k} by the measure $\bar{\mu}_{t_k} = \frac{1}{n} \sum_{i=1}^{n} \delta_{Y_{t_k}^i}$.

In a similar way, we approximate $V(t_k, .)$ by the cumulative distribution function of $\bar{\mu}_{t_k}$

$$\bar{V}_{t_k}(x) = \frac{1}{n} \sum_{i=1}^{n} H(x - Y_{t_k}^i). \tag{2.12}$$

Then Bossy-Talay in [5] prove that

Theorem 2.6 *Assume that b is a bounded Lipschitz continuous function on $\mathbb{R}^2$, that $\sigma \in C_b^1(\mathbb{R}^2)$ and that there exists a constant $s > 0$ such that $\sigma(x,y) \geq s$ for every x,y. Assume moreover that P_0 satisfies (1) or (2). Then there exist strictly positive constants C_1 and C_2 depending on σ, b, V_0, T, and C depending on M, η and α, such that for all $k \in \{1, ..., K\}$, one gets*

$$\|V_0 - \bar{V}_0\|_{L^1} \leq \frac{C\sqrt{\log n}}{n}$$

$$E(\|V(t_k, .) - \bar{V}_{t_k}(.)\|_{L^1}) \leq C_1 \left(\|V_0 - \bar{V}_0\|_{L^1} + \frac{1}{\sqrt{n}} + \sqrt{\Delta t} \right). (2.13)$$

3. The Vortex Equation in the whole Plane

We will now adapt these results to the specific case of the two-dimensional Navier-Stokes equation.

3.1. THE DETERMINISTIC FRAMEWORK

Let us consider the velocity flow $u(t,x) \in \mathbb{R}^2, t \in \mathbb{R}_+, x \in \mathbb{R}^2$ of a viscous and incompressible fluid in the whole plane. The equation governing this motion is the Navier-Stokes equation given by

$$\frac{\partial u}{\partial t}(t,x) + (u.\nabla)u(t,x) = \nu\Delta u(t,x) - \nabla p \; ;$$

$$\nabla.u(t,x) = 0 \; ; \; u(x,t) \to 0 \text{ as } |x| \to +\infty, \text{ for } 0 \leq t < +\infty, (3.1)$$

where p is the pressure function and $\nu > 0$ the viscosity (assumed to be constant).

The first step in the probabilistic approach consists in considering the equation satisfied by the vorticity flow $w(t,x) = \text{curl } u(t,x)$. Heuristically, since the divergence of u is zero, w is solution of the nonlinear partial differential equation, called **vortex equation**

$$\frac{\partial w}{\partial t}(t,x) + (u \cdot \nabla)w(t,x) = \nu\Delta w(t,x) \; ;$$

$$w(t,x) \to 0 \text{ as } |x| \to +\infty, \text{ for } 0 \leq t < +\infty; w_0(x) = w(0,x). (3.2)$$

This second equation is a one-dimensional equation in which the unknown pressure term has disappeared. To get it closed, we need to write u in function of w, which is possible since $\nabla.u = 0$. Indeed, the function u can then be formally written as the orthogonal gradient $\nabla^\perp\psi$ of a courant function ψ ($\nabla^\perp = (\partial_{x_2}, -\partial_{x_1})$), and then $w(t,x) = \text{curl } u(t,x)$ satisfies

$\Delta \psi = w$. The function ψ is then equal to $G \star w$, where $\star$ denotes the convolution and G is the fundamental solution of the Poisson equation in dimension two, defined for each $r > 0$ by $G(r) = -\frac{1}{2\pi}\ln r$.

Then, for every $t \geq 0$,

$$u(t,x) = \int_{\mathbb{R}^2} \nabla^\perp G(|x - y|)w(t,y)dy = \int_{\mathbb{R}^2} K(x - y)w(t,y)dy. \qquad (3.3)$$

The Biot-Savart kernel $K(x) = \nabla^\perp G$ is given for all $x = (x_1, x_2) \in \mathbb{R}^2$ by

$$K(x) = \frac{1}{2\pi} \frac{1}{(x_1^2 + x_2^2)}(-x_2, x_1). \qquad (3.4)$$

Note that $\nabla \cdot K = 0$. The main difficulty is the explosion at 0.

However, The function K is bounded at infinity and integrable near zero. So if $w \in L^\infty \cap L^1(\mathbb{R}^2)$, (3.3) makes sense. Indeed, denoting $K_1 = \int_{B(0,1)} |K(y)|dy$ and $K_\infty = \sup_{y \in B(0,1)^c} |K(y)|$, we get for all $g \in L^1 \cap L^\infty$ and $x \in \mathbb{R}^2$,

$$|K * g(x)| \leq K_1 \|g\|_{L^\infty} + K_\infty \|g\|_{L^1}, \qquad (3.5)$$

which leads to work in the space $L^1 \cap L^\infty$, well adapted to a probabilistic point of view. In all the following, we will assume that

$$w_0 \in L^1 \cap L^\infty. \qquad (3.6)$$

We are now interested in weak solutions of the equation (3.2). At this level, there are two probabilistic approaches in the literature. The first one consists in interpreting classical solutions of the vortex equation, using a Feynman-Kac approach, as the expectations of some stochastic processes. This approach has been developed by Busnello in [8] and allows to prove an existence and uniqueness theorem, but not to obtain a constructive method to simulate the solutions of the equation. So in the following, we will present an approach which consists in looking for weak solutions of the vortex equation, considered as a Fokker-Planck equation. This approach follows the precursive ideas of Chorin and has been firstly mathematically developed by Marchioro and Pulvirenti [26] and then by Méléard [29].

3.2. A CUT-OFF EQUATION

The drift kernel K explodes at 0 and we approximate it by bounded kernels.

Let us consider as in [26] the cut-off kernel K_ε defined in the following way. We denote by $G(r) = -\frac{1}{2\pi}\ln r$ the fundamental solution of the Poisson equation. One knows that for $x \in \mathbb{R}^2$, $K(x) = \nabla^\perp G(|x|)$. For each

$\varepsilon > 0$, we consider G_ε defined as $G_\varepsilon(r) = G(r)$ if $|r| \geq \varepsilon$ and arbitrarily extended to an even $C^2(\mathbb{R})$ function such that $|G'_\varepsilon(r)| \leq |G'(r)|$ and $|G''_\varepsilon(r)| \leq |G''(r)|$. Then we define

$$K_\varepsilon(x) = \nabla^\perp G_\varepsilon(|x|).$$

The function K_ε is bounded by a real M_ε and Lipschitz continuous with a Lipschitz constant L_ε.

Let us also remark that by construction, $\nabla.K_\varepsilon = 0$, and that K_ε satisfies a similar inequality as (3.5): for every $\varepsilon > 0$ and every function $g \in L^1 \cap L^\infty$

$$\|K_\varepsilon * g\|_\infty \leq K_1 \|g\|_{L^\infty} + K_\infty \|g\|_{L^1}. \tag{3.7}$$

For each fixed $\varepsilon > 0$, we consider the cut-off equation

$$\partial_t w^\varepsilon = \nu \Delta w^\varepsilon - (K_\varepsilon \star w^\varepsilon.\nabla)w^\varepsilon \ ; \ w_0 \in L^1 \cap L^\infty. \tag{3.8}$$

We are in the McKean-Vlasov context, except that the initial condition is not a probability measure. To overpass this problem, we use a trick due to Jourdain [20] to pass from a density initial function to any $w_0 \in L^1 \cap L^\infty$.

We define the bounded function h by

$$\forall x \in \mathbb{R}^2 \ , \quad h(x) = \frac{w_0(x)\|w_0\|_1}{|w_0(x)|}, \text{ with the convention } \frac{0}{0} = 0. \tag{3.9}$$

Let us remark that for each $x \in \mathbb{R}^2$,

$$-\|w_0\|_1 \leq h(x) \leq \|w_0\|_1, \tag{3.10}$$

and that $w_0(x)dx = h(x)\frac{|w_0(x)|}{\|w_0\|_1}dx$, where $\frac{|w_0|}{\|w_0\|_1}$ is thus a probability density.

Now, for Q a probability measure on $C([0, +\infty), \mathbb{R}^2)$, we define the family $(\tilde{Q}_t)_{t \geq 0}$ of weighted signed measures on $\mathbb{R}^2$ by

$$\forall B \text{ Borel subset of } \mathbb{R}^2 \ , \quad \tilde{Q}_t(B) = E^Q(1_B(X_t)h(X_0)), \tag{3.11}$$

where X denotes the canonical process on $C([0, +\infty), \mathbb{R}^2)$. (One associates with each sample-path a signed weight depending on the initial position).

Remark 3.1 Since the function h is bounded by $\|w_0\|_1$ and using (3.11), one observes that for each $t \geq 0$, the signed measure $\tilde{Q}_t$ is bounded, and its total mass is less than $\|w_0\|_1$, and that if Q_t is absolutely continuous with respect to the Lebesgue measure, then the same holds for $\tilde{Q}_t$.

The equation (3.8) understood in its weak form leads naturally to the following nonlinear stochastic differential equation.

Definition 3.2 Let us consider a random $\mathbb{R}^2$-valued variable X_0 with distribution $\frac{|w_0(x)|}{\|w_0\|_1}dx$ and a 2-dimensional Brownian motion B independent of X_0. A solution $Z^\varepsilon \in C(\mathbb{R}_+, \mathbb{R}^2)$ of the nonlinear stochastic differential equation satisfies $\forall t \in \mathbb{R}_+$

$$Z_t^\varepsilon = X_0 + \sqrt{2\nu}B_t + \int_0^t K_\varepsilon \star \tilde{P}^\varepsilon{}_s(Z_s^\varepsilon)ds,$$

$$P_s^\varepsilon = \mathcal{L}(Z_s^\varepsilon) \text{ and } \tilde{P}_s^\varepsilon \text{ is related to } P_s^\varepsilon \text{ by (3.11).} \qquad (3.12)$$

Since K_ε is Lipschitz continuous and bounded, we adapt the results of the previous section to show the existence and pathwise uniqueness of the solution Z^ε of (3.12). Girsanov's theorem implies the existence, for each $t > 0$, of a density p_t^ε (w.r.t. the Lebesgue measure) for P_t^ε, and then the existence of a density $\tilde{p}_t^\varepsilon$ for the weighted measure $\tilde{P}_t^\varepsilon$.

By Itô's formula, one proves that for any $T > 0$, the probability measure P^ε on $\mathcal{P}(C([0,T], \mathbb{R}^2))$ is solution of the nonlinear martingale problem $(\mathcal{M}^\varepsilon)$: for any $\phi \in C_b^2(\mathbb{R}^2)$ and $t \leq T$,

$$\phi(X_t) - \phi(X_0) - \int_0^t K^\varepsilon * \tilde{P}_s(X_s).\nabla\phi(X_s)ds - \nu\int_0^t \Delta\phi(x_s)ds \qquad (3.13)$$

is a P^ε-martingale, where X is the canonical process on $C([0,T], \mathbb{R}^2)$, P_0 is equal to $\frac{|w_0(x)|}{\|w_0\|_1}dx$, and $\tilde{P}_s^\varepsilon$ is related to $P_s^\varepsilon = P^\varepsilon \circ X_s^{-1}$ by (3.11).

Multiplying all terms of (3.13) by $h(X_0)$, taking expectations and using that $\nabla.K_\varepsilon = 0$, we show that $(\tilde{p}_t^\varepsilon)_t$ is a weak solution of the equation (3.8), and further for each $t > 0$,

$$\|\tilde{p}_t^\varepsilon\|_{L^1} \leq E^{P^\varepsilon}(|h(X_0)|) \leq \|w_0\|_{L^1}. \qquad (3.14)$$

That implies in particular that the drift coefficient $K_\varepsilon \star \tilde{p}^\varepsilon$ is bounded, uniformly in time, by $\|w_0\|_{L^1}$. One can further apply analysis results due to Friedman [12] to show that the solution $\tilde{p}^\varepsilon$ of (3.8) is continuous on $[0,T] \times \mathbb{R}^d$ and belongs to $C^{1,2}((0,T] \times \mathbb{R}^d)$. Since $\nabla.K_\varepsilon = 0$, one proves that $\tilde{p}^\varepsilon$ is also a strong solution of the equation

$$\partial_t v = \nu\Delta v - K_\varepsilon \star v.\nabla v \; ; \; v_0 = w_0$$

Then one knows (cf. [13] or [22]) that $\tilde{p}^\varepsilon$ satisfies a Feynman-Kac formula and can be written for any $t \leq T$ as $E(w_0(Y_t^{x,\varepsilon}))$, where the process $(Y_t^{x,\varepsilon})$ is defined by

$$Y_t^{x,\varepsilon} = x + \sqrt{2\nu}W_t + \int_0^t K_\varepsilon \star \tilde{p}_s^\varepsilon(Y_s^{x,\varepsilon})ds$$

from which we deduce that

$$\sup_{t\leq T} \|\tilde{p}_t^\varepsilon\|_{L^\infty} \leq \|w_0\|_{L^\infty}. \qquad (3.15)$$

Hence the good space to define the solutions is the space

$$\mathcal{H} = \{q \in L^\infty([0,T], L^1 \cap L^\infty); \ \sup_{t\leq T} \|q_t\|_{L^1} \leq \|w_0\|_{L^1}$$
$$\text{and } \sup_{t\leq T} \|q_t\|_{L^\infty} \leq \|w_0\|_{L^\infty}\}$$

and we define, for $q \in V$ and $t \leq T$, the norm

$$\|\|q_t\|\| = \|q_t\|_{L^1} + \|q_t\|_{L^\infty}.$$

Let us now prove the uniqueness of the solution of (3.8) in this space. We use to this aim the mild form of the equation and firstly remark that the heat kernel G_t^ν on $\mathbb{R}^2$ defined by $G_t^\nu(x) = \frac{1}{4\pi t\nu} e^{-\frac{|x|^2}{4t\nu}}$ satisfies

Lemma 3.3

$$\|\nabla_x G_t^\nu\|_{L^1} \leq \frac{C}{\sqrt{\nu t}} \qquad (3.16)$$

where C is a real constant, and then $\int_0^t \|\nabla_x G_{t-s}^\nu\|_{L^1} ds < +\infty$.

Proposition 3.4 1) *Each weak solution $w^\varepsilon \in \mathcal{H}$ of (3.8) is a.s. solution of the evolution equation*

$$w^\varepsilon(t,x) = G_t^\nu \star w_0(x) + \int_0^t \int_{\mathbb{R}^2} \nabla_x G_{t-s}^\nu(x-y).K_\varepsilon \star w_s^\varepsilon(y)w_s^\varepsilon(y)dyds. \qquad (3.17)$$

2) *There exists a unique weak solution of (3.8) in $\mathcal{H}$.*

Proof.

1) Using Fubini's theorem (allowed by Lemma 3.16 and (3.7)), we easily prove that for every function $\psi(t,x) \in C_b^{1,2}(\mathbb{R}_+ \times \mathbb{R}^2)$,

$$\int_{\mathbb{R}^2} \psi(t,x)w_t^\varepsilon(x)dx = \int_{\mathbb{R}^2} \psi(0,x)w_0(x)dx + \nu \int_0^t \int_{\mathbb{R}^2} \Delta\psi(s,x)w_s^\varepsilon(x)dxds$$

$$+ \int_0^t \int_{\mathbb{R}^2} \partial_s\psi(s,x)w_s^\varepsilon(x)dxds + \int_0^t \int_{\mathbb{R}^2} (K_\varepsilon \star w_s^\varepsilon(x).\nabla_x\psi(s,x))w_s^\varepsilon(x)dxds.$$

Then by choosing for a fixed time t, $\psi(s,x) = \int_{\mathbb{R}^2} G_{t-s}^\nu(x-y)\phi(y)dy$ for $\phi \in C_b^2(\mathbb{R}^2)$ and thanks again to Fubini's theorem, we obtain the mild equation (3.17).

2) Let q and q' be two solutions of (3.8) belonging to $\mathcal{H}$. Then

$$q_t(x) - q_t'(x) = \int_0^t \int \nabla_x G_{t-s}^\nu(x-y)(K_\varepsilon \star q_s(y))(q_s(y) - q_s'(y))$$

$$+ K_\varepsilon \star (q_s(y) - q_s'(y))q_s'(y))dyds.$$

Thus by (3.16) and (3.7), a simple computation gives (the constant C may change from line to line),

$$\||q_t - q_t'\||$$

$$\leq \int_0^t \|| \int \nabla_x G_{t-s}^\nu(x-y)(K_\varepsilon \star q_s(y))(q_s(y) - q_s'(y))$$

$$+ K_\varepsilon \star (q_s(y) - q_s'(y))q_s'(y))dy\||ds$$

$$\leq C(K_1 + K_\infty)(\|w_0\|_{L^1} + \|w_0\|_{L^\infty}) \int_0^t \|\nabla_x G_{t-s}^\nu\|_{\mathbf{L}^1} \||q_s - q_s'\||ds$$

$$\leq \frac{C}{\sqrt{\nu}} \int_0^t \frac{\||q_s - q_s'\||}{\sqrt{t-s}}ds.$$

By an iteration, we obtain

$$\||q_t - q_t'\|| \leq C \int_0^t \frac{1}{\sqrt{t-s}} \int_0^s \frac{\||q_u - q_u'\||}{\sqrt{s-u}}duds$$

$$\leq C \int_0^t \||q_u - q_u'\|| \int_u^t \frac{1}{\sqrt{t-s}\sqrt{s-u}}dsdu$$

$$\leq C \int_0^t \||q_u - q_u'\||du.$$

Therefore, by Gronwall's lemma,

$$\sup_{t\in[0,T]} \||q_t - q_t'\|| = 0$$

and the uniqueness in (3.8) is proved.

$\square$

3.3. EXISTENCE AND UNIQUENESS OF THE VORTEX EQUATION

We are now able to study the existence and uniqueness of a solution of the vortex equation in $\mathcal{H}$.

Theorem 3.5 *Assume $w_0 \in L^1 \cap L^\infty$. There exists a unique weak solution $w \in \mathcal{H}$ to the vortex equation (3.2). This solution is moreover solution of the evolution equation*

$$w_t(x) = G_t^\nu \star w_0(x) + \int_0^t \int_{\mathbb{R}^2} \nabla_x G_{t-s}^\nu(x-y).K \star w_s(y)w_s(y)dyds. \quad (3.18)$$

Proof. The uniqueness is proved following a similar proof as the one in Proposition 3.2, using (3.5) instead of (3.7).

Let us now show the existence. Let us first remark that the space $L^1 \cap L^\infty$ endowed with the norm $|||.|||$ is a complete space.

For each $\varepsilon > 0$, we have constructed a solution $(\tilde{p}_t^\varepsilon)_t$ of (3.8). We are going to prove that this family is Cauchy (in ε) and that the limiting point satisfies the vortex equation. Let us fix $\varepsilon > 0, \varepsilon' > 0$ and consider the two families $(\tilde{p}_t^\varepsilon)_t$ and $(\tilde{p}_t^{\varepsilon'})_t$ previously defined. They are solution of the corresponding mild equations (3.17) and

$$\tilde{p}_t^\varepsilon(x) - \tilde{p}_t^{\varepsilon'}(x) = \int_0^t \int \nabla_x G_{t-s}^\nu(x-y)$$

$$\left(K_\varepsilon \star \tilde{p}_s^\varepsilon(y)\tilde{p}_s^\varepsilon(y) - K_{\varepsilon'} \star \tilde{p}_s^{\varepsilon'}(y)\tilde{p}_s^{\varepsilon'}(y) \right) dy ds$$

$$= \int_0^t \int \nabla_x G_{t-s}^\nu(x-y) \left(K_\varepsilon \star (\tilde{p}_s^\varepsilon(y) - \tilde{p}_s^{\varepsilon'}(y))\tilde{p}_s^\varepsilon(y) \right.$$

$$+ (K_\varepsilon \star \tilde{p}_s^{\varepsilon'}(y) - K_{\varepsilon'} \star \tilde{p}_s^{\varepsilon'}(y))\tilde{p}_s^\varepsilon(y)$$

$$\left. + K_{\varepsilon'} \star \tilde{p}_s^{\varepsilon'}(y)(\tilde{p}_s^\varepsilon(y) - \tilde{p}_s^{\varepsilon'}(y)) \right) dy ds.$$

Since the functions $(\tilde{p}_t^\varepsilon)_t$ and $(\tilde{p}_t^{\varepsilon'})_t$ belong to $\mathcal{H}$ and by (3.16) and (3.7), we will do similar calculations as before to estimate the first and third terms of the right hand side. Now, to control the second term, we need the following lemma.

Lemma 3.6 For each $t \leq T$, for each $x \in \mathbb{R}^2$, for $\varepsilon' > \varepsilon$,

$$\left| \int_{\mathbb{R}^2} (K_{\varepsilon'}(x-y) - K_\varepsilon(x-y))\tilde{p}_t^{\varepsilon'}(y)dy \right| \leq 2\varepsilon' \|w_0\|_\infty. \tag{3.19}$$

Proof. Let us assume that $\varepsilon' > \varepsilon$. Since $K_{\varepsilon'}$ and K_ε coincide for $|x| \geq \varepsilon'$, we have

$$\left| \int_{\mathbb{R}^2} (K_{\varepsilon'}(x-y) - K_\varepsilon(x-y))\tilde{p}_t^{\varepsilon'}(y)dy \right|$$

$$\leq \int_{|x-y| \leq \varepsilon'} |K_{\varepsilon'}(x-y) - K_\varepsilon(x-y)||\tilde{p}_t^{\varepsilon'}(y)|dy$$

$$\leq \int_{|x-y| \leq \varepsilon'} (|K_{\varepsilon'}(x-y)| + |K_\varepsilon(x-y)|)|\tilde{p}_t^{\varepsilon'}(y)|dy$$

$$\leq 2 \int_{|x-y| \leq \varepsilon'} |K(x-y)| |\tilde{p}_t^{\varepsilon'}(y)| dy \leq 2\|w_0\|_\infty \int_{|z| \leq \varepsilon'} |K(z)| dz \quad \text{by (3.15)}$$

$$\leq 2\varepsilon' \|w_0\|_\infty \quad \text{by an easy computation.}$$

We have used that for every $\varepsilon > 0$, $|K_\varepsilon(x-y)| \leq |K(x-y)|$. □

Let us now come back to the proof of Theorem 3.5. We apply Lemma 3.6 to show that

$$\||\tilde{p}_t^\varepsilon - \tilde{p}_t^{\varepsilon'}\|| \leq C_1 \varepsilon' + C_2 \int_0^t \frac{1}{\sqrt{t-s}} \||\tilde{p}_s^\varepsilon - \tilde{p}_s^{\varepsilon'}\|| ds \tag{3.20}$$

and by iteration of this inequality and Gronwall's lemma, we finally obtain that for $\varepsilon' > \varepsilon$,

$$\sup_{t \leq T} \||\tilde{p}_t^\varepsilon - \tilde{p}_t^{\varepsilon'}\|| \leq C\varepsilon' \exp^{C'T}. \tag{3.21}$$

The family $\varepsilon \to \tilde{p}^\varepsilon$ is then a Cauchy family in $L^\infty([0,T], L^1 \cap L^\infty)$.

Hence, there exists a function $w \in L^\infty([0,T], L^1 \cap L^\infty)$, such that for each $t \leq T$

$$\lim_{\varepsilon \to 0} \sup_{t \leq T} \||\tilde{p}_t^\varepsilon - w_t\|| = 0$$

That implies that

$$\sup_{t \leq T} \|w_t\|_{L^1} \leq \|w_0\|_{L^1} \ ; \ \sup_{t \leq T} \|w_t\|_{L^\infty} \leq \|w_0\|_{L^\infty}. \tag{3.22}$$

Next, with similar arguments as before, one can pass to the limit in ε in the mild equation (3.17) and show that w is solution of (3.18). We deduce that w is a weak solution of the vortex equation. The theorem is proved. □

3.4. THE NONLINEAR PROCESS ASSOCIATED WITH THE VORTEX EQUATION

The equation (3.2) leads naturally to a nonlinear martingale problem. Let us fix $T > 0$.

Definition 3.7 The probability measure $P \in \mathcal{P}(C([0,T], \mathbb{R}^2))$ is solution of the nonlinear martingale problem $(\mathcal{M})$ if for each $\phi \in C_b^2(\mathbb{R}^2)$ and $t \leq T$,

$$\phi(X_t) - \phi(X_0) - \int_0^t K * \tilde{P}_s(X_s).\nabla\phi(X_s) ds - \nu \int_0^t \Delta\phi(x_s) ds$$

is a P-martingale, where X is the canonical process on $C([0,T], \mathbb{R}^2)$, P_0 is equal to $\frac{|w_0(x)|}{\|w_0\|_1} dx$, and $\tilde{P}_s$ is related by (3.11) to $P_s = P \circ X_s^{-1}$.

This nonlinear martingale problem is related to the following nonlinear stochastic differential equation.

Definition 3.8 Let us consider a $\mathbb{R}^2$-valued random variable X_0 with distribution $\frac{|w_0(x)|}{\|w_0\|_1}dx$ and a 2-dimensional Brownian motion B independent of X_0. A solution $X \in C(\mathbb{R}_+, \mathbb{R}^2)$ of the nonlinear stochastic differential equation satisfies $\forall t \in \mathbb{R}_+$

$$X_t = X_0 + \sqrt{2\nu}B_t + \int_0^t K * \tilde{P}_s(X_s)ds,$$

$$P_s \text{ is the marginal at time } s \text{ of the law of } X_s. \qquad (3.23)$$

Notations: We denote by $\hat{\mathcal{P}}_\infty(C([0,T], \mathbb{R}^2))$ the space of probability measures on $C([0,T], \mathbb{R}^2)$ such that for each $s \leq T$, the time-marginal P_s (and then $\tilde{P}_s$) has a density with respect to the Lebesgue measure, belonging to $L^\infty(\mathbb{R}^2)$. Then there exists (cf. Meyer [31] p.194) a measurable version $(s,x) \rightarrow \tilde{p}(s,x)$ in L^∞ such that for $s \in [0,T]$, $\tilde{P}_s(dx) = \tilde{p}(s,x)dx$.

We will prove the following theorem.

Theorem 3.9 *Let us consider $w_0 \in L^1 \cap L^\infty$. Then there exists a unique solution $P \in \hat{\mathcal{P}}_\infty(C([0,T], \mathbb{R}^2))$ to the martingale problem $(\mathcal{M})$ such that $P_0(dx) = \frac{|w_0(x)|}{\|w_0\|_1}dx$.*

Moreover, for each $t \in [0,T]$, the "weighted" density $\tilde{p}(t,x)$ is a.s. equal to w_t defined in Theorem 3.5.

Proof.

1) If $P \in \hat{\mathcal{P}}_\infty(C([0,T], \mathbb{R}^2))$ is a solution of $(\mathcal{M})$ and $\tilde{p}_t$ a measurable version of the densities of $\tilde{P}_t$, then multiplying by $h(X_0)$ and taking the expectation in $(\mathcal{M})$, we get that $\tilde{p}$ is a weak solution of (3.2) with initial condition w_0. Then by the uniqueness given in Theorem 3.5, for each $t \in [0,T]$, $\tilde{p}_t(x) = w_t(x)$ a.s.

2) Let us consider this unique weak solution w of (3.2) issued from w_0. Then w_t is for each t a bounded integrable function. We say that $P^w \in \mathcal{P}(C([0,T], \mathbb{R}^2))$ is solution of the classical martingale problem $(\mathcal{M}^w)$ if for each $\phi \in C_b^2(\mathbb{R}^2)$,

$$\phi(X_t) - \phi(X_0) - \nu \int_0^t \Delta\phi(X_s)ds - \int_0^t K * w_s(X_s).\nabla\phi(X_s)ds$$

is a P^w-martingale and $P_0^w(dx) = \frac{|w_0(x)|}{\|w_0\|_1}dx$. This martingale problem is well-posed. Indeed, by (3.5) and (3.22), for each $s \geq 0$,

$$\|K * w_s\|_\infty \leq K_1\|w_s\|_\infty + K_\infty\|w_s\|_1 \leq K_1\|w_0\|_\infty + K_\infty\|w_0\|_1 \ (3.24)$$

so the drift coefficient is bounded, and by Girsanov's Theorem, we get the existence and uniqueness of the solution of $(\mathcal{M}^w)$. Moreover, every time marginal of P^w admits a density p_s^w, and multiplying by $h(X_0)$ all the terms of the martingale problem and taking expectations, we obtain immediately that $(\tilde{p}_s^w)_{s\in[0,T]}$ is solution of the weak equation: for each $\phi \in C_b^2(\mathbb{R}^2)$,

$$\int_{\mathbb{R}^2} \phi(x)\tilde{p}_t^w(x)dx \;=\; \int_{\mathbb{R}^2} \phi(x)w_0(x)dx + \nu \int_0^t \int_{\mathbb{R}^2} \Delta\phi(x)\tilde{p}_s^w(x)dxds$$

$$+ \int_0^t \int_{\mathbb{R}^2} (K * w_s(x).\nabla\phi(x)\tilde{p}_s^w(x)dxds. \quad (3.25)$$

Then the flow $\tilde{p}_t$ is solution of

$$\partial_t \tilde{p}^w + (K * w.\nabla)\tilde{p}^w = \nu\Delta\tilde{p}^w \;; \quad \tilde{p}_0^w = w_0.$$

Adapting what we have done in Section 3.2, since the divergence of K is zero and w_0 belongs to $L^\infty \cap L^1$, we are able to prove that for each $t \in [0, T]$,

$$\|\tilde{p}_t^w\|_\infty \leq \|w_0\|_\infty. \quad (3.26)$$

3) Let us now prove the existence and uniqueness of a solution of (3.25) in $L^\infty([0,T], L^1 \cap L^\infty)$. As before, by Fubini's theorem and thanks to (3.5), we show that $\tilde{p}^w$ is solution of the evolution equation

$$\tilde{p}^w(t,x) = G_t^\nu * w_0(x) + \int_0^t \int \nabla_x G_{t-s}^\nu(x-y) K * w_s(y)\tilde{p}^w(s,y)dy. \quad (3.27)$$

We can easily prove the uniqueness of the solution of (3.27) in $L^\infty([0,T], L^\infty)$. Since $\tilde{p}^w$ and w are solutions of (3.27),

$$\sup_{t\in[0,T]} \|\tilde{p}^w(t,.) - w(t,.)\|_\infty = 0.$$

So the probability measure P^w is solution of the nonlinear martingale problem $(\mathcal{M})$.

4) Let us now prove the uniqueness of a solution of this martingale problem. Let P and Q be two solutions. The same reasoning as above implies that $(\tilde{P}_t)_{t\geq 0}$ and $(\tilde{Q}_t)_{t\geq 0}$ are equal to $(w(t,x)dx)_{t\geq 0}$. Hence P and Q are solutions of the classical well-posed martingale problem $(\mathcal{M}^w)$ and are then equal, and Theorem 3.9 is proved. □

3.5. THE PARTICLE APPROXIMATIONS

3.5.1. The Cut-off Case
Let us firstly adapt to the cut-off case the results of Section 2.

Definition 3.10 Consider a sequence $(B^i)_{i \in \mathbb{N}}$ of independent Brownian motions on $\mathbb{R}^2$ and a $\mathbb{R}^2$-valued sequence of independent variables $(Z_0^i)_{i \in \mathbb{N}}$ distributed according $\frac{|w_0(x)|}{\|w_0\|_1} dx$, and independent of $(B^i)_{i \in \mathbb{N}}$. For a fixed ε, for each $n \in \mathbb{N}^*$, and $1 \leq i \leq n$, let us consider the interacting processes defined by

$$Z_t^{in,\varepsilon} = Z_0^i + \sqrt{2\nu} B_t^i + \int_0^t K_\varepsilon * \tilde{\mu}_s^{n,\varepsilon}(Z_s^{in,\varepsilon}) ds \qquad (3.28)$$

where $\tilde{\mu}^{n,\varepsilon} = \frac{1}{n} \sum_{j=1}^n h(Z_0^j) \delta_{Z^{jn,\varepsilon}}$ is the weighted empirical measure of the system. (That is a random finite measure on $C(\mathbb{R}_+, \mathbb{R}^2)$).

We also define the limiting independent processes by

$$\bar{Z}_t^{i,\varepsilon} = Z_0^i + \sqrt{2\nu} B_t^i + \int_0^t K_\varepsilon * \tilde{P}_s^\varepsilon(\bar{Z}_s^{i,\varepsilon}) ds, \quad P_s^\varepsilon \text{ is the law of } \bar{Z}_s^{i,\varepsilon} \quad (3.29)$$

Proposition 3.11   *1) For each $T > 0$ and for each n, there exits a unique (pathwise) solution to the interacting particle system (3.28) in $C([0,T], \mathbb{R}^{2n})$ and a unique (pathwise) solution to the nonlinear equation (3.29) in $C([0,T], \mathbb{R}^2)$.*
2) For each $T > 0$,

$$E(\sup_{t \leq T} |Z_t^{in,\varepsilon} - \bar{Z}_t^{i,\varepsilon}|) \leq \frac{M_\varepsilon}{\sqrt{n} L_\varepsilon} \exp(\|w_0\|_1 T L_\varepsilon). \qquad (3.30)$$

Proof. The proof of the first assertion is standard. The second assertion is obtained by an adaptation of the proof of Theorem 2.5. Since the diffusion coefficient is a constant, one can lead computations in L^1.   □

3.5.2. *The Approximating Interacting Particle System*
We now define the interacting particle system we are interested in.
     We consider $T > 0$ and a sequence (ε_n) tending to 0 such that

$$\lim_n \frac{M_{\varepsilon_n}}{\sqrt{n} L_{\varepsilon_n}} \exp(\|w_0\|_1 T L_{\varepsilon_n}) = 0. \qquad (3.31)$$

For each n and given independent Brownian motions $(B^i)_{1 \leq i \leq n}$, we consider a coupling between the n particle system $(Z^{in} = Z^{in,\varepsilon_n})$ defined with the drift K_{ε_n} as in (3.28), and the corresponding independent limiting processes $(\bar{Y}^{in} = \bar{Z}^{i,\varepsilon_n})_{1 \leq i \leq n}$ defined for each $t \leq T$ and n by

$$\bar{Y}_t^{in} = Z_0^i + \sqrt{2\nu} B_t^i + \int_0^t K_{\varepsilon_n} * \tilde{P}_s^n(\bar{Y}_s^{in}) ds, \qquad (3.32)$$

where P_s^n is the law of the $\bar{Y}_s^{in}$.

By similar arguments as before, P_s^n admits a density function p_s^n and then $\tilde{P}_s^n$ admits a density function $\tilde{p}_s^n$ belonging to $\mathcal{H}$ and weak solution of the equation

$$\partial_t \tilde{p}^n = \nu \Delta \tilde{p}^n - (K_{\varepsilon_n} * \tilde{p}^n \cdot \nabla)\tilde{p}^n \; ; \; \tilde{p}_0^n = w_0. \tag{3.33}$$

and solution of the mild equation

$$\tilde{p}_t^n(x) = G_t^\nu * w_0(x) + \int_0^t \nabla_x G_{t-s}^\nu * (\tilde{p}_s^n . K_{\varepsilon_n} * \tilde{p}_s^n)(x)ds. \tag{3.34}$$

Let us now introduce for each n the coupling of processes $(Z^{in}, \bar{Y}^{in}, \bar{X}^i)_{1 \le i \le n}$, where $(\bar{X}^i)$ are independent copies of X defined as in (3.23) on a certain probability space and $Z^{in}, \bar{Y}^{in}$ are driven, for each i respectively, following the same Brownian motion as $\bar{X}^i$. We will now compare the two processes $\bar{Y}^{in}$ and $\bar{X}^i$. So we need to estimate $w - \tilde{p}^n$. Using (3.18) and (3.34), we obtain

$$\tilde{p}_t^n(x) - w_t(x)$$
$$= \int_0^t \int_{\mathbb{R}^2} \nabla_x G_{t-s}^\nu(x - y) . \left(K_{\varepsilon_n} * \tilde{p}_s^n(y)\tilde{p}_s^n(y) - K * w_s(y)w_s(y) \right) dyds \tag{3.35}$$

We will prove the

Theorem 3.12 *There exist positive real constants C_1 and C_2 such that*

$$\sup_{t \le T} |||\tilde{p}_t^n - w_t||| \le \frac{C_1}{\sqrt{\nu}}\varepsilon_n |||w_0|||^2 \sqrt{T} \exp(C_2 |||w_0|||T), \tag{3.36}$$

The proof begins with

Lemma 3.13 *For each $t \le T$, for each $x \in \mathbb{R}^2$,*

$$\left| \int_{\mathbb{R}^2} (K_{\varepsilon_n}(x - y) - K(x - y))\tilde{p}_t^n(y)dy \right| \le 2\varepsilon_n \|w_0\|_\infty \tag{3.37}$$

$$\|K_{\varepsilon_n} * \tilde{p}_t^n - K * w_t\|_\infty \le 2\varepsilon_n \|w_0\|_\infty + (K_1 + K_\infty) ||| \tilde{p}_t^n - w_t |||. \tag{3.38}$$

Proof.

1) (3.37) is proved as (3.19).
2) For $x \in \mathbb{R}^2$,

$$|K_{\varepsilon_n} * \tilde{p}_t^n(x) - K * w_t(x)| \le 2\varepsilon_n \|w_0\|_\infty + \int_{\mathbb{R}^2} |K(x-y)||\tilde{p}_t^n(y) - w_t(y)|dy$$
$$\le 2\varepsilon_n \|w_0\|_\infty + K_\infty \|\tilde{p}_t^n - w_t\|_1 + K_1 \|\tilde{p}_t^n - w_t\|_\infty.$$

$\square$

Let us now prove Theorem 3.12.

Proof. We consider (3.35). Then,

$$|\tilde{p}_t^n(x) - w_t(x)|$$

$$\leq |\int_0^t \int_{\mathbb{R}^2} \nabla_x G_{t-s}^\nu(x-y) \cdot$$

$$\left(\tilde{p}_s^n(y)(K_{\varepsilon_n} * \tilde{p}_s^n(y) - K * w_s(y)) + K * w_s(y)(\tilde{p}_s^n(y) - w_s(y)) \right) dyds|$$

$$\leq \int_0^t \int_{\mathbb{R}^2} |\nabla_x G_{t-s}^\nu(x-y)|$$

$$\left(|\tilde{p}_s^n(y)|(2\varepsilon_n \|w_0\|_\infty + 2(K_\infty + K_1)|||w_0|||.|||\tilde{p}_s^n - w_s|||) \right) dyds$$

by (3.38), (3.24) and (3.5)

$$\leq \frac{C}{\sqrt{\nu}} \int_0^t \frac{1}{\sqrt{t-s}} \left(\|w_0\|_\infty(2\varepsilon_n \|w_0\|_\infty + 2(K_\infty + K_1)|||w_0|||.|||\tilde{p}_s^n - w_s|||) \right) ds$$

$$\leq \frac{C}{\sqrt{\nu}} \left(4\varepsilon_n \|w_0\|_\infty^2 \sqrt{T} + 2(K_\infty + K_1)|||w_0||| \int_0^t \frac{1}{\sqrt{t-s}} |||\tilde{p}_s^n - w_s|||ds \right).$$

Consider now the L^1-norm of $\tilde{p}_t^n - w_t$ and by similar computations,

$$\int_{\mathbb{R}^2} |\tilde{p}_t^n(x) - w_t(x)|dx$$

$$\leq \int_0^t \frac{1}{\sqrt{t-s}} \|\tilde{p}_s^n\|_1(2\varepsilon_n \|w_0\|_\infty + (K_\infty + K_1)|||\tilde{p}_s^n - w_s|||)ds$$

$$+ \int_0^t \frac{1}{\sqrt{t-s}} \|K * w_s\|_\infty \|\tilde{p}_s^n - w_s\|_1 ds$$

$$\leq \frac{A}{\sqrt{\nu}} \left(4\varepsilon_n |||w_0|||^2 \sqrt{T} + 2(K_\infty + K_1)|||w_0||| \int_0^t \frac{1}{\sqrt{t-s}} |||\tilde{p}_s^n - w_s|||ds \right).$$

Combining the two previous results gives

$$|||\tilde{p}_t^n - w_t||| \leq \frac{C}{\sqrt{\nu}} \left(8\varepsilon_n |||w_0|||^2 \sqrt{T} + 2|||w_0||| \int_0^t \frac{1}{\sqrt{t-s}} |||\tilde{p}_s^n - w_s|||ds \right).$$

We iterate twice this inequality and obtain finally by Gronwall's lemma that

$$\sup_{t \leq T} |||\tilde{p}_t^n - w_t||| \leq \frac{C_1}{\sqrt{\nu}} \varepsilon_n |||w_0|||^2 \sqrt{T} \exp(C_2|||w_0|||T).$$

$$\square$$

Adding now (3.38) and (3.36), we deduce the

Corollary 3.14 *For each* $t \leq T$

$$\|K_{\varepsilon_n} * \tilde{p}_t^n - K * w_t\|_\infty \leq A_T \varepsilon_n, \tag{3.39}$$

where $A_T = 2\|w_0\|_\infty + (K_\infty + K_1)\frac{C_1}{\sqrt{\nu}}\varepsilon_n\||w_0\||^2\sqrt{T}\exp(C_2\||w_0\||T).$

We are now able to obtain our main theorem.

Theorem 3.15 *Let us consider independent processes, solutions of the stochastic differential equations defined on* $[0, T]$, $T > 0$ *by*

$$\bar{X}_t^i = Z_0^i + \sqrt{2\nu}B_t^i + \int_0^t K * \tilde{p}_s(\bar{X}_s^i)ds, \tag{3.40}$$

where $(B^i)_{i \in \mathbb{N}}$ *are independent Brownian motions on* $\mathbb{R}^2$ *and* $(Z_0^i)_{i \in \mathbb{N}}$ *are* $\mathbb{R}^2$-*valued iid random variables independent of* $(B^i)_{i \in \mathbb{N}}$ *with law* $P_0(dx) = \frac{|w_0(x)|}{\|w_0\|_1}dx$, $w_0 \in L^1 \cap L^\infty$. *The function* $\tilde{p}_s$ *is the density of the signed measure* $\tilde{P}_s$ *associated with the law* P_s *of* $\bar{X}_s^i$ *by (3.11).*

Let $(\varepsilon_n)_{n \in \mathbb{N}}$ *be a sequence of positive numbers tending to* 0 *and such that*

$$\lim_n \frac{M_{\varepsilon_n}}{\sqrt{n}L_{\varepsilon_n}} \exp(\|w_0\|_1 T L_{\varepsilon_n}) = 0.$$

We consider the coupled n-*particle system* $(Z^{in})_{1 \leq i \leq n}$ *defined by*

$$Z_t^{in} = Z_0^i + \sqrt{2\nu}B_t^i + \int_0^t \frac{1}{n}\sum_{j=1}^n h(Z_0^j)K_{\varepsilon_n}(Z_s^{in} - Z_s^{jn})ds. \tag{3.41}$$

Then, for each $1 \leq i \leq n$,

$$\lim_{n \to +\infty} E\left(\sup_{t \leq T}|Z_t^{in} - \bar{X}_t^i|\right) = 0, \tag{3.42}$$

(in the precise asymptotics given by (3.46)).

That implies the convergence in law (uniformly in time), of the weighted empirical measures $\tilde{\mu}_s^n = \frac{1}{n}\sum_{i=1}^n h(Z_0^{in})\delta_{Z_s^{in}}$ to $\tilde{P}_s$ and $\tilde{P}_s = w_s(x)dx$ where w is solution of the vortex equation with initial datum w_0.

Proof.

$$E\left(\sup_{t \leq T}|Z_t^{in} - \bar{X}_t^i|\right) \leq E\left(\sup_{t \leq T}|Z_t^{in} - \bar{Y}_t^{i,n}|\right) + E\left(\sup_{t \leq T}|\bar{Y}_t^{i,n} - \bar{X}_t^i|\right)$$

$$\leq \frac{M_{\varepsilon_n}}{nL_{\varepsilon_n}}\exp(\|w_0\|_1 T L_{\varepsilon_n}) + E\left(\sup_{t \leq T}|\bar{Y}_t^{i,n} - \bar{X}_t^i|\right). \tag{3.43}$$

But

$$|\bar{Y}_t^{i,n} - \bar{X}_t^i| \leq \int_0^t |K_{\varepsilon_n} * \tilde{p}_s^n(\bar{Y}_s^{i,n}) - K_{\varepsilon_n} * \tilde{p}_s^n(\bar{X}_s^i)|ds$$
$$+ \int_0^t |K_{\varepsilon_n} * \tilde{p}_s^n(\bar{X}_s^i) - K * w(\bar{X}_s^i)|ds.$$

The second right hand-side term is controled by Corollary 3.14. It remains to study the first right hand-side term, what we do following [26] Lemma 3.1 and Theorem 3.1. It is proved therein that for x and z in $\mathbb{R}^2$,

$$|K_{\varepsilon_n} * \tilde{p}_s^n(x) - K_{\varepsilon_n} * \tilde{p}_s^n(z)| \leq C_0(\|w_0\|_1 + \|w_0\|_\infty)\phi(x,z),$$

where $\phi(x,z) = \tilde{\phi}(|x - z|)$, and $\tilde{\phi}(r) = r(1 - \ln r)$ if $0 < r < 1$ and $\tilde{\phi}(r) = 1$ if $r \geq 1$. Let us remark that the function $\tilde{\phi}$ is non-decreasing and concave. Then, by noting $C = C_0(\|w_0\|_1 + \|w_0\|_\infty)$, one deduces that

$$|\bar{Y}_t^{i,n} - \bar{X}_t^i| \leq A_T\varepsilon_n + C \int_0^t \phi(\bar{Y}_s^{i,n}, \bar{X}_s^i)ds.$$

Thus,

$$E\left(\sup_{u \leq t} |\bar{Y}_u^{i,n} - \bar{X}_u^i|\right) \leq A_T\varepsilon_n + C \int_0^t E\left(\sup_{u \leq s} \phi(\bar{Y}_u^{i,n}, \bar{X}_u^i)\right)ds$$
$$\leq A_T\varepsilon_n + C \int_0^t E\left(\sup_{u \leq s} \phi(\bar{Y}_u^{i,n}, \bar{X}_u^i)\right)ds$$
$$\leq A_T\varepsilon_n + C \int_0^t E\left(\tilde{\phi}(\sup_{u \leq s} |\bar{Y}_u^{i,n} - \bar{X}_u^i|)\right)ds$$

since $\tilde{\phi}$ is non decreasing

$$\leq A_T\varepsilon_n + C \int_0^t \tilde{\phi}(E\left(\sup_{u \leq s} |\bar{Y}_u^{i,n} - \bar{X}_u^i|\right))ds$$

by concavity of $\tilde{\phi}$.

Let us denote by $H(t) = E\left(\sup_{u \leq t} |\bar{Y}_u^{i,n} - \bar{X}_u^i|\right)$. Then by the previous computations,

$$H(t) \leq A_T\varepsilon_n + C \int_0^t \tilde{\phi}(H(s))ds. \tag{3.44}$$

As in [26], we introduce the solution $h(x_0, t)$ of the equation

$$z'(t) = C\tilde{\phi}(z(t)); \quad z(0) = x_0 > 0.$$

Then, if $x_0 < 1$, and if $t_0 = \inf\{t, h(x_0, t) > 1\}$,

$$
\begin{aligned}
h(x_0, t) &= x_0^{\exp(-Ct)} \exp(1 - e^{-Ct}) \text{ if } h(x_0, t) < 1 \text{ , } t < t_0 \\
&= 1 + C(t - t_0) \text{ if } h(x_0, t) \geq 1, \; t \geq t_0
\end{aligned}
$$

and if $x_0 \geq 1$, $h(x_0, t) = x_0 + Ct$. Hence, by (3.44), we get $H(t) \leq h(A_T \varepsilon_n, t)$. But since ε_n tends to 0 as n tends to infinity, for n sufficiently large, we deduce that

$$
H(t) \leq (A_T \varepsilon_n)^{\exp(-Ct)} \exp(1 - e^{-Ct})
$$

and finally

$$
E\left(\sup_{u \leq T} |\bar{Y}_u^{i,n} - \bar{X}_u^i| \right) \leq (A_T \varepsilon_n)^{\exp(-CT)} \exp(1 - e^{-CT}). \tag{3.45}
$$

Now, by (3.43), and (3.45), we finally conclude that

$$
\begin{aligned}
E\left(\sup_{t \leq T} |Z_t^{in} - \bar{X}_t^i| \right) \leq \; & \frac{M_{\varepsilon_n}}{\sqrt{n} L_{\varepsilon_n}} \exp(\|w_0\|_1 T L_{\varepsilon_n}) \\
& + (A_T \varepsilon_n)^{\exp(-CT)} \exp(1 - e^{-CT}),
\end{aligned} \tag{3.46}
$$

where $C = C_0(\|w_0\|_1 + \|w_0\|_\infty)$ and then tends to 0 when n tends to infinity. $\qquad\square$

3.5.3. *Numerical Results*
We finally deduce from this study an algorithm for the simulation of the solution of the vortex equation. To approximate numerically this solution, one discretizes in time the particle system with an Euler scheme, as it has been described in Section 2 Theorem 2.6. The result obtained by Bossy-Talay [3] for the Burgers equation suggests us that if $\bar{\mu}_{l\Delta t}^n$ denotes the weighted empirical measure of the discretized system, then $K_{\varepsilon_n} \star \bar{\mu}_{l\Delta t}^n$ will converge to $K \star w_{l\Delta t}$ in L^1 with rate $\mathcal{O}(\sqrt{\Delta t} + \frac{1}{\sqrt{n}})$, where Δt denotes the time-step. The simulations realized with specific kernels K_ε confirm this behaviour.

3.6. GENERALIZATION TO A FINITE MEASURE INITIAL CONDITION

It is possible to generalize what we have done to a finite measure initial condition, instead of $w_0 \in L^1 \cap L^\infty$. We refer to [30] for details. In that case, the study of the vortex equation exposed by Giga-Miyakawa-Osada in [16] involves analytical results, in particular on generators of generalized divergence form. The solutions of the vortex equation live in L^q-spaces, with $1 < q < 2$. This induces new difficulties, even if the particle approximation method is essentially similar to the previous one. One looses the estimates

(3.45) and only obtains the convergence in law for the particle systems, instead of a L^1-convergence as in the bounded case.

4. The case of a Bounded Domain

We now consider a Navier-Stokes equation in a bounded domain Θ of $\mathbb{R}^2$ satisfying the no-slip boundary condition:

$$\partial_t u(t,x) + (u.\nabla)u(t,x) = \nu\Delta u(t,x) - \nabla p \quad \text{in } \Theta;$$
$$\nabla.u(t,x) = 0 \quad \text{in } \Theta \ ; \ u(0,x) = u_0(x) \text{ for } x \in \Theta;$$
$$u(t,x) = (0,0) \text{ for } x \in \partial\Theta,$$

where p is the pressure and $\nu > 0$ the viscosity coefficient. A probabilistic approach of this equation, based on branching processes, has already been developed by Benachour, Roynette, Vallois [1] and generalized in dimension 3 by Giet [15]. But even if the authors propose some particle approximations, the convergence of the method is not shown and the particle systems they describe are not for use in practice. Our purpose is to construct some easily simulable particle systems, and to rigorously show the convergence of some associated weighted empirical measures to a deterministic finite measure associated with the solution of the Navier-Stokes equation.

As in Section 3, we wish to associate a vortex equation with this Navier-Stokes equation. This approach would consist in replacing the Biot and Savart kernel by the orthogonal gradient K of the Green function of the Dirichlet problem in the domain. But one then only obtains the nullity of the normal component of the velocity on the boundary. To obtain in addition the nullity of the tangential component, we are inspired by Cottet [10], who proves that by adding a Neumann condition to the vortex equation, one obtains an admissible vorticity field in the sense that the associated velocity satisfies *a posteriori* the no-slip condition.

This Neumann condition badly depends on the vorticity and is really hard to take into account. So we will mainly deal in the following with a vortex equation in a bounded domain of R^2 with a given Neumann condition at the boundary. We obtain the existence and uniqueness of the solution of this vortex equation in an appropriate space. We are interested in proving the convergence of Monte-Carlo approximations to this solution. To our knowledge, there was so far no proof of convergence of deterministic or stochastic particle methods in this simplified case.

We associate with the vortex equation a nonlinear diffusive and reflected process, with random births at the boundary governed by the Neumann condition. We construct interacting normally reflected particle systems with space-time random births at the boundary and prove the propagation of chaos to the law of the nonlinear process associated with the vortex equation. We are inspired by the paper of Sznitman [36], which concerns interacting and reflected McKean-Vlasov particle systems living in a bounded domain. Some additional difficulties appear here, due to the singular interacting kernel K and to the space-time random births. Moreover, since w_0 and g are not probability densities, we introduce as in Section 3 some weights associated with the initial position.

We will be then able to describe the simulation algorithm, and we will finally explain how to adapt this numerical approach if we assume the Cottet condition (cf. [10]) to come back to the Navier-Stokes case.

Notation: If Θ is a bounded domain of $\mathbb{R}^2$, the Sobolev space $H^1(\Theta)$ consists in functions which belong together with their first order distribution derivatives to $L^2(\Theta)$.

4.1. THE MODEL

Let $T > 0$. Let us consider a function g defined on $\partial\Theta$. We are interested in the equation

$$\partial_t w(t, x) + \nabla.(wKw)(t, x) = \nu \Delta w(t, x) \text{ in }]0, T] \times \Theta;$$
$$w(0, x) = w_0(x) \text{ in } \Theta ; \ \partial_n w = \nabla w.n = g \text{ on }]0, T] \times \partial\Theta \quad (4.1)$$

where $n(x)$ denotes the outward normal to $\partial\Theta$ at the point x and $Kw(t, x) = \int_\Theta K(x, y)w(t, y)dy$. The kernel $K(x, y)$ is equal to $\nabla_x^\perp G(x, y) = (-\partial_{x_2} G(x, y), \partial_{x_1} G(x, y))$ where $G(x, y)$ is the fundamental solution of the Poisson equation

$$\Delta_x G(x, y) = \delta_y(x), \ x \in \Theta ; \ G(x, y) = 0, \ x \in \partial\Theta \quad (4.2)$$

Let us remark the important properties of the kernel K:

$$\forall x \neq y \in \emptyset , \ \nabla_x.K(x, y) = 0 ; \quad \forall x \in \partial\Theta, \forall y \in \emptyset, \ K(x, y).n(x) = 0 \quad (4.3)$$

In all the following, we will moreover assume

Hypotheses (H):
The domain Θ of $\mathbb{R}^2$ is bounded, simply connected and of class C^4.

$$w_0 \in L^2(\Theta) \ ; \quad g(t, x) \in L_t^2([0, T], L_x^2(\partial\Theta, d\sigma)), \quad (4.4)$$

where $d\sigma(x)$ denotes the surface measure on the boundary.

Thanks to the assumptions made on Θ, the following properties hold for the Green function G and the kernel $K = (K_1, K_2)$:

Lemma 4.1 $\exists C_0 > 0$, $\forall x \neq y \in \bar{\Theta}$,

$$|G(x,y)| \leq C_0(1 + |\ln|x - y||); \quad |K(x,y)| \leq \frac{C_0}{|x - y|}$$

$$|\nabla_x K_i(x,y)| + |\nabla_y K_i(x,y)| \leq \frac{C_0}{|x - y|^2} \text{ for } i = 1, 2.$$

Proof. For $y = (y_1, y_2) \in \mathbb{R}^2$, let $y^\perp = (-y_2, y_1)$ and $y^* = y/|y|^2$ if $y \neq (0,0)$. In case Θ is the unit disk $B(0,1)$ of $\mathbb{R}^2$, one has the following explicit expression for the Green function (see [17] p.19)

$$G_0(x,y) = \frac{1}{2\pi} \ln\left(\frac{|x - y|}{|y||x - y^*|}\right). \tag{4.5}$$

We remark that

$$\forall x, y \in \bar{B}(0,1),$$
$$|x - y^*| \geq |y||x - y^*| = \sqrt{|x - y|^2 + (|x|^2 - 1)(|y|^2 - 1)} \geq |x - y|. \tag{4.6}$$

As a consequence,

$$|2\pi G_0(x,y)| \leq -\ln|x - y| 1_{\{|x-y|\leq 1\}} + \ln(|y||x - y^*|) 1_{\{|y||x-y^*|\geq 1\}}.$$

As $|y||x - y^*| = |x|y| - y/|y|| \leq 2$, we conclude that $|2\pi G_0(x,y)| \leq |\ln|x - y|| + \ln(2)$. We also deduce from (4.6) the bound on the corresponding kernel

$$K(x,y) = \frac{1}{2\pi}\left(\frac{(x - y)^\perp}{|x - y|^2} - \frac{(x - y^*)^\perp}{|x - y^*|^2}\right) = \frac{1}{2\pi}\left(((x - y)^*)^\perp - ((x - y^*)^*)^\perp\right).$$

To estimate ∇K_i, we combine (4.6) and the fact that each term of the Jacobian matrix of $z \to z^*$ is bounded by $1/|z|^2$. When Θ is a general bounded and simply connected domain of class C^3, according to [33], there is a conformal mapping from $B(0,1)$ onto Θ which extends to a one-to-one C^2 mapping from $\bar{B}(0,1)$ to $\bar{\Theta}$ denoted by f and such that Df, $(Df)^{-1}$ and $D^2 f$ are bounded on $\bar{B}(0,1)$. Since the Green function for Θ is given by

$$G(x,y) = G_0(f^{-1}(x), f^{-1}(y)),$$

the estimates on G, K and ∇K_i follow from those obtained for the unit disk and the just mentionned properties of f. $\qquad\square$

We are interested in weak solutions of (4.1) defined in the following sense

Definition 4.2 We say that $w : [0,T] \times \Theta \to \mathbb{R}$ is a weak solution of (4.1) if $w(0,.) = w_0$ and

(i) $w \in L_t^\infty(L_x^2) \cap L_t^2(H_x^1)$ where $L_t^\infty(L_x^2)$ and $L_t^2(H_x^1)$ stand respectively for $L^\infty([0,T], L^2(\Theta))$ and $L^2([0,T], H^1(\Theta))$

(ii) for any $v \in H^1(\Theta)$, $\frac{d}{dt} \int_\Theta w_t v + \nu \int_\Theta \nabla w_t . \nabla v = \int_\Theta w_t K w_t . \nabla v + \nu \int_{\partial\Theta} g_t v d\sigma$.

Before stating the existence of a unique weak solution to (4.1), we are going to check the following Lemma which prepares the study of the non-linear term in (4.1).

Lemma 4.3

$$\forall 2 < p \le +\infty,$$
$$\exists C > 0, \forall w \in L^p(\Theta), \; Kw \in \mathcal{C}(\emptyset) \text{ and } \|Kw\|_{L^\infty} \le C\|w\|_{L^p} \tag{4.7}$$

$$\exists C > 0, \forall w \in L^2(\Theta), \; \|Kw\|_{L^2} \le C\|w\|_{L^2} \tag{4.8}$$

Proof. For $\alpha > 0$, let $K_\alpha(x,y) = \mathbf{1}_{\{|x-y|>\alpha\}} K(x,y)$. By Lebesgue's theorem and using the continuity of K away from the diagonal, we obtain the continuity of $x \in \emptyset \mapsto K_\alpha(x,.) \in L_y^q$, for each $q \ge 1$. When in addition $q < 2$, according to Lemma 3.3, $K_\alpha(x,.)$ converges to $K(x,.)$ in L_y^q uniformly on $\bar\Theta$, when α tends to 0. We deduce that $K(x,.)$ is continuous in L_y^q and obtain (4.7) by Hölder inequality. Let $w \in L^2(\Theta)$. Using Lemma 3.3 and Cauchy-Schwarz inequality, we get

$$\|Kw\|_{L^2}^2 \le \int_\Theta \left(\int_\Theta \frac{C_0}{|x-y|} dy \right) \left(\int_\Theta \frac{C_0 w^2(y)}{|x-y|} dy \right) dx$$

$$\le \left(\sup_{x \in \bar\Theta} \int_\Theta \frac{C_0}{|x-y|} dy \right)^2 \|w\|_{L^2}^2.$$

$\square$

We can now state the following existence and uniqueness theorem, and refer to [21] for the proof. It consists in obtaining energy estimates and in adapting the Galerkin approximation method.

Theorem 4.4 *Under hypotheses* **(H)**, *equation (4.1) has a unique solution w in the sense of Definition 4.2. In addition, $w \in C([0,T], L_x^2) \cap L_t^4(L_x^4)$.*

In order to give a probabilistic interpretation to this weak solution of (4.1), we introduce the semi-group $P_t^\nu(x,y)$ associated with $\sqrt{2\nu}$ times the Brownian motion normally reflected on the boundary and prove the following mild representation

Proposition 4.5 *Let w denote the weak solution of (4.1) given by Theorem 4.4. Then $\forall t \in [0, T]$, dx a.e. in Θ,*

$$w_t(x) = P_t^\nu w_0(y) + \int_0^t \nabla P_{t-s}^\nu . (w_s K w_s)(x) ds$$

$$\tag{4.9}$$

$$+ \ \nu \int_0^t \int_{\partial\Theta} P_{t-s}^\nu(y, x) g(s, y) d\sigma(y) ds$$

where $\nabla P_{t-s}^\nu . (w_s K w_s)(x) = \int_\Theta \nabla_y P_{t-s}^\nu(y, x) . w_s(y) K w_s(y) dy.$

Proof. Let $t \in]0, T]$ and φ be a smooth function on $\bar{\Theta}$ with a vanishing normal derivative at the boundary: $\partial_n \varphi(x) = 0$ for $x \in \partial\Theta$. According to [23] Theorem 5.3 p.320, the boundary value problem

$$\partial_s \psi + \nu \Delta \psi = 0 \ \text{ on } [0, t] \times \Theta; \ \partial_n \psi = 0 \ \text{ on } [0, t] \times \partial\Theta; \ \psi(t, .) = \varphi(.) \text{ on } \Theta$$

admits a classical solution $\psi(s, x)$ which is $C^{1,2}$ on $[0, t] \times \bar{\Theta}$. By the Feynman-Kac approach, this solution has the following representation: $\psi(s, x) = P_{t-s}^\nu \varphi(x)$. Clearly $\psi \in L^\infty([0, t], H^1(\Theta))$ and $\partial_s \psi \in L^2([0, t], (H^1)'_x(\Theta))$. By [39], Lemma 1.2 p. 261, we deduce that in $\mathcal{D}'(]0, t[)$,

$$\frac{d}{ds} \int_\Theta w_s \psi(s, .) = \int_\Theta w_s \partial_s \psi(s, .) - \nu \int_\Theta \nabla w_s . \nabla \psi(s, .)$$

$$+ \int_\Theta w_s K w_s . \nabla \psi(s, .) + \nu \int_{\partial\Theta} g_s \psi(s, .) d\sigma.$$

By the equation satisfied by ψ, the sum of the two first terms of the r.h.s. is nil. Hence

$$\int_\Theta w_t(x) \varphi(x) dx = \int_\Theta w_0(x) \psi(0, x) dx + \int_0^t \int_\Theta w_s K w_s(x) . \nabla \psi(s, x) dx ds$$

$$+ \ \nu \int_0^t \int_{\partial\Theta} \psi(s, x) g(s, x) d\sigma(x) ds.$$

By the symmetry of P^ν and hypotheses **(H)**, $\int_0^t \int_{\partial\Theta} \int_\Theta P_{t-s}^\nu(x, y) |\varphi(y)| \, |dy| g(s, x)| d\sigma(x) ds \leq \sup |\varphi| \|g\|_{L_t^1(L_x^1(\partial\Theta))} < +\infty$. Hence, by Fubini's theorem the last term of the r.h.s. is equal to $\nu \int_\Theta \varphi(x) \int_0^t \int_{\partial\Theta} P_{t-s}^\nu(y, x) g(s, y) d\sigma(y) ds dx$. We conclude the proof by applying similarly Fubini's theorem to the other terms of the r.h.s. and remarking that the derived equality holds for any smooth function φ with vanishing normal derivative.

To justify the use of Fubini's theorem in the second term, we need the following estimates given by [35] (a.13) and (a.14) p.600:

$$\forall x \in \bar{\Theta}, \; \forall y \in \bar{\Theta},$$
$$|\nabla_x P_t^\nu(x, y)| \le C_1/t^{3/2} \text{ and } \|\nabla_x P_t^\nu(x, y)\|_{L_y^1(\Theta)} \le C_1/\sqrt{t}. \tag{4.10}$$

Indeed the first one ensures that $\nabla \psi(s, x) = \int_\Theta \nabla_x P_{t-s}^\nu(x, y)\varphi(y)dy$. By the second one and (4.8),

$$\int_0^t \int_\Theta |w_s K w_s|(x) \int_\Theta |\nabla_x P_{t-s}^\nu(x, y)||\varphi(y)|dydxds \le$$
$$C \sup |\varphi| \|w\|_{L_t^\infty(L_x^2)}^2 \int_0^t (t - s)^{-1/2}ds.$$

$\square$

4.2. THE PROBABILISTIC INTERPRETATION OF THE VORTEX EQUATION WITH A NEUMANN BOUNDARY CONDITION

We are again in a McKean-Vlasov context and will associate a nonlinear martingale problem.

To bypass the difficulty due to the Neumann condition involving g, we essentially follow Fernandez-Méléard [11], proving that this term is related to space-time random births located at the boundary. We have also to take into account to the bounded domain instead of the whole space and the diffusion processes we consider will be reflected on the boundary. There are also births inside the domain at time 0 and the functions w_0 and g are not probability densities. As in Section 3, we follow Jourdain [20] to overpass this problem.

Let $\|w_0\|_1 = \int_\Theta |w_0|$ and $\|g\|_1 = \int_{[0,T] \times \partial \Theta} |g| d\sigma dt$. To govern the times and positions of births we introduce on $[0, T] \times \bar{\Theta}$ the probability measure

$$P_0(dt, dx) = \mathbf{1}_{\{x \in \Theta\}} \delta_{\{0\}}(dt) \frac{|w_0(x)|}{\|w_0\|_1 + \nu\|g\|_1} dx$$

$$+ \mathbf{1}_{\{x \in \partial \Theta\}} \frac{\nu|g(t, x)|}{\|w_0\|_1 + \nu\|g\|_1} dt d\sigma(x). \tag{4.11}$$

We also consider for $t \in [0, T]$ and $x \in \bar{\Theta}$ the measurable function

$$h(t, x) = \mathbf{1}_{\{t=0, x \in \Theta\}} \frac{w_0(x)}{|w_0(x)|} (\|w_0\|_1 + \nu\|g\|_1)$$

$$+ \mathbf{1}_{\{x \in \partial \Theta\}} \frac{g(t, x)}{|g(t, x)|} (\|w_0\|_1 + \nu\|g\|_1) \tag{4.12}$$

with values in $\{-(\|w_0\|_1 + \nu\|g\|_1), 0, \|w_0\|_1 + \nu\|g\|_1\}$. Let us remark that if φ a bounded measurable function on $[0, T] \times \bar{\Theta}$, then

$$\int_{[0,T] \times \bar{\Theta}} \varphi(t, x) h(t, x) P_0(dt, dx) = \int_{\Theta} \varphi(0, x) w_0(x) dx$$

(4.13)

$$+\nu \int_{[0,T] \times \partial\Theta} \varphi(t, x) g(t, x) dt d\sigma(x)$$

Let $(\tau, (X_t)_{t \leq T}, (k_t)_{t \leq T})$ denote the canonical process on $[0, T] \times \mathcal{C}([0, T], \emptyset) \times \mathcal{C}([0, T], \mathbb{R}^2)$. For a probability measure Q on this space, we define the family $(\tilde{Q}_t)_{t \in [0,T]}$ of signed measures on $\emptyset$ by

$$\forall B \in \mathcal{B}(\emptyset), \quad \tilde{Q}_t(B) = E^Q(h(\tau, X_0) \mathbf{1}_{\{\tau \leq t\}} \mathbf{1}_B(X_t)), \qquad (4.14)$$

It is easy to check that for each $t \in [0, T]$, the signed measure $\tilde{Q}_t$ is bounded with a total mass less than $\|w_0\|_1 + \nu\|g\|_1$.

To give a probabilistic interpretation to the equation, we are inspired by Sznitman [36] and Bossy-Jourdain [6] for the reflected contribution and by Fernandez-Méléard [11] for the space-time random births.

Definition 4.6 Let $T > 0$. We denote by $\mathcal{P}_T$ the space of probability measures Q on $[0, T] \times \mathcal{C}([0, T], \emptyset) \times \mathcal{C}([0, T], \mathbb{R}^2)$ such that for each $t \in [0, T]$, the signed measure $\tilde{Q}_t$ has a density $\tilde{q}_t$ with respect to the Lebesgue measure on Θ and the measurable version $\tilde{q}$ belongs to $L_t^\infty(L_x^2) \cap L_t^2(H_x^1)$.

Definition 4.7 A probability measure $P \in \mathcal{P}_T$ is solution of the nonlinear martingale problem $(\mathcal{M}_B)$ if

1) $P \circ (\tau, X_0, k_0)^{-1} = P_0 \otimes \delta_{(0,0)}$
2) for each $\phi \in \mathcal{C}_b^2(\mathbb{R}^2)$,

$$M_t^\phi = \phi(X_t + k_t) - \phi(X_0)$$

$$- \int_0^t \mathbf{1}_{\{\tau \leq s\}} \left(K\tilde{p}_s(X_s) . \nabla\phi(X_s + k_s) + \nu\Delta\phi(X_s + k_s) \right) ds$$

is a P-martingale, for the filtration $\mathcal{F}_t = \sigma(\tau, (X_s, k_s), s \leq t)$ ($\tilde{p}(s, x)$ denotes a measurable version of the densities of $\tilde{P}_s$).
3) P a.s., $\forall t \in [0, T]$,

$$\int_0^t d|k|_s < +\infty, \quad |k|_t = \int_0^t \mathbf{1}_{\{X_s \in \partial\Theta\}} \mathbf{1}_{\{\tau \leq s\}} d|k|_s, \quad k_t = \int_0^t n(X_s) d|k|_s.$$

The following lemma states the link between $(\mathcal{M}_B)$ and the vortex equation (4.1).

Lemma 4.8 If $P \in \mathcal{P}_T$ solves $\mathcal{M}_B$ then $\tilde{p}$ is a weak solution of (4.1).

Proof. By Definition 4.7 1), (4.14) and (4.13), $\tilde{p}_0 = w_0$.

According to Definition 4.7 2), $\beta_t = X_t - X_0 - \int_0^t \mathbf{1}_{\{\tau \leq s\}} K \tilde{p}_s(X_s) ds + k_t$ is a P-continuous martingale with bracket $< \beta >_t = 2\nu(t-\tau)^+ I_2$ where I_2 denotes the 2×2 identity matrix, which implies that $\beta_t = 0$ for $t \in [0, \tau]$. Using moreover Definition 4.7 3), we deduce that $X_t = X_0$ for $t \in [0, \tau]$. Hence for $\psi \in C^{1,2}([0,T] \times \bar{\Theta})$,

$$\int_0^T \partial_s \psi(s, X_s) ds + \psi(0, X_0) = \psi(\tau, X_0) + \int_0^T \mathbf{1}_{\{\tau \leq s\}} \partial_s \psi(s, X_s) ds.$$

If moreover $\forall (s, x) \in [0, T] \times \partial\Theta$, $\partial_n \psi(s, x) = 0$, by Itô's formula, we deduce that

$$\psi(T, X_t) = \psi(\tau, X_0) + \int_0^T \nabla\psi(s, X_s).d\beta_s$$

$$+ \int_0^T \mathbf{1}_{\{\tau \leq s\}} (\partial_s \psi(s, X_s) + K \tilde{p}_s(X_s).\nabla\psi(s, X_s) + \nu\Delta\psi(s, X_s)) ds$$

Multiplying by the $\mathcal{F}_0$-measurable variable $h(\tau, X_0)$, taking expectations and using the definition of $\tilde{p}$ and (4.13), we deduce that

$$\int_\emptyset \psi(T, x)\tilde{p}(T, x) dx = \int_\emptyset \psi(0, x) w_0(x) dx + \nu \int_0^T \int_{\partial\Theta} \psi(s, x) g(s, x) d\sigma(x) ds$$

$$+ \int_0^T \int_\emptyset (\partial_s \psi(s, x) + K \tilde{p}_s(x).\nabla\psi(s, x) + \nu\Delta\psi(s, x))\tilde{p}(s, x) dx ds,$$

For the choice $\psi(s, x) = \varphi(s)v(x)$ where v is a C^2 function on $\bar{\Theta}$ such that $\partial_n v = 0$ on $\partial\Theta$ and $\varphi \in \mathcal{D}(]0, T[)$, we obtain

$$\int_0^T \left(\varphi'(s) \int_\emptyset \tilde{p}_s v + \varphi(s) \left(\int_\Theta \tilde{p}_s K \tilde{p}_s.\nabla v + \nu \int_\Theta \tilde{p}_s \Delta v + \nu \int_{\partial\Theta} g_s v d\sigma \right) \right) ds = 0.$$

As $P \in \mathcal{P}_T$, $\tilde{p} \in L_t^2(H_x^1)$. By Green's formula for functions in $H^1(\Theta)$ ([7] p.197) and since $\partial_n v$ vanishes on the boundary, ds a.e. in $[0, T]$, $\int_\Theta \tilde{p}_s \Delta v = - \int_\Theta \nabla\tilde{p}_s.\nabla v$. Since Θ is C^4, the $C^2(\bar{\Theta})$-functions with a vanishing normal derivative are dense in $H^1(\Theta)$ (see [7]), and we conclude that $\tilde{p}$ satisfies Definition 4.2 (ii). $\qquad\square$

Theorem 4.9 *Under Hypotheses* **(H)**, *the martingale problem* $(\mathcal{M}_B)$ *has a unique solution P. In addition, the corresponding $\tilde{p}$ is a weak solution of (4.1) and satisfies (4.9).*

Proof. 1) Uniqueness

Let P^1 and P^2 be two solutions of $(\mathcal{M}_B)$. Then according to Lemma 4.8, $\tilde{p}^1$ and $\tilde{p}^2$ are weak solutions of (4.1). According to Theorem 4.4, $\tilde{p}_1 = \tilde{p}_2 = w$. Hence P^1 and P^2 both solve the martingale problem defined like $(\mathcal{M}_B)$ but with known drift coefficient Kw_s replacing $K\tilde{p}_s$ in Definition 4.7 2). Since $w \in L^4_t(L^4_x)$, by (4.7), $\|Kw_s\|_{L^\infty_x} \in L^4_t$.

Let Γ denote the first marginal of the probability measure P_0 on $[0, T] \times \bar{\Theta}$ and for $i = 1, 2$ and $u \in [0, T]$, $p^i(u, .)$ be a regular conditional probability on $[0, T] \times \mathcal{C}([0, T], \emptyset) \times \mathcal{C}([0, T], \mathbb{R}^2)$ endowed with P^i given $\tau = u$.

Then $d\Gamma(u)$ a.e., $p^i(u, .)$ a.s., $\tau = u$, Definition 4.7 3) is satisfied and $p^i(u, .) \circ (X_0, k_0)^{-1}$ is equal to

$$\mathbf{1}_{\{u=0\}}\frac{|w_0(x)|dx}{\|w_0\|_1} \otimes \delta_{(0,0)} + \mathbf{1}_{\{u>0\}}\frac{|g(u, x)|d\sigma(x)}{\int_{\partial\Theta}|g(u, y)|d\sigma(y)} \otimes \delta_{(0,0)} \qquad (4.15)$$

and $\forall \phi \in C^2_b(\mathbb{R}^2)$,

$$\phi(X_t + k_t) - \phi(X_0) - \int_0^t \mathbf{1}_{\{u\leq s\}}\left(Kw_s(X_s).\nabla\phi(X_s + k_s) + \nu\Delta\phi(X_s + k_s)\right)ds$$

is a $p^i(u, .)$-martingale.

Reasoning like in the proof of Lemma 4.8, we obtain that $d\Gamma(u)$ a.e., $p^i(u, .)$ a.s., $X_t = X_0$ and $k_t = (0, 0)$ for $t \in [0, u]$. With (4.15), we deduce that $d\Gamma(u)$ a.e., $p^1(u, .) \circ (X_u, k_u)^{-1} = p^2(u, .) \circ (X_u, k_u)^{-1}$ and that for $i = 1, 2$, $p^i(u, .)$ is equal to the image of $p^i(u, .) \circ ((X_{t+u}, k_{t+u})_{t\in[0,T-u]})^{-1}$ by the mapping

$$(X_t, k_t)_{t\geq 0} \in \mathcal{C}([0, T-u], \emptyset \times \mathbb{R}^2) \rightarrow (X_{(t-u)^+}, k_{(t-u)^+})_{t\in[0,T]} \in \mathcal{C}([0, T], \emptyset \times \mathbb{R}^2).$$

Moreover $d\Gamma(u)$ a.e. , $W_t = \frac{1}{\sqrt{2\nu}}\left(X_{t+u} - X_u - \int_u^{t+u} Kw_s(X_s)ds + k_{t+u}\right)$ is a $p^i(u, .)$ Brownian motion. Since $s \rightarrow \|Kw_s\|_{L^\infty}$ is square integrable, combining trajectorial uniqueness for the Brownian motion normally reflected at the boundary of Θ (see [25]), Girsanov's theorem and the equality $p^1(u, .) \circ (X_u, k_u)^{-1} = p^2(u, .) \circ (X_u, k_u)^{-1}$ which holds $d\Gamma(u)$ i a.e., we deduce that $d\Gamma(u)$ a.e.,

$$p^1(u, .) \circ ((X_{t+u}, k_{t+u})_{t\in[0,T-u]})^{-1} = p^2(u, .) \circ ((X_{t+u}, k_{t+u})_{t\in[0,T-u]})^{-1}.$$

Hence $d\Gamma(u)$ p.p. $p^1(u, .) = p^2(u, .)$ and $P^1 = P^2$.

2) Existence.

Let w be the solution of the vortex equation given by Theorem 4.4. We recall that $\|Kw_s\|_{L^\infty} \in L^4_t$. We construct a solution to the linear martingale problem defined like $(\mathcal{M}_B)$ but with known drift coefficient $Kw_s(.)$ replacing $K\tilde{p}_s$ in Definition 4.7 2) and we check that this probability measure solves $(\mathcal{M}_B)$.

Let (τ, X_0) be a random variable with law P_0 independent from a two-dimensional Brownian motion $(W_t)_{t\in[0,T]}$. Existence and trajectorial uniqueness hold for the stochastic differential equation with normal reflection

$$X_t = X_0 + \sqrt{2\nu} \int_0^t \mathbf{1}_{\{\tau \le s\}} dW_s - k_t \ ;$$

$$|k|_t = \int_0^t \mathbf{1}_{\{X_s \in \partial\Theta\}} \mathbf{1}_{\{\tau \le s\}} d|k|_s \ ; \quad k_t = \int_0^t n(X_s) d|k|_s.$$

Moreover $\forall t \in [0,T]$, the law of X_t admits

$$x \to \frac{1}{\|w_0\|_1 + \nu\|g\|_{L^1([0,t]\times\partial\Theta)}}$$

$$\left(|w_0| P_t^\nu(x) + \nu \int_0^t \int_{\partial\Theta} |g|(s,y) P_{t-s}^\nu(y,x)\sigma(dy)ds \right)$$

as density w.r.t. the Lebesgue measure on ø. Since $\|Kw_s\|_{L^\infty}$ is square integrable, we deduce by Girsanov's theorem that the martingale problem defined like $(\mathcal{M}_B)$, with Kw_s replacing $K\tilde{p}_s$, admits a solution P such that for each $t \in [0,T]$, the measure $\tilde{P}_t$ has a density. Let $\tilde{p}$ denote a measurable version of the densities.

We set $t \in [0,T]$. Reasoning like in the proof of Lemma 4.8, we obtain that for $\psi \in \mathcal{C}^{1,2}([0,t] \times \bar{\Theta})$ such that $\forall(s,x) \in [0,t] \times \partial\Theta$, $\partial_n\psi(s,x) = 0$,

$$\int_\emptyset \psi(t,x)\tilde{p}(t,x)dx = \int_\emptyset \psi(0,x)w_0(x)dx + \nu \int_0^t \int_{\partial\Theta} \psi(s,x)g(s,x)d\sigma(x)ds$$

$$+ \int_0^t \int_\emptyset (\partial_s\psi(s,x) + Kw_s(x).\nabla\psi(s,x) + \nu\Delta\psi(s,x))\tilde{p}(s,x)dxds.$$

Choosing $\psi(s,x) = P_{t-s}^\nu\varphi(x)$ like in the proof of Proposition 4.5 and remarking that because of (4.10) and the uniform in time bound $\|\tilde{p}_t\|_{L^1} \le \|w_0\|_1 + \nu\|g\|_1$,

$$\int_0^t \int_{\Theta^2} |\nabla_x P_{t-s}^\nu(x,y)| |\varphi(y)| |\tilde{p}_s(x)| |Kw_s(x)| dxdyds$$

$$\le C \int_0^t \frac{\|Kw_s\|_{L^\infty} ds}{\sqrt{t-s}} < +\infty,$$

we deduce by Fubini's theorem that

$$dx \ a.e., \ \tilde{p}_t(x) = P_t^\nu w_0(x) \ + \ \int_0^t \nabla P_{t-s}^\nu \cdot (\tilde{p}_s K w_s)(x) ds$$

$$+ \ \nu \int_{(0,t] \times \partial \Theta} P_{t-s}^\nu(x, y) g(s, y) d\sigma(y) ds.$$

Now, using the mild equation (4.9) satisfied by w and (4.10), we obtain

$$\exists C > 0, \ \forall t \in [0, T], \ \|\tilde{p}_t - w_t\|_{L^1} \leq C \int_0^t \|\tilde{p}_s - w_s\|_{L^1} \frac{\|K w_s\|_{L^\infty}}{\sqrt{t-s}} ds. \ (4.16)$$

By iterating this bound, then using Hölder's inequality, we obtain

$$\|\tilde{p}_t - w_t\|_{L^1} \ \leq \ C \int_0^t \|\tilde{p}_s - w_s\|_{L^1} \|K w_s\|_{L^\infty} \int_s^t \frac{\|K w_u\|_{L^\infty} \ du}{\sqrt{t-u}\sqrt{u-s}} ds$$

$$\leq \ C \int_0^t \|\tilde{p}_s - w_s\|_{L^1} \|K w_s\|_{L^\infty} \|K w\|_{L_t^4(L_x^\infty)}$$

$$\left(\int_s^t ((t-u)(u-s))^{-2/3} du \right)^{3/4} ds.$$

Hence (4.16) holds with $(t-s)^{-1/2}$ replaced by $(t-s)^{-1/4}$ in the r.h.s. After the next iteration we obtain that (4.16) holds with $(t-s)^{-1/2}$ replaced by 1 and conclude by Gronwall's lemma that $\forall t \in [0, T], \ \tilde{p}_t = w_t$. □

4.3. STOCHASTIC APPROXIMATIONS OF THE SOLUTION OF THE VORTEX EQUATION

4.3.1. *The Case of a Cut-off Kernel*

As in Section 3, we introduce a cut-off kernel K_ε preserving the properties (4.3). More precisely we consider an increasing C^2-function η from $\mathbb{R}_+$ to $\mathbb{R}_+$, such that $\eta(x) = x$ for $x \leq \frac{1}{2}$ and $\eta(x) = 1$ for $x \geq 1$. For $\varepsilon \leq 1$, we set

$$G_\varepsilon(x, y) \ = \ \eta\left(\frac{|x-y|^3}{\varepsilon^3}\right) G(x, y); \tag{4.17}$$

$$K_\varepsilon(x, y) \ = \ \nabla_x^\perp G_\varepsilon(x, y)$$

$$= \ \eta\left(\frac{|x-y|^3}{\varepsilon^3}\right) K(x, y)$$

$$+ \eta'\left(\frac{|x-y|^3}{\varepsilon^3}\right) \frac{3(x-y)^\perp |x-y|}{\varepsilon^3} G(x, y) \tag{4.18}$$

The following Lemma states usefull properties of this cutoff kernel:

Lemma 4.10   1) There exists a constant C independent of ε, such that

$$\nabla_x.K_\varepsilon(x,y) = 0 \quad ; \quad K_\varepsilon(x,y) \cdot n(x) = 0 \text{ for } x \in \partial\Theta,$$
$$K_\varepsilon(x,y) = K(x,y) \text{ if } |x-y| \geq \varepsilon$$
$$\forall x,y \in \bar\Theta \ , \ |K_\varepsilon(x,y)| \leq \frac{C(1+|\ln|x-y||)}{|x-y|} \qquad (4.19)$$

2) $\sup_{x \in \emptyset} \|K(x,.) - K_\varepsilon(x,.)\|_{L_y^p}$ tends to 0 as ε tends to 0 as soon as $p < 2$.

3) For ε sufficiently small, the kernel K_ε is bounded by $M_\varepsilon \leq \frac{C|\ln\varepsilon|}{\varepsilon}$ and Lipschitz continuous in both variables with constant $L_\varepsilon \leq \frac{C|\ln\varepsilon|}{\varepsilon^2}$ where C does not depend on ε.

Proof. The two first properties in 1) are obvious and 2) is an easy consequence of (4.19).

By Lemma 4.1 and the above definition of η, the norm of first term of the r.h.s. of (4.18) is smaller than $C_0(\frac{1}{|x-y|} \wedge \sup_{r \in [0,\varepsilon 2^{-1/3}]} \frac{r^2}{\varepsilon^3}) \leq C_0(\frac{1}{|x-y|} \wedge \frac{1}{\varepsilon})$. By the estimate of G in Lemma 3.3 and since $\eta'(x) = 0$ for $x > 1$, the second term of the r.h.s. of (4.18) is smaller than $3C_0\|\eta'\|_\infty$ times

$$\left(\frac{1+|\ln|x-y||}{|x-y|} \wedge \sup_{r \in [0,\varepsilon]} \frac{r^2(1+|\ln(r)|)}{\varepsilon^3} \right) \leq \left(\frac{1+\ln|x-y|}{|x-y|} \wedge \frac{1+|\ln(\varepsilon)|}{\varepsilon} \right)$$

as $\varepsilon \leq 1$. We deduce both (4.19) and the upper-bound in $C|\ln(\varepsilon)|/\varepsilon$. To prove that K_ε is Lipschitz continuous, we use in a similar way Lemma 3.3 combined with the definition of η to check that the gradient of each coordinate of K_ε w.r.t. either x or y is bounded by $C|\ln(\varepsilon)|/\varepsilon^2$ (the contribution of the first term of the r.h.s. of (4.18) is C/ε^2 whereas the one of the second term is $C|\ln(\varepsilon)|/\varepsilon^2$). $\qquad\square$

With a slight adaptation of Sznitman [36] to take into account the random births on the boundary, we obtain the existence and pathwise uniqueness of the following interacting particle systems.

Definition 4.11 Consider a sequence $(B^i)_{i \in \mathbb{N}}$ of independent Brownian motions on $\mathbb{R}^2$ and a sequence of independent variables $(\tau^i, Z_0^i)_{i \in \mathbb{N}}$ with values in $[0,T] \times \emptyset$ distributed according to P_0, and independent of the Brownian motions. For a fixed ε, for each $n \in \mathbb{N}^*$, and $1 \leq i \leq n$, let us consider the interacting processes defined by

$$Z_t^{in,\varepsilon} \in \emptyset, \forall t \in [0,T]$$
$$Z_t^{in,\varepsilon} = Z_0^i + \sqrt{2\nu} \int_0^t \mathbf{1}_{\{\tau^i \leq s\}} dB_s^i + \int_0^t \mathbf{1}_{\{\tau^i \leq s\}} K_\varepsilon \tilde\mu_s^{n,\varepsilon}(Z_s^{in,\varepsilon}) ds - k_t^{in,\varepsilon};$$
$$|k^{in,\varepsilon}|_t = \int_0^t \mathbf{1}_{\{Z_s^{in,\varepsilon} \in \partial\Theta\}} \mathbf{1}_{\{\tau^i \leq s\}} d|Z^{in,\varepsilon}|_s \ ; \ k_t^{in,\varepsilon} = \int_0^t n(Z_s^{in,\varepsilon}) d|k^{in,\varepsilon}|_s$$

$$(4.20)$$

where $\tilde{\mu}_s^{n,\varepsilon} = \frac{1}{n}\sum_{j=1}^n h(\tau^j, Z_0^j)\mathbf{1}_{\{\tau^j \leq s\}}\delta_{Z_s^{jn,\varepsilon}}$ is the weighted empirical measure of the system.

Let us remark that the particles either have birth at time 0 inside the domain and evolve as diffusive particles with normal reflecting boundary conditions, or have birth at a random time on the boundary of the domain, and evolve after birth as the other ones. Moreover, all particles, as soon as they are born, interact together following a mean field depending on the parameter ε.

Again according to [36], we also get the existence and pathwise uniqueness of the limiting processes (when n tends to infinity and ε is fixed), coupled with the interacting processes as follows.

Definition 4.12 We define $\bar{Z}^{i,\varepsilon}$ by

$$\bar{Z}_t^{i,\varepsilon} \in \emptyset, \forall t \in [0,T]$$

$$\bar{Z}_t^{i,\varepsilon} = Z_0^i + \sqrt{2\nu}\int_0^t \mathbf{1}_{\{\tau^i \leq s\}}dB_s^i + \int_0^t \mathbf{1}_{\{\tau^i \leq s\}}K_\varepsilon \tilde{Q}_s^\varepsilon(\bar{Z}_s^{i,\varepsilon})ds - \bar{k}_t^{i,\varepsilon} ;$$

$$|\bar{k}^{i,\varepsilon}|_t = \int_0^t \mathbf{1}_{\{\bar{Z}_s^{i,\varepsilon}\in\partial\Theta\}}\mathbf{1}_{\{\tau^i\leq s\}}d|\bar{k}^{i,\varepsilon}|_s ; \quad \bar{k}_t^{i,\varepsilon} = \int_0^t n(\bar{Z}_s^{i,\varepsilon})d|\bar{k}^{i,\varepsilon}|_s \quad (4.21)$$

where Q^ε is the common law of $(\tau^i, \bar{Z}^{i,\varepsilon}, \bar{k}^{i,\varepsilon})$, and $\tilde{Q}_s^\varepsilon$ is defined from Q^ε by (4.14).

Sznitman also proves a propagation of chaos result, but without precise estimates on the rate of convergence. In order to get such estimates, we denote by H a $C_b^2(\bar{\Theta})$-extension of the distance-function $d(.,\partial\Theta)$ (defined on a restriction to Θ of a neighbourhood of $\partial\Theta$). The function H satisfies (see [17])

$$\nabla H = -n \text{ on } \partial\Theta. \quad (4.22)$$

We also recall that the domain Θ (since C^4) satisfies the uniform "exterior sphere" condition:

$$\exists C_{sp} \geq 0 , \forall x \in \partial\Theta , \forall x' \in \bar{\Theta} , C_{sp}|x - x'|^2 + n(x).(x - x') \geq 0. \quad (4.23)$$

Proposition 4.13 For $t \leq T$, for each $i \in \{1, ..., n\}$,

$$E(\sup_{s\leq t}|Z_s^{in,\varepsilon} - \bar{Z}_s^{i,\varepsilon}|^2) \leq 2d(\Theta)\sqrt{\frac{A_\varepsilon}{n}}$$

$$E(\sup_{s\leq t}|k_s^{in,\varepsilon} - \bar{k}_s^{i,\varepsilon}|) \leq E(\sup_{s\leq t}|Z_s^{in,\varepsilon} - \bar{Z}_s^{i,\varepsilon}|) \overset{\exp(K_H(1 + (\|w_0\|_1 + \nu\|g\|_1)(M_\varepsilon/2 + L_\varepsilon)t))}{}$$

$$+2t(\|w_0\|_1 + \nu\|g\|_1)\left(L_\varepsilon E(\sup_{s\leq t}|Z_s^{in,\varepsilon} - \bar{Z}_s^{i,\varepsilon}|) + \frac{M_\varepsilon}{\sqrt{n}}\right)$$

*where K_H is only depends on the upper-bounds of the function H and its
derivatives, $d(\Theta)$ is the diameter of Θ and $A_\varepsilon = \frac{4(\|w_0\|_1 + \nu\|g\|_1)^2 M_\varepsilon^2}{2 + (\|w_0\|_1 + \nu\|g\|_1)(M_\varepsilon + 2L_\varepsilon)}$.*

Remark 4.14 The convergence rate given above is not optimal in n.
Indeed one can check that $E(\sup_{s \leq t} |Z_s^{in,\varepsilon} - \bar{Z}_s^{i,\varepsilon}|^4)$ is smaller than
$\frac{C_1 M_\varepsilon^4 t}{n^2 C_2} \exp(C(M_\varepsilon^2 + 4L_\varepsilon^2)t^2)$, but in the next section, we will choose $\varepsilon = \varepsilon_n$
depending on n and converging to 0 in such a way that $E(\sup_{s \leq t} |Z_s^{in,\varepsilon_n} - \bar{Z}_s^{i,\varepsilon_n}|^2)$ tends to 0. The estimation given in the proposition allows a quicker
(but still very slow) convergence of ε_n to 0 than the previous one.

Proof. We compare the two processes $Z^{in,\varepsilon}$ and $\bar{Z}^{i,\varepsilon}$. We denote for sim-
plicity Z, k, $\bar{Z}$ and $\bar{k}$ instead of $Z^{in,\varepsilon}$, $k^{in,\varepsilon}$, $\bar{Z}^{i,\varepsilon}$ and $\bar{k}^{i,\varepsilon}$, $h_t = H(Z_t)$,
$\bar{h}_t = H(\bar{Z}_t)$, $h'_t = \nabla H(Z_t)$, $\bar{h}'_t = \nabla H(\bar{Z}_t)$, $h''_t = \Delta H(Z_t)$, $\bar{h}''_t = \Delta H(\bar{Z}_t)$,
$b_t = K_\varepsilon \tilde{\mu}_s^{n,\varepsilon}(Z_t)$ and $\bar{b}_t = K_\varepsilon \tilde{Q}_t^\varepsilon(\bar{Z}_t)$. Computing $d \exp(-2C_{sp}(h_t + \bar{h}_t))|Z_t - \bar{Z}_t|^2$ by Itô's formula, we get

$$1_{\{\tau_i \leq t\}} \exp(-2C_{sp}(h_t + \bar{h}_t)) \times \left[2(Z_t - \bar{Z}_t).(d\bar{k}_t - dk_t) \right.$$

$$- 2C_{sp}|Z_t - \bar{Z}_t|^2(d|k|_t + d|\bar{k}|_t) - 2C_{sp}|Z_t - \bar{Z}_t|^2 \left(\sqrt{2\nu}(h'_t + \bar{h}'_t)dB_t^i \right.$$

$$+ \left. \left\{ h'_t b_t + \bar{h}'_t \bar{b}_t + \nu(-2C_{sp}|h'_t + \bar{h}'_t|^2 + h''_t + \bar{h}''_t) \right\} dt \right)$$

$$+ \left. 2(Z_t - \bar{Z}_t).(b_t - \bar{b}_t)dt \right] \tag{4.24}$$

Because of the "exterior sphere" condition, the local time terms of the first
line have a non-positive contribution after integration over time. We deduce
that for a computable constant K_H which only depends on upper-bounds
of the function H and its derivatives,

$$E(|Z_t^{in,\varepsilon} - \bar{Z}_t^{i,\varepsilon}|^2) \leq K_H \left((1 + M_\varepsilon(\|w_0\|_1 + \nu\|g\|_1)) \int_0^t E(|Z_s^{in,\varepsilon} - \bar{Z}_s^{i,\varepsilon}|^2)ds \right.$$

$$+ \left. \int_0^t E(|Z_s^{in,\varepsilon} - \bar{Z}_s^{i,\varepsilon}||K_\varepsilon \tilde{\mu}_s^{n,\varepsilon}(Z_s^{in,\varepsilon}) - K_\varepsilon \tilde{Q}_s^\varepsilon(\bar{Z}_s^{i,\varepsilon})|)ds \right) \tag{4.25}$$

Using the Lipschitz continuity of K_ε, the boundedness of h and the exchangeability of the processes $(Z^{in,\varepsilon}, \bar{Z}^{i,\varepsilon})$, $1 \leq i \leq n$, we obtain

$$E \left(|Z_s^{in,\varepsilon} - \bar{Z}_s^{i,\varepsilon}||K_\varepsilon \tilde{\mu}_s^{n,\varepsilon}(Z_s^{in,\varepsilon}) - K_\varepsilon \tilde{Q}_s^\varepsilon(\bar{Z}_s^{i,\varepsilon})| \right)$$

$$\leq (\|w_0\|_1 + \nu\|g\|_1)L_\varepsilon E(|Z_s^{in,\varepsilon} - \bar{Z}_s^{i,\varepsilon}|(|Z_s^{in,\varepsilon} - \bar{Z}_s^{i,\varepsilon}| + \frac{1}{n}\sum_{j=1}^n |Z_s^{jn,\varepsilon} - \bar{Z}_s^{j,\varepsilon}|))$$

$$+E(|Z_s^{in,\varepsilon} - \bar{Z}_s^{i,\varepsilon}| \| \frac{1}{n} \sum_{j=1}^{n} h(\tau_j, Z_0^j)\mathbf{1}_{\{\tau^j \leq s\}} K_\varepsilon(\bar{Z}_s^{i,\varepsilon}, \bar{Z}_s^{j,\varepsilon}) - K_\varepsilon \tilde{Q}_s^\varepsilon(\bar{Z}_s^{i,\varepsilon})|)$$

$$\leq (1 + 2(\|w_0\|_1 + \nu\|g\|_1)L_\varepsilon)E(|Z_s^{in,\varepsilon} - \bar{Z}_s^{i,\varepsilon}|^2)$$

$$+E(|\frac{1}{n} \sum_{j=1}^{n} h(\tau_j, Z_0^j)\mathbf{1}_{\{\tau^j \leq s\}} K_\varepsilon(\bar{Z}_s^{i,\varepsilon}, \bar{Z}_s^{j,\varepsilon}) - K_\varepsilon \tilde{Q}_s^\varepsilon(\bar{Z}_s^{i,\varepsilon})|^2)$$

After expansion of $E(|\frac{1}{n} \sum_{j=1}^{n} h(\tau_j, Z_0^j)\mathbf{1}_{\{\tau^j \leq s\}} K_\varepsilon(\bar{Z}_s^{i,\varepsilon}, \bar{Z}_s^{j,\varepsilon}) - K_\varepsilon \tilde{Q}_s^\varepsilon(\bar{Z}_s^{i,\varepsilon})|^2)$, many terms disappear by independence of the variables which are centered conditionnally to $\bar{Z}^{i,\varepsilon}$ and it only remains n bounded terms. We deduce that

$$E(|Z_t^{in,\varepsilon} - \bar{Z}_t^{i,\varepsilon}|^2) \leq K_H \left((2 + (\|w_0\|_1 \right.$$

$$+ \nu\|g\|_1)(M_\varepsilon + 2L_\varepsilon)) \int_0^t E(|Z_s^{in,\varepsilon} - \bar{Z}_s^{i,\varepsilon}|^2)ds$$

$$\left. + \frac{4(\|w_0\|_1 + \nu\|g\|_1)^2 M_\varepsilon^2 t}{n} \right) \qquad (4.26)$$

Using Gronwall's Lemma, we obtain that both sides of (4.25) and (4.26) are smaller than

$$f(t) = \frac{4(\|w_0\|_1 + \nu\|g\|_1)^2 M_\varepsilon^2}{n(2 + (\|w_0\|_1 + \nu\|g\|_1)(M_\varepsilon + 2L_\varepsilon))}$$

$$\exp(K_H(2 + (\|w_0\|_1 + \nu\|g\|_1)(M_\varepsilon + 2L_\varepsilon))t).$$

Integrating (4.24) w.r.t. time, dealing with the stochastic integral thanks to Doob's inequality and using that the r.h.s. of (4.25) is smaller than $f(t)$, we get

$$E(\sup_{s \leq t} |Z_s^{in,\varepsilon} - \bar{Z}_s^{i,\varepsilon}|^2) \leq \left(K_H \int_0^t E(|Z_s^{in,\varepsilon} - \bar{Z}_s^{i,\varepsilon}|^4)ds \right)^{1/2} + f(t)$$

$$\leq d(\Theta) \left(K_H \int_0^t E(|Z_s^{in,\varepsilon} - \bar{Z}_s^{i,\varepsilon}|^2)ds \right)^{1/2} + f(t)$$

$$\leq d(\Theta)\sqrt{f(t)} + f(t)$$

since the r.h.s. of (4.26) is smaller than $f(t)$

The l.h.s. being smaller than $d(\Theta)^2$, it is smaller than $2d(\Theta)\sqrt{f(t)}$ when $f(t) \geq d(\Theta)^2$ and the r.h.s. is smaller than $2d(\Theta)\sqrt{f(t)}$ otherwise. We

deduce the desired estimate for $E(\sup_{s\le t}|Z_s^{in,\varepsilon} - \bar{Z}_s^{i,\varepsilon}|^2)$.
Now remarking that

$$\sup_{s\le t}|k_s^{in,\varepsilon} - \bar{k}_s^{i,\varepsilon}| \le \int_0^t |K_\varepsilon \tilde{\mu}_s^{n,\varepsilon}(Z_s^{in,\varepsilon}) - K_\varepsilon \tilde{Q}_s^\varepsilon(\bar{Z}_s^{i,\varepsilon})|ds + \sup_{s\le t}|Z_s^{in,\varepsilon} - \bar{Z}_s^{i,\varepsilon}|$$

and using arguments developed above we obtain the second assertion. $\square$

Remark 4.15 Let us remark that if ø is a convex region then the rate of convergence is easier to obtain. Indeed the constant C_{sp} defined in (4.23) can be chosen equal to 0:

$$\forall x \in \partial\Theta \ , \ \forall x' \in \text{ø} \ , \ n(x).(x - x') \ge 0. \tag{4.27}$$

In the expression of $|Z_t^{in,\varepsilon} - \bar{Z}_t^{i,\varepsilon}|^2$ given by Itô's formula, the local time terms are non-positive and therefore

$$E(\sup_{s\le t}|Z_s^{in,\varepsilon} \quad \bar{Z}_s^{i,\varepsilon}|^2) \le (1 + 2(\|w_0\|_1 + \nu\|g\|_1)L_\varepsilon))$$

$$\int_0^t E(\sup_{u\le s}|Z_u^{in,\varepsilon} - \bar{Z}_u^{i,\varepsilon}|^2)ds + \frac{4(\|w_0\|_1 + \nu\|g\|_1)^2 M_\varepsilon^2 t}{n}$$

and we conclude by Gronwall's Lemma.

4.3.2. Convergence of the Limiting Laws

We will prove that the laws Q^ε of $(\tau^1, \bar{Z}^{1,\varepsilon}, \bar{k}^{1,\varepsilon})$ converge to the unique solution P of $(\mathcal{M}_B)$ as ε tends to 0. We are first going to check that the drift coefficient $K_\varepsilon \tilde{Q}_s^\varepsilon$ converges to $K\tilde{p}_s$.
By Girsanov's theorem, it turns out that $\forall s > 0$, the measure $\tilde{Q}_s^\varepsilon$ admits a density function q_s^ε. Moreover, reasoning like in the proof of Theorem 4.9 and using the boundedness of K_ε, we show that q^ε is the unique solution in $L_T^1 = \{p_t, \sup_{t\le T}\|p_t\|_{L^1} < +\infty\}$ of the equation

$$q_t^\varepsilon(x) = P_t^\nu w_0(x) \ + \ \int_0^t \nabla_x P_{t-s}^\nu.(q_s^\varepsilon K_\varepsilon q_s^\varepsilon)(x)ds$$

$$\tag{4.28}$$

$$+ \ \nu \int_0^t \int_{\partial\Theta} P_{t-s}^\nu g(s,y)d\sigma(y)ds.$$

On the other hand, thanks to Lemma 4.19 1), we can apply to the equation

$$\partial_t w(t,x) + \nabla.(wK_\varepsilon w)(t,x) = \nu\Delta w(t,x) \ \text{in} \ \Theta;$$

$$w(x,0) = w_0 \ \text{in} \ \Theta \ ; \ \partial_n w = \nabla w.n = g \ \text{on} \ \partial\Theta \tag{4.29}$$

all what we have done for the equation (4.1). Then there exists a unique weak solution w^ε belonging to $L_t^\infty(L_x^2) \cap L_t^2(H_x^1)$. Now, like in Proposition 4.5, we obtain that w^ε is also solution of (4.28). Since it belongs to L_T^1 (Θ is bounded), we conclude that $w^\varepsilon = q^\varepsilon$. Thanks to (4.19), one can check that

$$\sup_{\varepsilon \in (0,1]} \left(\|q^\varepsilon\|_{L_t^\infty(L_x^2)} + \|q^\varepsilon\|_{L_t^2(H_x^1)} + \|\partial_t q^\varepsilon\|_{L_t^2(H_x^{1'})} + \|q^\varepsilon\|_{L_t^4(L_x^4)} \right) < +\infty. \quad (4.30)$$

Remark 4.16 Similarly the non-negative measures $B \in \mathcal{B}(\emptyset) \to E^{Q^\varepsilon}(1_{\{\tau \leq t\}} 1_B(X_t))$ have densities p_t^ε w.r.t. the Lebesgue measure which are the unique solution in L_T^1 of the mild equation obtained by replacing respectively w_0 and g by $|w_0|/(\|w_0\|_1 + \nu\|g\|_1)$ and $|g|/(\|w_0\|_1 + \nu\|g\|_1)$ in (4.28).

Identifying p^ε with the unique weak solution of the problem obtained from (4.29) by replacing w_0 and g in the same way, we check that (4.30) holds for p^ε.

We can now prove the convergence of q^ε to w.

Proposition 4.17

$$\lim_{\varepsilon \to 0} \|q^\varepsilon - w\|_{L_t^2(L_x^2)} = 0 \; ; \; \lim_{\varepsilon \to 0} \|K_\varepsilon q^\varepsilon - Kw\|_{L_t^2(L_x^2)} = 0.$$

Proof. Thanks to (4.30), one can extract from each sequence q^{ε_n} with ε_n tending to 0, a sub-sequence (still denoted q^{ε_n} for simplicity), which converges strongly in $L_t^2(L_x^2)$ and in $L_t^2(H_x^1)$ and weakly* in $L_t^\infty(L_x^2)$ to $\tilde{w}$. We can show that $\tilde{w}$ is a weak solution of (4.1) and conclude that $\tilde{w} = w$ by uniqueness for this equation.

Let $1 < p < 2$. Combining the Sobolev inequality $\|q_s^{\varepsilon_n}\|_{L^{\frac{p}{p-1}}} \leq C\|q_s^{\varepsilon_n}\|_{H^1}$, Lemma 4.10 2) and (4.30), we obtain that the term $(K_{\varepsilon_n} - K)q_s^{\varepsilon_n}$ converges to 0 in $L_t^2(L_x^\infty)$. Now, writing

$$\|K_\varepsilon q^\varepsilon - Kw\|_{L_t^2(L_x^2)} \leq \|K(q^\varepsilon - w)\|_{L_t^2(L_x^2)} + \|(K_\varepsilon - K)q^\varepsilon\|_{L_t^2(L_x^2)},$$

and using (4.8), one easily deduces the second assertion.   □

Theorem 4.18 *The probability measures Q^ε on $[0,T] \times C([0,T],\emptyset) \times C([0,T],\mathbb{R}^2)$ converge weakly to the unique solution P of $(\mathcal{M}_B)$, as ε tends to 0.*

Proof. The weak convergence topology being metrizable, we check that $(Q^n = Q^{\varepsilon_n})_{n \in \mathbb{N}}$ converges weakly to P when the sequence ε_n tends to 0 as n tends to $+\infty$. Let us firstly prove the uniform tightness of the sequence $(Q^n)_n$, next identify the limiting points.

1) By (4.19) and (4.30), we easily obtain that

$$\sup_{n} \|K_{\varepsilon_n} q_s^{\varepsilon_n}\|_{L_t^4(L_x^\infty)} < +\infty. \tag{4.31}$$

Then the Kolmogorov tightness criterion is satisfied for the laws of

$$\bar{Y}_t^{1,\varepsilon_n} = Z_0^1 + \sqrt{2\nu} \int_0^t \mathbf{1}_{\{\tau^1 \le s\}} dB_s^1 + \int_0^t \mathbf{1}_{\{\tau^1 \le s\}} K_{\varepsilon_n} q_s^n (\bar{Z}_s^{1,\varepsilon_n}) ds.$$

The uniform tightness of the laws Q^n of the processes $(\tau^1, \bar{Z}^{1,\varepsilon_n}, \bar{k}^{1,\varepsilon_n})$ is thus a simple consequence of the fact that the application sending $y \in C([0,T], \mathbb{R}^2)$ on the solution $(x, k) \in C([0,T], \emptyset) \times C([0,T], \mathbb{R}^2)$ of the Skorohod problem is continuous (See [25]).

   2) Let us now denote by Q^∞ a limiting value of a convergent subsequence still denoted by (Q^n) for simplicity and prove by arguments inspired from [36] that $Q^\infty = P$.

   If as usual (τ, X, k) denotes the canonical process on $[0,T] \times C([0,T], \emptyset) \times C([0,T], \mathbb{R}^2)$, let us define, for $p \in \mathbb{N}^*$, $0 \le s_1 \le \dots \le s_p \le s < t \le T$, $\phi \in C_b^2(\mathbb{R}^2)$, $g \in C_b([0,T], (\emptyset \times \mathbb{R}^2)^p)$ the function

$$G_n(\tau, X, k) = g(\tau, X_{s_1}, k_{s_1}, \dots, X_{s_p}, k_{s_p})\Big(\phi(X_t + k_t) - \phi(X_s + k_s)$$

$$- \int_s^t \mathbf{1}_{\{\tau \le u\}}\Big(\nu \Delta\phi(X_u + k_u) + K_{\varepsilon_n} q_u^{\varepsilon_n}(X_u) . \nabla\phi(X_u + k_u)\Big) du\Big)$$

Then $E^{Q^n}(G_n(\tau, X, k)) = 0$. Now if we define the function G by replacing $K_{\varepsilon_n} q_s^{\varepsilon_n}$ by Kw_s in (4.3.2), we wish to prove that $E^{Q^\infty}(G(\tau, X, k)) = 0$.

$$E^{Q^\infty}(G(\tau, X, k)) = E^{Q^\infty}(G(\tau, X, k)) - E^{Q^n}(G(\tau, X, k))$$

$$+ E^{Q^n}(G(\tau, X, k) - G^n(\tau, X, k)).$$

Since $w \in L_t^4(L_x^4)$, by (4.7), ds a.e. in $[0,T]$ $x \in \emptyset \to Kw_s(x)$ is continuous and $Kw_s \in L_t^4(L_x^\infty)$. We deduce that $G(\tau, X, k)$ is a continuous function on the path space, and the first term of the r.h.s. tends to 0 as n tends to infinity. On the other hand, using Remark 4.16 and Proposition 4.17, we obtain

$$E^{Q^n}|G^n(\tau, X, k) - G(\tau, X, k)|$$

$$\le CE\Big(\int_0^t \mathbf{1}_{\{\tau^1 \le s\}}|K_{\varepsilon_n} q_s^{\varepsilon_n}(\bar{Z}_s^{1,\varepsilon_n}) - Kw_s(\bar{Z}_s^{1,\varepsilon_n})|ds\Big)$$

$$\le C\|p^{\varepsilon_n}\|_{L_t^2(L_x^2)}\|K_{\varepsilon_n} q^{\varepsilon_n} - Kw\|_{L_t^2(L_x^2)} \to 0 \text{ as } n \to +\infty.$$

Hence $E^{Q^\infty}(G(\tau, X, k)) = 0$. Since $\forall n$, $Q^n \circ (\tau, X_0, k_0)^{-1} = P_0 \otimes \delta_{(0,0)}$, $Q^\infty \circ (\tau, X_0, k_0)^{-1} = P_0 \otimes \delta_{(0,0)}$. We are now going to prove that Q^∞-almost surely,

$$|k|_T < \infty \text{ and } \forall t \in [0, T], \ |k|_t = \int_0^t \mathbf{1}_{\{X_s \in \partial\Theta\}} \mathbf{1}_{\{\tau \leq s\}} d|k|_s \ ;$$

$$k_t = \int_0^t n(X_s) d|k|_s.$$

Since according to the proof of Theorem 4.9, P is the unique solution of the linear martingale problem defined like $\mathcal{M}_T$ but with known drift coefficient Kw_s, we will conclude that $Q^\infty = P$. According to the following Lemma the proof of which is postponed,

Lemma 4.19 For any $A \geq 0$, the following subset of $[0, T] \times C([0, T], \emptyset) \times C([0, T], \mathbb{R}^2)$

$$F_A = \left\{ (u, x, k) : \ |k|_T = \int_0^T \mathbf{1}_{\{u \leq s\}} \ \mathbf{1}_{\{x_s \in \partial\Theta\}} d|k|_s \leq A \right.$$

$$\left. \text{and } \forall t \in [0, T], \ k_t = \int_0^t n(x_s) d|k|_s \right\}$$

is closed.

We have

$$Q^\infty \left(\bigcup_{A>0} F_A \right) \geq 1 - \lim_{A \to +\infty} \liminf_{n \to +\infty} Q^n(F_A^c) \geq 1 - \lim_{A \to +\infty} \frac{\sup_{n \in \mathbb{N}} E^{Q^n} |k|_T}{A}.$$

Therefore it is enough to check that $\sup_{n \in \mathbb{N}} E|\bar{k}^{1,\varepsilon_n}|_T < +\infty$ to conclude the proof. Since $\nabla H = -n$ on $\partial\Theta$, applying Itô's formula to compute $H(\bar{Z}_T^{1,\varepsilon_n})$, we get that $|\bar{k}^{1,\varepsilon_n}|_T$ is equal to

$$H(\bar{Z}_T^{1,\varepsilon_n}) - H(Z_0^1)$$

$$- \int_0^T \mathbf{1}_{\{\tau^1 \leq s\}} \left((\nu \Delta H + K_{\varepsilon_n} q_s^{\varepsilon_n} . \nabla H)(\bar{Z}_s^{1,\varepsilon_n}) ds + \sqrt{2\nu} \nabla H(\bar{Z}_s^{1,\varepsilon_n}) . dB_s^1 \right).$$

Taking expectations and using (4.31), we obtain the desired result. □

Proof of Lemma 4.19 Let $(u^n, x^n, k^n) \in F_A$ converge to (u, x, k) as $n \to +\infty$. Since $\sup_n |k^n|_T \leq A$, by extraction of a subsequence, we can suppose that the measure $d|k^n|$ (resp. dk^n) converges weakly to a positive measure da with mass smaller than A (resp. to db_s). Of course $db_s = \lambda(s) da_s$

for some measurable function $\lambda : [0, T] \rightarrow \mathbb{R}^2$ and since k^n converges uniformly on $[0, T]$ to k, $db_s = dk_s$. Since $d(x_s^n, \partial\Theta)$, where $d(., \partial\Theta)$ denotes the (continuous) distance from the boundary function, converges uniformly on $[0, T]$ to $d(x_s, \partial\Theta)$,

$$\int_0^T d(x_s, \partial\Theta) da_s = \lim_n \int_0^T d(x_s^n, \partial\Theta) d|k|_s^n = 0.$$

We deduce that da_s a.e. and therefore $d|k|_s$ a.e., $x_s \in \partial\Theta$. Since the functions k^n which are equal to $(0,0)$ on $[0, u^n]$ converge uniformly to k, this function is equal to $(0,0)$ on $[0, u]$ and $|k|_u = 0$. To check the only lacking property: $dk_s = n(x_s)d|k|_s$, we remark that

$$\forall f \in \mathcal{C}([0, T], \mathbb{R}_+), \ \forall g \in \mathcal{C}([0, T], \emptyset),$$

$$\int_0^T f(s)\left((x_s - g(s)).dk_s + C_{sp}|x_s - g(s)|^2 da_s\right) \geq 0$$

by taking the limit $n \rightarrow +\infty$ in the similar inequalities satisfied with (x, dk, da) replaced by $(x^n, dk^n, d|k^n|)$ according to the uniform "exterior sphere" condition (4.23). We deduce that $dk_s = |\lambda(s)|n(x_s)da_s$ which implies the desired property.   □

4.4. THE CONVERGENCE THEOREM

We now consider a sequence (ε_n) tending to 0 as n tends to infinity, in such a way that

$$\lim_{n \rightarrow +\infty} L_{\varepsilon_n}^2 \sqrt{\frac{A_{\varepsilon_n}}{n}} \exp(K_H(1 + (\|w_0\|_1 + \nu\|g\|_1)(M_{\varepsilon_n}/2 + L_{\varepsilon_n})T))$$

$$+ \frac{M_{\varepsilon_n}}{\sqrt{n}} = 0.$$
(4.32)

The convergence of ε_n to 0 is very slow and this choice is certainly not optimal.

Let us now consider for each n the system of processes (τ^i, Z^{in}, k^{in}) where $Z^{in} = Z^{in,\varepsilon_n}$ and $k^{in} = k^{in,\varepsilon_n}$ are defined as in (4.35) but with K_{ε_n} replacing K_ε. We are now able to obtain our main theorem.

Theorem 4.20 *1) The laws of the n-particle system* $(\tau^i, Z^{in}, k^{in})_{1 \leq i \leq n}$, *are P-chaotic (where P is the solution of the problem* $(\mathcal{M}_B)$):

$$\forall p \in \mathbb{N}^*, \ \mathcal{L}((\tau^1, Z^{1n}, k^{1n}), ..., (\tau^p, Z^{pn}, k^{pn})) \overset{\text{weakly}}{\Longrightarrow} P^{\otimes p} \ \text{as } n \rightarrow +\infty. \ (4.33)$$

2) The approximate velocity field converges to Kw:

$$\lim_{n \to +\infty} E(\|K_{\varepsilon_n} \tilde{\mu}_t^{n,\varepsilon_n}(x) - Kw_t(x)\|_{L_t^2(L_x^2)}^2) = 0. \tag{4.34}$$

Proof. 1) Since the processes $(\tau^i, \bar{Z}^{i,\varepsilon_n}, \bar{k}^{i,\varepsilon_n})_i$ are independent, Theorem 4.18 implies that for every fixed $p \in \mathbb{N}^*$, the law of $((\tau^1, \bar{Z}^{1,\varepsilon_n}, \bar{k}^{1,\varepsilon_n}), ...,$ $(\tau^p, \bar{Z}^{p,\varepsilon_n}, \bar{k}^{p,\varepsilon_n}))$ converges weakly to $P^{\otimes p}$. Let $\mathcal{C}_T = [0,T] \times C([0,T], \emptyset) \times C([0,T], \mathbb{R}^2)$. We endow $\mathcal{C}_T^p$ with the uniform metric and $\mathcal{P}(\mathcal{C}_T^p)$ with the Vaserstein metric

$$\rho(\mu, \nu) = \inf \left\{ \int_{\mathcal{C}_T^p \times \mathcal{C}_T^p} d(x,y) \wedge 1) R(dx, dy); R \text{ has marginals } \mu \text{ and } \nu \right\}$$

compatible with the topology of the weak convergence. Hence

$$\rho(\mathcal{L}((\tau^1, \bar{Z}^{1,\varepsilon_n}, \bar{k}^{1,\varepsilon_n}), ..., (\tau_p, \bar{Z}^{p,\varepsilon_n}, \bar{k}^{p,\varepsilon_n})), P^{\otimes p}) \to 0 \text{ as } n \to +\infty.$$

By Proposition 4.13, and (4.32)

$$\lim_{n \to +\infty} E\left(d\left(((\tau^1, Z^{1n}, k^{1n}), ..., (\tau^p, Z^{pn}, k^{pn})), \right. \right.$$

$$\left. \left. ((\tau^1, \bar{Z}^{1,\varepsilon_n}, \bar{k}^{1,\varepsilon_n}), ..., (\tau_p, \bar{Z}^{p,\varepsilon_n}, \bar{k}^{p,\varepsilon_n})) \right) \right) = 0$$

which ensures that

$$\lim_{n \to +\infty} \rho\left(\mathcal{L}\left((\tau^1, Z^{1n}, k^{1n}), ..., (\tau^p, Z^{pn}, k^{pn}) \right), \right.$$

$$\left. \mathcal{L}\left((\tau^1, \bar{Z}^{1,\varepsilon_n}, \bar{k}^{1,\varepsilon_n}), ..., (\tau^p, \bar{Z}^{p,\varepsilon_n}, \bar{k}^{p,\varepsilon_n}) \right) \right) = 0.$$

We conclude that $\rho(\mathcal{L}((\tau^1, Z^{1n}, k^{1n}), ..., (\tau^p, Z^{pn}, k^{pn})), P^{\otimes p})$ converges to 0.

2) On the other hand,

$$E(|K_{\varepsilon_n} \tilde{\mu}_t^{n,\varepsilon_n}(x) - Kw_t(x)|^2) \le$$

$$3E\left(\left| K_{\varepsilon_n} \tilde{\mu}_t^{n,\varepsilon_n}(x) - \frac{1}{n} \sum_{i=1}^{n} 1_{\{\tau^i \le t\}} h(\tau^i, Z_0^i) K_{\varepsilon_n}(x, \bar{Z}_t^{i,\varepsilon_n}) \right|^2 \right.$$

$$+ \left| \frac{1}{n} \sum_{i=1}^{n} 1_{\{\tau^i \le t\}} h(\tau^i, Z_0^i) K_{\varepsilon_n}(x, \bar{Z}_t^{i,\varepsilon_n}) \right.$$

$$\left. - K_{\varepsilon_n} \tilde{Q}_t^{\varepsilon_n}(x) \right|^2 + |K_{\varepsilon_n} \tilde{Q}_t^{\varepsilon_n}(x) - Kw_t(x)|^2 \right)$$

$$\le 3\left((\|w_0\|_1 + \nu\|g\|_1)^2 (L_{\varepsilon_n}^2 E(\sup_{s \le t} |Z_s^{in} - \bar{Z}_s^{i,\varepsilon_n}|^2) + \frac{4M_{\varepsilon_n}^2}{n}) \right.$$

$$\left. + |K_{\varepsilon_n} q_t^{\varepsilon_n}(x) - Kw_t(x)|^2 \right).$$

We conclude using (4.32), Proposition 4.13 and Proposition 4.17.   □

Remark 4.21 Since the laws $\mathcal{L}((\tau^1, Z^{1n}, k^{1n}), ..., (\tau^n, Z^{nn}, k^{nn}))$ are exchangeable, the propagation of chaos is equivalent to the convergence in probability of the empirical measures to P, as probability measures on the path space (cf. [37]). As a consequence, if the space of finite measures on ø is endowed with the weak convergence topology, the random finite measures $\tilde{\mu}_t^{n,\varepsilon n} = \frac{1}{n} \sum_{i=1}^n \mathbf{1}_{\{\tau^i \le t\}} h(\tau^i, Z_0^i) \delta_{Z_t^{in}}$ converge in measure to $w_t(x)dx$ for any $t \in [0, T]$, w being the unique solution of the vortex equation.

4.4.1. *Numerical Comments*

We finally deduce from this study a simulation algorithm for the solution of the equation (4.1). To approximate numerically this solution necessitates to discretize in time the particle system. This can be achieved thanks to the Euler scheme for reflected diffusions proposed by Gobet [18]. Adapting these results to our case with identity diffusion matrix and normal reflection, one could hope (as in Bossy-Jourdain [6]), that if $\bar{\mu}_{l\Delta t}^n$ denotes the weighted empirical measure of the discretized system, $K_{\varepsilon_n} \bar{\mu}_{l\Delta t}^n$ converges to $K w_{l\Delta t}$ in $L^\infty(\Theta)$, with the rate $\mathcal{O}(\Delta t + \frac{1}{\sqrt{n}})$, where Δt denotes the time step.

4.5. SOME COMMENTS ON THE GENERALIZATION TO THE NAVIER-STOKES CASE

In [10], the Neumann condition obtained by Cottet has the form

$$\partial_n w = \int_{\partial\Theta} B w(x) d\sigma(x)$$

on $\partial\Theta$, with a specific kernel $B(x, y)$ which is a sophisticated derivation operator with a bad behaviour when $x = y$. The dependence of the Neumann condition on w makes things more complicated, since this condition can not be interpreted as the law of some births on the boundary. We consider a particle system with a cut-off bounded kernel B_ε instead of B, who creates new particles on the boundary with a rate depending on the empirical measure of the alive particles. So the number of particles does not stay constant, and we will consider at each time t the empirical measure of the particles alive at t, which is a finite (point) measure on $\bar{\Theta}$.

This work is in progress. So, let us summarily describe this sytem and explain the numerical algorithm, without proof of convergence.

Let $n \in \mathbb{N}^*$ and $(Z_0^{in,\varepsilon})_{1 \le i \le n}$ denote independent initial random variables with law $\frac{|w_0|(x)}{\|w_0\|_1} dx$ independent from a sequence $(B^i)_{i \ge 1}$ of two-dimensional Brownian motions. For $1 \le i \le n$, we assign the weight $s_i =$

$\frac{w_0}{|w_0|}(Z_0^{in,\varepsilon})$ to the i-th particle. Let also $(\tau_k, Z_k, U_k)_{k\geq 1}$ be a sequence of independent random variables with law

$$C_\varepsilon|\partial\Theta|1_{\{t\geq 0\}}e^{-C_\varepsilon|\partial\Theta|t}dt \otimes \frac{d\sigma(z)}{|\partial\Theta|} \otimes 1_{[0,1]}(u)du,$$

where C_ε is an upper-bound of the kernel B_ε and $|\partial\Theta| = \int_{\partial\Theta} d\sigma(x)$.

We set $T_0 = 0$. The system with $N_0 = n$ initial particles is constructed inductively for $k \geq 1$ as follows:

- $T_k = T_{k-1} + \frac{\tau_k}{N_{k-1}}$.
- On the time-interval $[T_{k-1}, T_k]$, the number of particles remains equal to N_{k-1} and their positions $Z_t^{in,\varepsilon}$, $1 \leq i \leq N_{k-1}$ evolve according to the following stochastic differential equation with normal reflection:

$$Z_t^{in,\varepsilon} \in \emptyset, \forall t \in [T_{k-1}, T_k];$$

$$Z_t^{in,\varepsilon} = Z_{T_{k-1}}^{in,\varepsilon} + \sqrt{2\nu}B_t^i + \frac{\|w_0\|_1}{n}\int_0^t \sum_{j=1}^{N_{k-1}} s_j K_\varepsilon(Z_s^{in,\varepsilon}, Z_s^{jn,\varepsilon})ds - k_t^{in,\varepsilon};$$

$$|k^{in,\varepsilon}|_t = \int_0^t 1_{\{Z_s^{in,\varepsilon}\in\partial\Theta\}}d|Z^{in,\varepsilon}|_s \; ; \; k_t^{in,\varepsilon} = \int_0^t n(Z_s^{in,\varepsilon})d|k^{in,\varepsilon}|_s$$

$$(4.35)$$

- At time T_k,

  • either $U_k \leq \dfrac{|\sum_{i=1}^{N_{k-1}} s_i B_\varepsilon(Z_k, Z_{T_k}^{in,\varepsilon})|}{N_{k-1}C_\varepsilon}$ and we create a new particle: $N_k = N_{k-1} + 1$, $Z_{T_k}^{N_k n,\varepsilon} = Z_k$, s_{N_k} equal to the sign of $\sum_{i=1}^{N_{k-1}} s_i B_\varepsilon(Z_k, Z_{T_k}^{in,\varepsilon})$.

  • or the converse inequality holds and no particle is created: $N_k = N_{k-1}$.

As $\forall k \in \mathbb{N}$, $N_k \leq n + k$, $\lim_{k\to+\infty} T_k \geq \sum_{k\in\mathbb{N}} \frac{\tau_{k+1}}{n+k}$ and

$$E(e^{-\lim_{k\to+\infty} T_k}) \leq E\left(\prod_{k\in\mathbb{N}} e^{-\frac{\tau_{k+1}}{n+k}}\right) = \prod_{k\in\mathbb{N}}\left(1 - \frac{1}{1 + C_\varepsilon|\partial\Theta|(n+k)}\right) = 0.$$

We deduce that a.s. $\lim_{k\to+\infty} T_k = +\infty$ and the particle system is defined on the time interval $[0, +\infty)$.

References

1. Benachour, S.; Roynette, B. and Vallois, P. (2001) *Branching Process Associated with 2d-Navier-Stokes Equation*, *Revista Mathematica Iberoamericana*, **Vol.17 no. 2**, pp. 331–373.

2. Bossy, M. (1995) *Vitesse de Convergence d'Algorithmes Particulaires Stochastiques et Application à l'Équation de Burgers*, Thèse, Université de Provence.
3. Bossy, M. and Talay, D. (1996) Convergence Rate for the Approximation of the Limit Law of Weakly Interacting Particles: Application to the Burgers Equation, *The Annals of Probability*, **Vol.6 no. 3**, pp. 818–861.
4. Bossy, M.; Fezoui, L. and Piperno, S. (1997) Comparison of a Stochastic Particle Method and a Finite Volume Deterministic Method Applied to the Burgers Equation, *Monte Carlo Methods Appl.*, **Vol.3 no. 2**, pp. 45–53.
5. Bossy, M. and Talay, D. (1997) A Stochastic Particle Method for the McKean-Vlasov and the Burgers Equation, *Mathematics of Computation*, **Vol.66 no. 217**, pp. 157–192.
6. Bossy, M. and Jourdain, B. (2002) Rate of Convergence of a Particle Method for the Solution of a 1d Viscous Scalar Conservation Law in a Bounded Interval, *Ann. Prob.*, **Vol.30 no. 4**, pp. 1797–1832.
7. Brezis, H. (1993) *Analyse Fonctionnelle*, Masson.
8. Busnello, B. (1999) A Probabilistic Approach to the Two-Dimensional Navier-Stokes Equations, *The Annals of Prob.*, **Vol.27 no. 4**, pp. 1750–1780.
9. Chorin, A.J. (1994) *Vorticity and Turbulence*, Applied Mathematical Sciences **Vol.103**, Springer-Verlag.
10. Cottet, P.H. (1989) Boundary Conditions and Deterministic Vortex Methods for the Navier-Stokes Equations, in *Mathematical Aspects of Vortex Dynamics*, SIAM, Philadelphia PA.
11. Fernandez, B. and Méléard, S. (2000) Asymptotic Behaviour for Interacting Diffusion Processes with Space-Time Random Birth, *Bernoulli*, **Vol.6 no. 1**, pp. 91–111.
12. Friedman, A. (1964) *Partial Differential Equations of Parabolic Type*, Prentice Hall, Englewoods Cliffs, N.J.
13. Friedman, A. (1975) *Stochastic Differential Equations and Applications*, **Vol.1**, Academic Press.
14. Gärtner, J. (1988) On the McKean-Vlasov Limit for Interacting Diffusions, *Math. Nachr.*, **Vol.137**, pp. 197–248.
15. Giet, J.S. (2000) *Processus Stochastiques: Application à l'Équation de Navier-Stokes; Simulation de la Loi de Diffusions Irrégulières; Vitesse du Schéma d'Euler pour des Fonctionnelles*, Thèse de Doctorat, Université Henri Poincaré, Nancy 1.
16. Giga, Y.; Miyakawa, T. and Osada, H. (1988) Two-Dimensional Navier-Stokes Flow with Measures as Initial Vorticity, *Arch. Rational Mech. Anal.*, **Vol.104**, pp. 223–250.
17. Gilbarg, D. and Trudinger, N.S. (1983) *Elliptic Partial Differential Equations of Second Order*, 2nd ed., Berlin-Heidelberg-New-York, Springer-Verlag.
18. Gobet, E. (2001) Euler Schemes and Half-Space Approximations for the Simulation of Diffusion in a Domain, *ESAIM P&S*, **Vol.5**, pp. 261–293.
19. Jourdain, B. and Méléard, S. (1998) Propagation of Chaos and Fluctuations for a Moderate Model with Smooth Initial Data, *Annales de l'IHP*, **Vol.34 no. 6**, pp. 727–767.
20. Jourdain, B. (2000) Diffusion Processes Associated with Nonlinear Evolution Equations for Signed Initial Measures, *Methodol. Comput. Appl. Probab.*, **Vol.2 no. 1**, pp. 69–91.
21. Jourdain, B. and Méléard, S. (2002) Probabilistic Interpretation and Particle Method for Vortex Equations with Neumann's Boundary Conditions, Prépub. 02/8, Université Paris 10.
22. Karatzas, I. and Shreve, S.E. (1991) *Brownian motion and Stochastic calculus*, 2nd ed., Springer-Verlag.
23. Ladyzenskaja, O.A., Solonnikov, V.A. and Ural'ceva, N.N. (1968) *Linear and Quasilinear Equations of Parabolic Type*, AMS.
24. Léonard, C. (1986) Une Loi des Garnds Nombres pour des Systèmes de Diffusion

avec Interaction et à Coefficients non Bornés, *Ann. I.H.P.*, **Vol.22**, pp. 237–262.

25. Lions, P.L. and Sznitman, A.S. (1984) Stochastic Differential Equations with Reflecting Boundary Conditions, *Communications on Pure and Applied Mathematics*, **Vol.37**, pp. 511–537.

26. Marchioro, C. and Pulvirenti, M. (1982) Hydrodynamics in Two Dimensions and Vortex Theory, *Commun. Math. Phys.*, **Vol.84**, pp. 483–503.

27. McKean, H.P. (1967) Propagation of Chaos for a Class of Nonlinear Parabolic Equations, *Lecture Series in Differential Equations*, **Vol.7**, pp. 41–57.

28. Méléard, S. (1996) Asymptotic Behaviour of some Interacting Particle Systems, McKean-Vlasov and Boltzmann Models, *CIME Lectures, L.N. in Math.*, Springer Verlag, **Vol.1627**, pp. 42–95.

29. Méléard, S. (2000) A Trajectorial Proof of the Vortex Method for the Two-Dimensional Navier-Stokes Equation, *Ann. Appl. Prob.*, **Vol.10 no. 4**, pp. 1197–1211.

30. Méléard, S. (2001) Monte-Carlo Approximations of the Solution of Two-Dimensional Navier-Stokes Equations with Finite Measure Initial Data, *P.T.R.F.*, **Vol.121 no. 3**, pp. 367–388.

31. Meyer, P.A. (1966) *Probabilités et Potentiel*, Hermann.

32. Osada, H. (1987) Propagation of Chaos for the Two Dimensional Navier-Stokes Equations, Probabilistic Methods in Math. Phys., (K. Itô and N. Ikeda Eds.), Tokyo: Kinokuniya, pp. 303–334.

33. Pommerenke, Ch. (1992) Boundary Behaviour of Conformal Maps, Springer-Verlag.

34. Rachev, S.T. (1991) Probability Metrics and the Stability of Stochastic Models, Chichester, John Wiley and Sons.

35. Sato, K. and Ueno, T.(1965) Multi-Dimensional Diffusion and the Markov Process on the Boundary, *J. Math. Kyoto Univ.*, **Vol.4 no. 3**, pp. 529–605.

36. Sznitman, A.S. (1984) Nonlinear Reflecting Diffusion Process, and the Propagation of Chaos and Fluctuations Associated, *J. of Functional Analysis*, **Vol.56**, pp. 311–336.

37. Sznitman, A.S. (1991) Topics in Propagation of Chaos, Ecole d'Été de Probabilités de Saint-Flour XIX - 1989, *L.N. in Math.*, **Vol.1464**, Springer-Verlag.

38. Tanaka, H. (1982) Limit Theorems for Certain Diffusion Processes with Interaction, Tanigushi Symp. on Stochastic Analysis, Katata, pp. 469–488.

39. Temam, R. (1979) *Navier-Stokes Equations*, North Holland.

RANDOM AND UNIVERSAL METRIC SPACES

ANATOLY M.VERSHIK

St. Petersburg branch of Steklov Institute of Mathematics of Russian Academy of Science
Fontanka 27, 119011 St. Petersburg, Russia
vershik@pdmi.ras.ru

Abstract. We introduce a model of the set of all Polish (=separable complete metric) spaces: the cone $\mathcal{R}$ of distance matrices, and consider geometric and probabilistic problems connected with this object. The notion of the universal distance matrix is defined and we proved that the set of such matrices is everywhere dense G_δ set in weak topology in the cone $\mathcal{R}$. Universality of distance matrix is the necessary and sufficient condition on the distance matrix of the countable everywhere dense set of so called universal Urysohn space which he had defined in 1924 in his last paper. This means that Urysohn space is generic in the set of all Polish spaces. Then we consider metric spaces with measures (metric triples) and define a complete invariant: its - matrix distribution. We give an intrinsic characterization of the set of matrix distributions, and using the ergodic theorem, give a new proof of Gromov's "reconstruction theorem". A natural construction of a wide class of measures on the cone $\mathcal{R}$ is given and for these we show that *with probability one a random Polish space is again the Urysohn space.* There is a close connection between these questions, metric classification of measurable functions of several arguments, and classification of the actions of the infinite symmetric group ([15, 16]).

Contents

A. Maass et al. (eds.), Dynamics and Randomness II, 199–228.

1. Introduction: The Cone $\mathcal{R}$ of Distance Matrices as a Set of all the Polish Spaces

Consider the set of all infinite real matrices

$$\mathcal{R} = \{\{r_{i,j}\}_{i,j=1}^{\infty} : r_{i,i} = 0, \ r_{i,j} \geq 0, \ r_{i,j}$$
$$= r_{j,i}, \ r_{i,k} + r_{k,j} \geq r_{i,j}, \ \text{for } i, j, k = 1, 2, \ldots\}$$

We will call the elements of $\mathcal{R}$ *distance matrices*. Each such matrix defines a semi-metric on the set of natural numbers $\mathbf{N}$. We allow zeros away from the principal diagonal, so in general ρ is only a semi-metric. If matrix has no zeros away from the principal diagonal we will call it *a proper distance matrix*.

The set of all distance matrices is a weakly closed convex cone in the real linear space $Mat_{\mathbf{N}}(\mathbf{R}) = \mathbf{R}^{\mathbf{N}^2}$ endowed with the ordinary weak topology. We always consider the cone $\mathcal{R}$ with this topology and will call it the cone of distance matrices. The subset of proper distance matrices is everywhere dense open subcone in the cone $\mathcal{R}$.

If the distance matrix r is proper then the completion of the metric space $(\mathbf{N}, r)$ is a complete separable metric space (=Polish space) (X_r, ρ_r) with a distinguished everywhere dense countable subset $\{x_i\}$ which is the image of the natural numbers in the completion. A general distance matrix (with possible zeros away from the diagonal) defines on the set of natural numbers structure of *semi-metric space*. By the completion of $(\mathbf{N}, r)$ in this case we mean the completion of the corresponding quotient metric space of the classes of points with zero distances. For example the zero matrix is a distance matrix on the natural numbers with zero distances between each two numbers and its "completion" is the singleton metric space. Thus finite metric spaces also could be considered in this setting.

Suppose now that we have some Polish space (X, ρ), equipped with the orderes everywhere dense countable set $\{x_i\}_{i=1}^{\infty}$. Defining the matrix $r = \{r_{i,j}\} \in \mathcal{R}$ by $r_{i,j} = \rho(x_j, x_j), i, j = 1\ldots$ we obtain a proper distance matrix. which we interpret as a *metric on the set of natural numbers*. Clearly this distance matrix analogously to structural constant in the algebraic situation, contains complete information about the original space (X, ρ) because (X, ρ) is the canonical completion of the set of naturals with this metric. Any invariant property of the metric space (topological and homological etc.) could be expressed in terms of the distance matrix for any dense countable subset of that space. We will study the theory of Polish spaces from this point of view and consider the cone of distance matrices $\mathcal{R}$ as *the universe of all separable complete metric spaces* with a fixed dense countable subset and study the properties of the metric spaces as well as properties of whole set of its using thuis cone. We can view $\mathcal{R}$ as a "fibering", whose base is the collection of all individual Polish spaces, and the fiber over a given space is the set of all countable ordered dense subsets in this metric space. Because of the universality of the Urysohn space $\mathcal{U}$ (see below) the set of all closed subsets of $\mathcal{U}$ could be considered as a base of this bundle. Thus the space $\mathcal{R}$ plays the role of a "tautology fibration" over the space of classes of isometric Polish spaces, analogous to common topological constructions of a tautology fiber bundles.

The question arises: what kind of distance matrix is "generic" in the sense of the topology of $\mathcal{R}$. One of the main results (section 3) is the Theorem 1, which is a generalization of Urysohn's results and which asserts that Urysohn space is generic (=dense G_δ set in $\mathcal{R}$) in this sense. The main tool is the notion of *universal distance matrices*, an example of such matrix was used in indirect way by P.S.Urysohn in his pioneer paper [13] for the proof of exitence of the universal metric space. An explicit formulation of the notion of universality of the distance matrix is given in Statement 1 (section 3.1). We give a new version of his main results and a new proof in the section 3. Related consideration of the Urysohn space can be found also in the papers [8, 2, 14, 11]. I want to point out that the fact that during almost 70 years Urysohn's paper [13] with this result was out of attention of the mathematicians is astonishimg; I do not know any text-book or monographs on general topology in which Urysohn universal metric space was mentioned!

Introduce a partition ξ of the $\mathcal{R}$ into the equivalence classes of distance matrices with isometric completions. The quotient space over the partition ξ (or space of the classes of equivalence) is the set of the *isometry classes* of the Polish spaces. As was conjectured in [15] and proved in the paper [4] the quotient by this equivalence relation is not "smooth", in the sense that it has no good Borel or topological structure and thus the problem of the classification of the Polish spaces up to isometry is "wild". At the same time the restricted problem for the case of compact Polish spaces is smooth (see [7]) and the space of all isometry classes of compact metric spaces has a natural topology. Surprisingly, if we consider the problem of classification of the Polish spaces with measure (metric triples) up-to isometry which preserves the measures, this classification is "smooth", and we will consider in detales a complete invariant ("matrix distribution") of metric triples (section 4). One direction -the completeness of this invariant - was proved in the book by M.Gromov [7]; we will give another proof of his reconstruction theorem based on ergodic methods and a new description of the invariant. Then we prove a theorem about the precise description of the matrix distribution of the metric triples (section 4) as a measure on $\mathcal{R}$. The section 2 is devoted to the elementary geometry of the space $\mathcal{R}$ which we use throughout all the paper, and especially in the section 5 in which we consider the various types of measures on the cone $\mathcal{R}$, and methods of the constructing of them. The measure on the above cone is nothing more than a random metric on the set of naturals numbers. Thus we can construct a "random" metric space as the result of completion of the random metric on the natural numbers. In this way we prove that loosely speaking, a Polish space randomly constructed, by a very natural procedure gives us with probability 1 again universal Urysohn space. We can say, that *the random space is universal space* (see [19]).

One of the previous simple analogy of such theorems is the theorem due to P.Erdös and A.Rényi about random graphs (see [6, 3]). The results of the paper about the genericity of the Urysohn space (Theorem 1) and probabilistic typicalness of its (Theorem 7) show that these two properties coincide in the category of the Polish spaces as well as in more simple case of the graphs. Perhaps, this coincidence has more a general and deep feature and takes place in the other categories. As a similar facts recall universality of Poulsen simplex ([9]) and of the Guraij's Banach space ([10]) (Y.Beniamini's remark), exsitence of the group which is universal in the class of finite groups homogeneous Hall's group.

Many questions about Urysohn space remain open, it is not clear if it is contractable or not there are no good realization of it; one of the main question is to construct a natural probability measures in the space. The group of the isometries of Urysohn space is also very intriguing object (see [14, 11]). We will discuss these questions elsewhere.

2. Geometry and Topology of the Cone $\mathcal{R}$

2.1. CONVEX STRUCTURE

Analogously to $\mathcal{R}$ let us define the finite dimensional cone $\mathcal{R}_n$ of distance matrices of order n. Cone $\mathcal{R}_n$ is a polyhedral cone inside the positive orthant in $Mat_n(\mathbf{R}) \equiv \mathbf{R}^{n^2}$. Denote by $\mathbf{M}_n^s(\mathbf{R}) \equiv \mathbf{M}_n^s$ of the space *symmetric matrices with zeros on the principal diagonal*. We have $\mathcal{R}_n \subset \mathbf{M}_n^s$ and the latter space is evidently the linear hull of the cone: $\mathrm{span}(\mathcal{R}_n) = \mathbf{M}_n^s$, because the interior of $\mathcal{R}_n$ is not empty. It is clear that $span(\mathcal{R}) \subset \mathbf{M}_\mathbf{N}^s$, where $\mathbf{M}_\mathbf{N}^s$ is the space of all real infinite symmetric matrices with zero principal diagonal, the geometry of the cone $\mathcal{R}$ is very complicated.

Each matrix $r \in \mathcal{R}_n$ defines a (semi)metric space X_r on the of n-point set.

Define the projection

$$p_{m,n} : \mathbf{M}_m^s \longrightarrow \mathbf{M}_n^s, m > n$$

which associates with the matrix r of order m its NW-corner of order n. The cones $\mathcal{R}_n$ are consistent with the projections i.e. $p_{n.m} : p_{m,n}(\mathcal{R}_m) = \mathcal{R}_n$. The projections $p_{n,m}$ extend naturally to the space of infinite symmetric matrices with zero diagonal - $p_n : M_N^s \longrightarrow M_n^s(\mathbf{R})$, and p_n also preserve the cones: $p_n(\mathcal{R}) = \mathcal{R}_n$. It is clear that $\mathcal{R}$ is the inverse limit as topological space (in weak topology) of the system $(\mathcal{R}_n, \{p_n\})$.

An important but evident property of the cone $\mathcal{R}_n$ is its invariance under the action of the symmetric group S_n simultaneously permuting the rows and columns of the matrices.

Let us consider the geometrical structure of $\mathcal{R}_n$ and $\mathcal{R}$.

For the first two dimensions we have $\mathcal{R}_1 = \{0\}), \mathcal{R}_2 = \mathbf{R}$. It is interesting to describe the extremal rays (in the sense of convex geometry) of the convex polyhedral cone $\mathcal{R}_n, n = 3, \ldots, \infty$. This is a well-known problem - see [5, 1] and the list of literature there. Each extremal ray in $\mathcal{R}_n, n \leq 4$ is of the type $\{\lambda \cdot l : \lambda \geq 0\}$, where l is a symmetric $0-1$- distance matrix which corresponds to the semi-metric space whose quotient metric space has just two points. For $n \geq 5$ there are extremal rays of other type. The complete description of the set extremal rays is rather a difficult and very interesting combinatorial problem. The most important question for us concerns to the asymptotic properties of cone $\mathcal{R}_n$ and especially the description of the set of extremal rays of the infinite dimensional cone $\mathcal{R}$. It happens that this set is a dense G_δ in $\mathcal{R}$ and some of the so called universal distance matrices (see par 3.) are extremal. This is in consistent with the estimation in [1] of the number of extremal rays of $\mathcal{R}_n$ which grows very rapidly. The algebro-geometric structure and stratification of the cones $\mathcal{R}_n$ as semi-algebraic sets. Are also very intriguing. In order to clarify topological and convex structure of the cones $\mathcal{R}_n$ we will use an *inductive description* of these cones and will study it in the next subsection.

2.2. ADMISSIBLE VECTORS AND STRUCTURE OF THE $\mathcal{R}$

Suppose $r = \{r_{i,j}\}_1^n$ is a distance matrix of order n ($r \in \mathcal{R}_n$), choose a vector $a \equiv \{a_i\}_{i=1}^n \in \mathbf{R}^n$ such that if we attaching to the matrix r with vector a as the last column and the last row then the new matrix of order $n + 1$ still belongs to $\mathcal{R}_{n+1}$. We will call such vector *an admissible vector* for fixed distance matrix r and denote the set of of all admissible vectors for r as $A(r)$. For given $a \in A(r)$ denote as (r^a), distance matrix of order $n + 1$ obtained from matrix r adding vector $a \in A(r)$ as the last row and column. It is clear that $p_n(r^a) = r$. The matrix r^a has the form

$$
r^a = \begin{pmatrix}
0 & r_{1,2} & \cdots & r_{1,n} & a_1 \\
r_{1,2} & 0 & \cdots & r_{2,n} & a_2 \\
\vdots & \vdots & \ddots & \vdots & \vdots \\
r_{1,n} & r_{2,n} & \cdots & 0 & a_n \\
a_1 & a_2 & \cdots & a_n & 0
\end{pmatrix}
$$

The (semi)metric space X_{r^a} corresponding to matrix r^a is an extension of X_r: we add one new point x_{n+1} and $a_i, i = 1 \ldots n$ is the distance between x_{n+1} and x_i. The admissibility of a is equivalent to the following set of inequalities: the vector $a = \{a_i\}_{i=1}^n$ must satisfy to the series of triangle inequalities for all $i, j, k = 1, 2 \ldots n$; (matrix $\{r_{i,j}\}_{i,j=1}^n$ is fixed):

$$|a_i - a_j| \leq r_{i,j} \leq a_i + a_j \tag{1}$$

So for given distance matrix r of order n the set of admissible vectors is $A(r) = \{\{a_i\}_{i=1}^n : |a_i - a_j| \leq r_{i,j} \leq a_i + a_j, i, j = 1 \ldots n\}$. It makes sense to mention that we can view on the vector $a = \{a_i\}$ as a *Lipshitz function* $f(.)$ on the space $X_r = \{1, 2 \ldots n\}$ with r as a metric: $f_a(i) = a_i$ with Lipshitz constant equal 1. This point of view helps to consider a general procedure of extension of metric space.

Geometrically the set $A(r)$ can be identified with the intersection of cone $\mathcal{R}_{n+1}$ and the affine subspace which consists of matrices of order $n+1$ with given matrix r as the NW-corner of order n. It is clear from the linearity of inequalities that the set $A(r)$ is an unbounded closed convex polyhedron in $\mathbf{R}^n$. If $r_{i,j} \equiv 0, i, j = 1 \ldots n \geq 1$, then $A(r)$ is diagonal: $A(0) = \Delta_n \equiv \{(\lambda, \ldots \lambda) : \lambda \geq 0\} \subset \mathbf{R}_+^n$. Let us describe the structure of $A(r)$ more carefully.

Lemma 1 *For each proper distance matrix r of order n the set of admissible vectors $A(r)$ is a closed convex polyhedron in the orthant $\mathbf{R}_+^n$, namely it is a Minkowski sum:*

$$A(r) = M_r + \Delta_n,$$

where Δ_n is the half-line of constant vectors in the space $\mathbf{R}_+^n$, and M_r is a compact convex polytope of dimension n. This polytope M_r is the convex hull of extremal points of the polyhedron $A(r)$: $M_r = conv(ext A(r))$.

Proof. The set $A(r) \subset \mathbf{R}^n$ is the intersection of finitely many closed half-spaces, and evidently it does not contain straight lines, so, by a general theorem of convex geometry $A(r)$ is a sum of the convex closed polytope and some cone (which does not contain straight lines) with the vertex at origin. This convex polytope is the convex hull of the extremal points of convex set $A(r)$. But this cone is half-line of the constant (diagonal) nonnegative vectors in $\mathbf{R}^n$ because if it contains half-line which is different from the diagonal then the triangle inequality is violated. The dimension of $A(r)$ equal to n for proper distance matrix; in general it depends on the matrix r and could be less than n for some matrix r; while the dimension of M_r is equal to $\dim A(r)$ or to $\dim A(r) - 1$. The assertion about topological structure of $A(r)$ follows from what was claimed above.   □

The next lemma asserts that this correspondence $r \to A(r)$ is covariant under the action of symmetric group in $\mathbf{R}^n$. The proof is evident.

Lemma 2 *For any $r \in \mathcal{R}_n$ we have coincideness of the sets: $A(grg^{-1}) = g(A(r))$, where $g \in S_n$ is element of symmetric group S_n which acts in a natural way on the space of matrices $M_N(\mathbf{R})$ as well as on the space of the convex subsets of the vector space $\mathbf{R}^n$.*

The convex structure of polytopes M_r, $A(r)$ is very interesting and seems to have not been studied before. For dimensions higher than 3 *combinatorial type* of the polytope M_r hardly depends on r but for dimension three the combinatorial type of polytopes M_r, and consequently the combinatorial structure of the sets $A(r)$ is the same for all proper distances matrices r.

Example For $n = 3$ the description of the set $A(r)$ and of its extremal points is the following. Let r be a matrix

$$r = \begin{pmatrix} 0 & r_{1,2} & r_{1,3} \\ r_{1,2} & 0 & r_{2,3} \\ r_{1,3} & r_{2,3} & 0 \end{pmatrix}$$

Denote $r_{1,2} = \alpha, r_{1,3} = \beta, r_{2,3} = \gamma$, then r^a:

$$r^a = \begin{pmatrix} 0 & \alpha & \beta & a_1 \\ \alpha & 0 & \gamma & a_2 \\ \beta & \gamma & 0 & a_3 \\ a_1 & a_2 & a_3 & 0 \end{pmatrix}$$

Denote $\delta = \frac{1}{2}(\alpha+\beta+\gamma)$ There are seven extremal points $a = (a_1, a_2, a_3)$ of $A(r)$: the first one is a vertex which is the closest to the origin: $(\delta - \gamma, \delta - \beta, \delta - \alpha)$, then another three non degenerated extremal points: $(\delta, \delta - \alpha, \delta - \gamma), (\delta - \beta, \delta, \delta - \alpha), (\delta - \gamma, \delta - \beta, \delta)$, and three degenerated extremal points $(0, \alpha, \beta), (\alpha, 0, \gamma), (\beta, \gamma, 0)$.

If $\alpha = \beta = \gamma = 1$ then those seven points are as follows

$$(1/2, 1/2, 1/2), (3/2, 1/2, 1/2), (1/2, 3/2, 1/2),$$
$$(1/2, 1/2, 3/2), (0, 1, 1), (1, 0, 1), (1, 1, 0).$$

Remark that all non-degenerated extremal points in the example defines the finite metric spaces which can not be isometrically embedded to Euclidean space.

2.3. PROJECTIONS AND ISOMORPHISMS

Let r be a distance matrix of order N and $p_n(r)$ its NW-corner of order $n < N$. Then we can define a projection χ_n^r of $A(r)$ to $A(p_n(r))$: $\chi_n^r : (b_1, \ldots b_n, b_{n+1}, \ldots b_N) \mapsto (b_1 \ldots b_n)$. (We omit the index N in the notation χ_n^r). The next simple lemma plays a very important role for our construction.

Lemma 3 *Let $r \in \mathcal{R}_n$ be a distance matrix of order n. For any two vectors $a = (a_1, \ldots a_n) \in A(r)$ and $b = (b_1, \ldots b_n) \in A(r)$ there exists a real nonnegative number $h \in \mathbf{R}$ such that vector $\bar{b} = (b_1, \ldots b_n, h) \in A(r^a)$ (and also $\bar{a} = (a_1, \ldots a_n, h) \in A(r^b)$).*

Corollary 1 *For each $r \in \mathcal{R}_n$ and $a \in A(r)$ the map $\chi^r_{n+1,n}$: $(b_1, \ldots b_n, b_{n+1}) \mapsto (b_1 \ldots b_n)$ of $A(r^a) \to A(r)$ is the epimorphism of $A(r^a)$ onto $A(r)$ (by definition $p_{n+1,n}(r^a) = r$)*

Proof. The assertion of this lemma as we will see, follows from simple geometrical observation: suppose we have two finite metric spaces $X = \{x_1, \ldots x_{n-1}, x_n\}$ with metric ρ_1 and $Y = \{y_1, \ldots y_{n-1}, y_n\}$ with metric ρ_2. Suppose the subspaces of the first $n - 1$ points $\{x_1, \ldots x_{n-1}\}$ and $\{y_1, \ldots y_{n-1}\}$ are isometric, i.e. $\rho_1(x_i, x_j) = \rho_2(y_i, y_j), i, j = 1, \ldots n - 1$. Then there exists a third space $Z = \{z_1, \ldots z_{n-1}, z_n, z_{n+1}\}$ with metric ρ and two isometries I_1, I_2 of both spaces X and Y to the space Z so that $I_1(x_i) = z_i, I_2(y_i) = z_i, i = 1, \ldots n - 1, I_1(x_n) = z_n, I_2(y_n) = z_{n+1}$. In order to prove existence of Z we need to show that it is possible to define only nonnegative number h which will be the distance $\rho(z_n, z_{n+1}) = h$ between z_n and z_{n+1} (images of x_n and y_n in Z correspondingly) such that all triangle inequalities are valid in the space Z. The existence of h follows from the inequalities:

$$\rho_1(x_i, x_n) - \rho_2(y_i, y_n) \le \rho_1(x_i, x_j) + \rho_1(x_j, x_n) - \rho_2(y_i, y_n) =$$

$$= \rho_1(x_j, x_n) + \rho_2(y_i, y_j) - \rho_2(y_i, y_n) \le \rho_1(x_j, x_n) + \rho_2(y_j, y_n)$$

for all $i, j = 1, \ldots n - 1$.
    Consequently

$$\max_i |\rho_1(x_i, x_n) - \rho_2(y_i, y_n)| \equiv M \le m \equiv \min_j (\rho_1(x_j, x_n) + \rho_2(y_j, y_n)).$$

Thus, a number h could be chosen as an arbitrary number from the nonempty closed interval $[M, m]$ and we set $\rho(z_n, z_{n+1}) \equiv h$; it follows from the definitions that all triangle inequalities are satisfied. Now suppose we have a distance matrix r of order $n - 1$ and an admissible vector $a \in A(r)$, so we have a metric space $\{x_1, \ldots x_{n-1}, x_n\}$ (the first $n - 1$ points correspond to the matrix r, and whole space - to the extended matrix r^a. Now suppose we choose another admissible vector $b \in A(r)$, and giving distance matrix r^b of the space $\{y_1, \ldots y_{n-1}, y_n\}$, where the subset of first $n - 1$ points is isometric to the space $\{x_1, \ldots x_{n-1}\}$. As we proved we can define space Z whose distance metric $\bar{r}$ of order $n + 1$ gives the required property.   □

Now we can formulate a general assertion about the projections χ^r.

Lemma 4 *For arbitrary natural numbers N and n, $N > n$, and any $r \in \mathcal{R}_N$ the map χ_n^r is epimorphism of $A(r)$ onto $A(p_n(r))$. In other words for each $a = (a_1, \ldots a_n) \in A(p_n(r))$ there exists a vector $(b_{n+1}, \ldots b_N)$ such that $b = (a_1, \ldots a_n, b_{n+1}, \ldots b_N) \in A(r)$.*

Proof. The above proof shows how to define the first number b_{n+1}. But the projection χ_n^r seen as a map from $A(r), r \in \mathcal{R}_N$ to $A(p_n(r))$ is the product of projections $\chi_n^r \cdots \chi_{N-1}^r$. Because each factor is epimorphism the product is epimorphism also. □

It is convenient for our goals to represent the infinite distance matrix $r \equiv \{r_{i,j}\} \in \mathcal{R}$ as a sequence of admissible vectors of increasing lengths:

$$r(1) = \{r_{1,2}\},$$
$$r(2) = \{r_{1,3}, r_{2,3}\}, \ldots r(k) = \{r_{1,k+1}, r_{2,k+1}, \ldots r_{k,k+1}\} \ldots, k = 1, 2 \ldots,$$
$$(2)$$

satisfying conditions $r(k) \in A(p_k(r))$, (recall that $p_k(r)$ is the NW-projection of matrix r on the space $\mathbf{M}_k^s$ defined above), so each vector $r(k)$ is admissible for the *previous distance matrix* $p_k(r)$. We can consider the following sequence of the cones and maps:

$$0 = \mathcal{R}_1 \xleftarrow{p_2} \mathcal{R}_2 = \mathbf{R}_+ \xleftarrow{p_3} \mathcal{R}_3 \longleftarrow \ldots \longleftarrow \mathcal{R}_{n-1} \xleftarrow{p_n} \mathcal{R}_n \longleftarrow \ldots \quad (3)$$

the projection p_n here is the restriction of the projection defined above onto the cone $\mathcal{R}_n$. The preimage of the point $r \in \mathcal{R}_{n-1}$ (fiber over r) is the set $A(r)$ which described in Lemma 1. Note that this is not fibration in the usual sense: the preimages of the points could even not be homeomorphic to each other for various r (even dimensions could be different). But that sequence defines allows to define cone $\mathcal{R}$ as an inverse limit of the cones $\mathcal{R}_n$. We will use the sequence (3) in order to define the measures on the cone $\mathcal{R}$ in the spirit of the theory of the Markov processes.

3. Universality and Urysohn Space

3.1. UNIVERSAL DISTANCE MATRICES

The following definition plays a crucial role.

Definition 1

1. An infinite proper distance matrix $r = \{r_{i,j}\}_{i,j=1}^\infty \in \mathcal{R}$ is called a UNIVERSAL distance matrix if the following condition is satisfied: for any $\epsilon > 0$, $n \in \mathbf{N}$ and for any vector $a = \{a_i\}_{i=1}^n \in A(p_n(r))$ there exists $m \in \mathbf{N}$ such that $\max_{i=1\ldots n} |r_{i,m} - a_i| < \epsilon$.

In another words: for each $n \in \mathbf{N}$ the set of vectors $\{\{r_{i,j}\}_{i=1}^{n}\}_{j=n+1}^{\infty}$ is everywhere dense in the set of admissible vectors $A(p_n(r))$.

2. An infinite proper distance matrix $r = \{r_{i,j}\}_{i,j=1}^{\infty} \in \mathcal{R}$ is called a weakly universal distance matrix if for any $n \in \mathbf{N}$ the set of all submatrices $\{r_{i_k,i_s}\}_{k,s=1}^{n}$ of the matrix r of order n over all n-tuples $\{i_k\}_{k=1}^{n} \subset \mathbf{N}$ is dense in the cone $\mathcal{R}_n$.

Let us denote the set of universal distance matrices by $\mathcal{M}$. We will prove that $\mathcal{M}$ is not empty but before we formulate some properties of universal matrices.

Lemma 5 *Each universal distance matrix is weakly universal. There exist nonuniversal but weakly universal distance matrices.*

Proof. Choose any distance matrix $q \in \mathcal{R}_n$; we will prove that for given positive ϵ it is possible to find a set $\{i_k\}_{k=1}^{n} \subset \mathbf{N}$ such that $\max_{k,s=1,\dots n} |r_{i_k,i_s} - q_{k,s}| < \epsilon$. Because $r^1 = \{r_{1,1} = 0\}$ then $A(r^1) = \mathbf{R}_+$ (see 2.2), and by universality of r the sequence $\{r_{1,n}\}_{n=2}^{\infty}$ must be dense in $\mathbf{R}_+$, so we can choose some i_1 such that $|r_{1,i_1} - q_{1,2}| < \epsilon$, then using density of the columns of length 2 which follows from the universality condition we can choose a natural number i_2 such that $|r_{1,i_2} - q_{1,3}| < \epsilon, |r_{2,i_2} - q_{2,3}| < \epsilon$, etc.

There are many examples of weakly universal but nonuniversal distance matrices. The distance matrix of the arbitrary countable everywhere dense set of any universal but not homogeneously universal (see below) Polish spaces (like $C([0,1])$) gives such a counterexample, but the simplest one is the distance matrix of the disjoint union of all finite metric spaces with the rational distance matrices (B.Weiss's example).   □

The following corollary of universality gives useful tool for tre studying of the Urysohn's space:

Corollary 2 *(ϵ-extension of the isometry) Suppose r is an infinite universal distance matrix and q is a finite distance matrix of order N such that for some $n < N, r_{i,j} = q_{i,j}, i,j = 1 \dots n$. Denote $i_k = k, k = 1 \dots n$. Then for any positive ϵ there exist the natural numbers $i_{n+1} \dots i_N$ such that $\max_{k,s=1 \dots N} |r_{i_k,i_s} - q_{k,s}| < \epsilon$. In another words, we can enlarge the set of the first n natural numbers with some set of $N - n$ numbers: $i_{n+1}, \dots, i_N$ in such a way that the distance matrix of whole set $i_1 = 1, i_2 = 2 \dots i_n = n, i_{n+1}, \dots i_N$ differs with the distance matrix q (in norm) less than ϵ.*

Conversely, if the infinite distance matrix r has property above for any finite distance submatrix q and for any positive ϵ then matrix r is a universal matrix.

Proof. For $N = n + 1$ the claim follows directly from the definition of universality of r, then we can use induction on N. The second claim follows from the definition. $\qquad\square$

Once more reformulation of the notion of universality uses the term of group action. Suppose $q \in \mathcal{R}_n$; denote $\mathcal{R}^n(q)$ the set of all $r \in \mathcal{R}$, with NW-corner equal to the matrix q. Consider the group S^n_∞ of permutations, which preserve as fixed the first n rows and columns of the matrices from $\mathcal{R}$ and consequently map the set $\mathcal{R}^n(q)$ to itself. The following criteria of the universality is a direct corollary of the definition:

Statement 1 *A matrix $r \in \mathcal{R}$ is universal iff for each n the orbit of r under the action of the group S^n_∞ is everywhere dense in $\mathcal{R}^n(r^n)$ in weak topology, here r^n is the NW-corner of matrix r of the order n. The matrix r is weakly universal iff its S_∞-orbit is everywhere dense in $\mathcal{R}$.*

From other side the existence of universal distance matrix as well as existence of Urysohn space is not evident. We simplify and a little bit strengthen Urysohn's existence theorem and prove the following

Theorem 1 *The set $\mathcal{M}$ of the universal matrices is nonempty. Moreover, this set is everywhere dense G_δ-set in the cone $\mathcal{R}$ in the weak topology.*

Proof. We will use the representation described in the lemmas in previous section for construction of at least one universal proper distance matrix in the cone $\mathcal{R}$.

Let us fix sequence $\{m_n\}_{n=1}^\infty$ of natural numbers in which each natural number occurs infinitely many times and for each $n, m_n \leq n; m_1 = 1$. For each proper finite distance matrix $r \in \mathcal{R}_n$ let us choose an ordered countable dense subset $\Gamma_r \subset A(r)$ of the vectors with positive coordinates: $\Gamma_r = \{\gamma_k^r\}_{k=1}^\infty \subset A(r) \subset \mathbf{R}^n$ and choose any metric in $A(r)$, say the Euclidean metric.

The first step consists of the choice of positive real number $\gamma_1^1 \in \Gamma_1 \subset A(0) = \mathbf{R}_+^1$ so that we define a distance matrix r of order 2 with element $r_{1,2} = \gamma_1^1$.

Our construction of the universal matrix r is inductive one, it used the representation of matrix as a sequence of admissible vectors $\{r(1), r(2), \ldots\}$ with increasing lengths (formula (2)). The conditions on the vectors are as follows $r(k) \in A(p_k(r_{k+1}))$. The sequence of the corresponding matrices $r_n, n = 1 \ldots$ stabilizes to the infinite matrix r. Suppose after $(n-1)$-th step we obtain a finite matrix r_{n-1}; then we choose a new admissible vector $r(n) \in A(r_{n-1})$. The choice of this vector (denote it a) is defined by the condition that the distance (in norm) between the projection $\chi^r_{m_n}(a)$ of the vector a onto the subspace of admissible vectors $A(r_{m_n})$ and the point

$\gamma_s^{m_n} \in \Gamma_{r_{m_n}} \subset A(r_{m_n})$ must be less than 2^{-n}, where $s = |i : m_i = m_n, 1 \leq i \leq n| + 1$:

$$\|\chi_{m_n}^r(a) - \gamma_s^{r_{m_n}}\| < 2^{-n}.$$

Recall now that the projection $\chi_{m_n}^r$ is an epimorphism from $A(r)$ to $A(p_{m_n}(r))$, (Lemma 4), hence a vector $a \in A(r_n)$ with these properties does exist. The number s is just the number of the points of $\Gamma_{r_{m_n}}$ which occur on the previous steps of the construction. After infinitely many steps we obtain the infinite distance matrix r.

Universality of r is evident, because for each n projection χ_n^r of vectors $r(k), k = n + 1 \ldots$ is a dense set in $A(r_n)$ by construction. This proves the existence of the universal matrix.

Now notice that the property of universality of the matrix are preserved under the action of finite simultaneous permutations of the rows and columns, and also under the NW-shift which cancels the first row and first column of the matrices. Also the set of universal matrices $\mathcal{M}$ is invariant under the changing of the finite part of the matrix. Consequently, $\mathcal{M}$ contains together with the given matrix also its permutations and shifts. But because of the weak universality of any universal distance matrices r, even the orbit of matrix r under the action of the group of permutations $S_{\mathbf{N}}$ is everywhere dense in $\mathcal{R}$ in weak topology. A fortiori $\mathcal{M}$ is everywhere dense in $\mathcal{R}$.

Finally, the formula which follows directly from the definition of universality shows us immediately that the set of all universal matrices $\mathcal{M}$ is a G_δ-set:

$$\mathcal{M} = \cap_{k \in \mathbf{N}} \cap_{n \in \mathbf{N}} \cap_{a \in A(r^n) \cap \mathbf{Q}^{n^2}} \cup_{m \in \mathbf{N}, m > n} \{r \in \mathcal{R} : \max_{i=1,\ldots n} |r_{i,m} - a_i| < \frac{1}{k}\}.$$

$\square$

Let us fix some infinite universal proper distance matrix r, and provide the set of all natural numbers $\mathbf{N}$ with metric r. Denote the completion of the space $(\mathbf{N}, r)$ with respect to metric r as a metric space $(\mathcal{U}_r, \rho_r)$. Evidently, it is a Polish space.

Lemma 6 *The distance matrix of any everywhere dense countable subset $\{u_i\}$ of the space $\mathcal{U}_r$ is a universal distance matrix.*

Proof. Let us identify the set $\mathbf{N}$ with $\{x_i\} \subset \mathcal{U}_r$. Then by definition $\rho(x_i, x_j) = r_{i,j}$ By definition the universality of r means that for any n the closure (Cl) in $\mathbf{R}^n$ of the set of vectors coincide with the set of the admissible vectors of NW-corner of matrix r of order n: $Cl(\cup_{j>n}\{\{\rho(x_i, x_j)\}_{i=1}^n\} = A(p_n(r))$. Because the set $\{u_i\}$ is also everywhere dense in $(\mathcal{U}_r, \rho_r$ we can replace the previous set by the following: $Cl(\cup_{j>n}\{\{\rho(x_i, u_j)\}_{i=1}^n\} = A(p_n(r))$.

But because $\{x_i\}$ is everywhere dense in (U_r, ρ_r) we can change this on $Cl(\cup_{j>n}\{\{\rho(u_i, u_j)\}_{i=1}^n\} = A(p_n(r'))$, where r' is distance matrix for $\{u_i\}$. $\square$

We will see that $(\mathcal{U}_r, \rho_r)$ is the so called Urysohn space which is defined below, and the universality of matrix is necessary, and sufficient condition to be a countable everywhere dense set in Urysohn space.

3.2. URYSOHN UNIVERSAL SPACE AND UNIVERSAL MATRICES

Now we introduce the remarkable Urysohn space. In his last papers [13] Pavel Samuilovich Urysohn (1898-1924) gave a concrete construction of the universal Polish space which is now called "Urysohn space". It was the answer on the question whihc was posed to him by M.Freshet about universal Banach space. Later Banach and Mazur have proved existence of the universal Banach spaces (f.e. $C([0,1])$), but Urysohn's answer was more deep because his space is homogeneuos in some sense. Actually Urysohn had proved several theorems which we summarize as the following theorem:

Theorem 2 *(Urysohn [13])*

  A. *There exists a Polish(=separable complete metric) space $\mathcal{U}$ with the properties:*
     1) (Universality) For each Polish space X there exists the isometric embedding of X to the space $\mathcal{U}$;
     2) (Homogeneity) For each two isometric finite subsets $A = (a_1 \ldots a_m)$ and $B = (b_1 \ldots b_m)$ of $\mathcal{U}$ there exists isometry J of the whole space $\mathcal{U}$ which brings A to B: $JX = Y$;
  B. *(Uniqueness) Two Polish spaces with the previous properties 1) and 2) are isometric.*

The condition 2) of the theorem could be strenghened: the finite subsets possible to change to the compact subsets. So the group of isometry of the space acts transitively on the isometric compacts. But to enlarge the compact sets onto closed subsets is impossible. Also we can in equivalent manner formulate this condition as condition of the extension of the isometries from compacts to the whole space, see below.

The main result of this section is the following theorem, which includes the previous theorem.

Theorem 3

  1. *The completion $(\mathcal{U}_r, \rho_r)$ of the set of natural numbers $(\mathbf{N}, r)$ with respect to any universal proper distance matrix r satisfies to the properties 1) and 2) of the above theorem and consequently is the Urysohn space.*

2. *(Uniqueness).* *For any two universal distance matrices* r *and* r' *the completions of the spaces* $(\mathbf{N}, r)$ *and* $(\mathbf{N}, r')$ *are isometric.*

Corollary 3 *The isometric type of the space* $(\mathcal{U}_r, \rho_r)$ *does not depend on the choice of universal matrix* r. *The universality of the distance matrix is necessary and sufficient condition to be the distance matrix of a countable everywhere dense subset of the Urysohn space.*

The corollary follows from the Theorem 3 and Lemma 6. The proof of the Theorem 3 which we give here repeats and simplifies some arguments of Urysohn but he did not use infinite distance matrices and the useful notion of the universal matrix.

Proof. Suppose that matrix $r = \{r(i, j)\}_{i,j=1}^{\infty} \in \mathcal{R}$ be an arbitrary universal proper distance matrix (it is convenient now to write $r(i, j)$ instead of $r_{i,j}$) and the space $\mathcal{U}_r$ is the completion of the countable metric space $(\mathbf{N}, r)$, denote the corresponding metric on $\mathcal{U}_r$ as ρ_r, but we will omit the index r.

1. First of all we will prove that the metric space $(\mathcal{U}, \rho)$ is universal in the sense of property 1) of the theorem 2, and then that it is homogeneous in the sense of property 2) of theorem 2.

Let (Y, q) be an arbitrary Polish space. In order to prove there is an isometric embedding of (Y, q) into $(\mathcal{U}, \rho)$ it is enough to prove that there exists an isometric embedding of some countable everywhere dense set $\{y_n\}_1^{\infty}$ of the space (Y, q) to $(\mathcal{U}, \rho)$. This means that we must prove that for any infinite proper distance matrix $q = \{q(i, j)\} \in \mathcal{R}$ there exists some countable set $\{u_i\} \subset \mathcal{U}$ with distance matrix equal to q. In it turn for this we need to construct a set of the fundamental sequences in the space $(\mathbf{N}, r)$, say, $N_i = \{n_i^{(m)}\}_{m=i}^{\infty} \subset \mathbf{N}, i = 1, 2 \ldots$ such that

$$\lim_{m \to \infty} r(n_i^{(m)}, n_j^{(m)}) = q(i, j), \quad i, j = m, \ldots$$
$$\text{and for all} \quad i, \lim_{m,k \to \infty} r(n_i^{(m)}, n_i^{(k)}) = 0.$$

The convergence of each sequence $N_i = \{n_i^{(m)}\}_{m=i}^{\infty}$ in the space $(\mathcal{U}, \rho)$ when $m \to \infty$ to some point $u_i \in \mathcal{U}, i = 1, 2 \ldots$ follows from the fundamentality of this sequence e.g. from the second equality above, and because of the first equality, the distance matrix of the limit set $\{u_i\}$ coincided with the matrix q. Now we we will construct the needed sequences $\{N_i\}_{i=1}^{\infty} \subset \mathbf{N}$ by induction. Choose arbitrarily a point $n_1^{(1)} \in \mathbf{N}$, and suppose that for given $m > 1$ we already have defined the finite fragments $L_k = \{n_i^{(k)}\}_{i=1}^{k} \subset \mathbf{N}$ of the first m sets $\{N_i\}_{i=1}^{m}$, for $k = 1, 2 \ldots m$ with property $\max_{i,j=1,\ldots k} |r(n_i^{(k)}, n_j^{(k)}) - q(i, j)| = \delta_k < 2^{-k}, k = 1, \ldots m$, and the sets L_k mutually do not intersect.

Our construction of the set L_{m+1} will depend only on the set L_m, so we can for simplicity change the notations and renumber L_m as follow: $n_i^{(m)} = i, i = 1, \ldots m$.

Now we will construct a new set $L_{m+1} = \{n_i^{(m+1)}\}_{i=1}^{m+1} \subset \mathbf{N}$ with the needed properties in the following way. Consider the finite metric space (V, d) with $2m + 1$ points $y_1, \ldots y_m; z_1, \ldots z_m, z_{m+1}$ with the distances: $d(y_i, y_j) = r_{i,j}, i, j = 1 \ldots m, d(z_i, z_j) = q(i, j), i, j = 1, \ldots m+1; d(y_i, z_j) = q(i, j), i = 1 \ldots m, j = 1 \ldots m+1; i \neq j, d(y_i, z_i) = \delta_m, i = 1 \ldots m,$ for some δ_m. It is easy to check that this is correct definition of the distances. Denote the distance matrix of the space (V, d) as q_m. Now apply corollary 2 (ϵ-extension of isometry) and enlarge the set $L_m = \{1, 2 \ldots m\}$ with the new set L_{m+1} with $m + 1$ natural numbers $\{n_i^{(m+1)}\}_{i=m+1}^{2m+1} \subset \mathbf{N}$ in such a way that the distance matrix of L_{m+1} differ from the NW-corner of the order $m + 1$ of the matrix q not more than δ_m which is less than $2^{-(m+1)}$:

$\max_{i,j} |r(n_i^{(m+1)}, n^{(m+1)}_j) - q_m(i, j)| = \delta_{m+1} < 2^{-(m+1)}$ (remember that NW corners of order m of matrices q_m and r are coincided by construction). We can see that for each i the sequences $\{n^{(m_i)}\}_{m=i}^{\infty}$ is fundamental and $\lim_{m \to \infty} r(n_i^{(m)}, n_j^{(m)}) = q(i, j)$. Thus we have proved that each Polish space can be isometrically embedded into $(\mathcal{U}_r, \rho_r)$.

We can essentially refine now the corollary 2 as follow.

Corollary 4 (extension of isometry). *The space $(\mathcal{U}_r, \rho_r)$ has the following property: for any finite set $A = \{a_i\}_{i=1}^n \in \mathcal{U}_r$ and distance matrix q of order $N, N > n$ with NW-corner of order n which is equal to the matrix $\{\rho(a_i, a_j)\}_{i,j=1}^n$ there exist points $a_{n+1} \ldots a_N$ such that distance matrix of whole set $\{a_i\}_{i=1}^N$ is equal to the matrix q.*

The proof follows to the proof of Corollary 2 and uses the arguments which we use above.

Remark 1 As we have mentioned there exist examples of universal but not homogeneous Polish spaces (f.e. Banach space $C([0, 1])$). The corollary above shows that the main difference between such universal spaces and Urysohn space is the following: we can isometrically embed in any universal space a given metric space; but in the case of Urysohn space we can do more: the points of the of embedded metric space have given distances from the points of a fixed finite (or even compact) set.

Let us continue the proof of the Theorem 3.

2. In order to prove homogeneity let us fix two finite n-point sets $A = \{a_i\}_{i=1}^n$ and $B = \{b_i\}_{i=1}^n$ of $(\mathcal{U}_r, \rho_r)$ and construct two isometric ordered countable subsets C and D each of which is everywhere dense in $\mathcal{U}$ and C

begins with A and D begins with B. The method of constrution is well-known and called "back and forth". First of all we fix some countable everywhere dense subset F in $(\mathcal{U}_r, \rho_r)$, $F \cap A = F \cap B = \emptyset$, and represent it as increasing union: $F = \cup F_n$. Put $C_1 = A \cup F_1$, and find a set $D_1 = B \cup F_1'$ so that the isometry of A and B extends to F_1 and F_1'. Thus D_1 is isometric to C_1. This is possible to define because of Corollary 4 (extension of isometry). Then, choose $D_2 = D_1 \cup F_2$ and $C_2 = C_1 \cup F_2'$ and again extend the isometry from the part on which it was defined before to whole set. So we construct an isometry between D_2 and C_2 and so on. The alternating process gives us two everywhere dense isometrical sets $\cup C_i$ and $\cup D_i$ and the isometry between them extends isometry of A and B.

3. Uniqueness. Let r and r' be two universal proper distance matrices and the spaces $(\mathcal{U}_r, \rho_r)$ and $(\mathcal{U}_r', \rho_r')$ their completions. We will construct repectively in the spaces two countable everywhere dense sets F_1 and F_2 so isometry between them will extend to the whole space. Denote by $\{x_i\}$ and $\{u_i\}$ everywhere dense subsets of $(\mathcal{U}_r, \rho_r)$ and $(\mathcal{U}_r', \rho_r')$ to which are generated respectively by matrices r and r'. Now we repeat the same arguments as in the proof of the first part of the theorem. We start with finite number of the points $\{x_i\}_{i=1}^{n_1}$ in $(\mathcal{U}_r, \rho_r)$ and append to them the set of points $\{u_i'\}_{i=1}^{m_1} \subset \mathcal{U}_r$ with the same distance matrix as the distance matrix of the set of points $\{u_i\}_{i=1}^{m_1}$; this is possible because of universality of the $(\mathcal{U}_r, \rho_r)$ (property 1) which had been already proved). Then append to the set $\{u_i\}_{i=1}^{m_2}$, $(m_2 > m_1)$ in the space $\mathcal{U}_r'$ the set of points $\{x_i'\}_{i=1}^{n_2}$, $(n_2 > n_1)$ in such a way that the distance matrix of the subset $\{u_i\}_{i=1}^{m_1} \cup \{x_i'\}_{i=1}^{n_1}$ of the set $\{u_i\}_{i=1}^{m_2} \cup \{x_i'\}_{i=1}^{n_2}$ coincides with the distance matrix of the set $\{u_i'\}_{i=1}^{m_1} \cup \{x_i\}_{i=1}^{n_1}$ etc. continuing this process ad infinity as the result of this construction we obtain two sets - the first is $\{x_i\}_{i=1}^{\infty} \cup \{u_i'\}_{i=1}^{\infty} \subset \mathcal{U}_r$ and the second is $\{u_i\}_{i=1}^{\infty} \cup \{x_i'\}_{i=1}^{\infty} \subset \mathcal{U}_r'$ - which are everywhere dense in their spaces and are isometric. Thus we have concluded the proof of the theorem. $\square$

Theorems 1 and 3 give us the following the remarkable fact:

Corollary 5 *A generic ("typical") distance matrix is a universal matrix, and consequently a generic Polish space (in the sense our model of the cone $\mathcal{R}$) is the Urysohn space $\mathcal{U}$.*

In his paper P. Urysohn gave an example of a countable space with rational distance matrix (indeed that was universal incomplete metric space over rationals $\mathbf{Q}$). Our method of construction is more general: we construct the universal matrix based on the geometry of the cone $\mathcal{R}$ and allows to give necessary and sufficient condition on the distance matrix of any countable everywhere dense sets. In the section 5 we apply it to the construction of Urysohn space in probabilistic terms. We will give also the measure theoretic versions of the universality of Urysohn space and prove some facts about metric spaces with measure.

Urysohn also pointed out that there exist universal metric space of the given diameter (say, equal to 1). If we define in the same spirit the notion of universal matrix with entries from interval $[0, 1]$, we obtain the universal metric space of diameter 1 and the assertions of all theorems of this paragraph take place for that space.

4. Matrix Distribution as Complete Invariant of the Metric Triple and its Characterization

4.1. MATRIX DISTRIBUTION AND UNIQUENESS THEOREM

Now we begin to consider the metric spaces with measure and the random metrics on the natural numbers.

Suppose (X, ρ, μ) is a Polish spaces with metric ρ and with borel probability measure μ. We will call *metric triple* (In [7] the author used term "mm-space" another term is "probability metric space"). Two triples (X_1, ρ_1, μ_1) and (X_2, ρ_2, μ_2) are *isomorphic* if there exists isometry V which preserve the measure:

$$\rho_2(Vx, Vy) = \rho_2(x, y), V\mu_1 = \mu_2.$$

As it was mentioned the classification of the Polish space (non compact) is non smooth problem. Surprisingly the classification of metric triple is "smooth" and has a good answer which connects with the action of the group S_∞ and S_∞-invariant measures on the cone $\mathcal{R}$.

For the metric triple $T = (X, \rho, \mu)$ define the infinite product with the Bernoulli measure $(X^{\mathbf{N}}, \mu^{\mathbf{N}})$ and the map $F : X^{\mathbf{N}} \to \mathcal{R}$ as follows

$$F_T(\{x_i\}_{i=1}^\infty) = \{\rho(x_i, x_j)\}_{i,j=1}^\infty \in \mathcal{R}$$

The F_T-image of the measure $\mu^{\mathbf{N}}$ which we denoted as D_T will be called *matrix distribution of the triple* T: $F_T \mu^\infty \equiv D_T$.

The group S_∞ of all finite permutations of the natural numbers (infinite symmetric group $S_{\mathbf{N}}$) acts on the $\mathbf{M_N(R)}$ as well as on the cone $\mathcal{R}$ of the distance matrices as the group of simultaneous permutations of rows and columns of matrix.

Lemma 7 *The measure D_T is a Borel probability measure on $\mathcal{R}$ which is invariant and ergodic with respect to the action of infinite symmetric group, and invariant and ergodic with respect to simultaneous shift in vertical and horizontal directions (shortly NW-shift): $(NW(r))_{i,j} = r_{i+1,j+1}; i, j = 1, 2 \ldots$).*

Proof. All facts follow from the same properties of the measure μ^∞, which is invariant under the shift and permutations of the coordinates, and because map F_T commutes with action of the shifts and permutations. □

Let us call a measure on the metric space non-degenerated if there are no nonempty open sets of zero measure.

Theorem 4 *Two metric triples $T_1 = (X_1, \rho_1, \mu_1)$ and $T_2 = (X_2, \rho_2, \mu_2)$ with non-degenerated measures are equivalent iff its matrix distributions coincide: $D_{T_1} = D_{T_2}$ as measures on the cone $\mathcal{R}$.*

Proof. The necessity of the coincidence of the matrix distributions is evident: if there exists an isometry $V : X_1 \to X_2$ between T_1 and T_2 which preserves measures then the infinite power V^∞ preserves the Bernoulli measures: $V^\infty(\mu_1^\infty) = \mu_2^\infty$ and because of equality $F_{T_2} X_2^\infty = F_{T_2}(V^\infty X_1^\infty)$, the images of these measures are the same: $D_{T_2} = D_{T_1}$. Suppose now that $D_{T_2} = D_{T_1} = D$. Then D-almost all distance matrices r are the images under the maps F_{T_1} and F_{T_2}, say $r_{i,j} = \rho_1(x_i, x_j) = \rho_2(y_i, y_j)$ but this means that the identification of $x_i \in X_1$ and $y_i \in X_2$ for all i is an isometry V between these countable sets. The crucial point of the arguments: by the ergodic (with respect to NW-shift) theorem μ_1-almost all sequences $\{x_i\}$ and μ_2-almost all sequences $\{y_i\}$ are uniformly distributed on X_1 and X_2 respectively. This means that the μ_1 measure of each ball $B^l(x_i) \equiv \{z \in X_1 : \rho_1(x_i, z) < l\}$ is equal to

$$\mu_1(B^l(x_i)) = \lim_{n \to \infty} \frac{1}{n} \sum_{k=1}^{n} 1_{[0,l]}(\rho_1(x_k, x_i)).$$

But because of the isometry V $(r_{i,j} = \rho_1(x_i, x_j) = \rho_2(y_i, y_j)$ - see above) the same quantity is a μ_2-measure of the ball: $B^l(y_i) \equiv \{u \in X_2 : \rho_2(y_i, u) < l\}$

$$\mu_2(B^l(y_i)) = \lim_{n \to \infty} \frac{1}{n} \sum_{k=1}^{n} 1_{[0,l]}(\rho_2(y_k, y_i)) = \mu_1(B^l(x_i)).$$

Finally, both measures are non-degenerated, consequently each of the sequences $\{x_i\}$ and $\{y_i\}$ is everywhere dense in its own space. Because both measures are Borel it is enough to conclude their coincidence if we establish that the measures of the all such balls are the same. □

Corollary 6 *Matrix distribution is complete invariant of the equivalence classes (up-to isometries which preserve the measure) of the of metric triples with non-degenerated measures.*

We can call this theorem the "Uniqueness Theorem" because it asserts the uniqueness up-to isomorphism of the metric triple with the given matrix distribution. Firstly this theorem as the "Reconstruction Theorem" in another formulation has been proved in the book [7] pp.117-123 by Gromov. He formulated it in the terms of finite dimensional distributions of what we called matrix distribution and proved it using analytical method. He asked me in 1997 about this theorem and I suggested the proof which is written here (see also in [15]) and which he had quoted (pp.122-123) in the book. Gromov had invited the readers to compare two proofs, one of which is rather analytical (Weierstrass approximations) and another (above) in fact uses only the reference to the ergodic theorem. The explanations of this difference is the same as in all applications of the ergodic theorem - it helps us to replace the methods of space approximation by operations with infinite (limit) orbits. In our case the consideration of infinite matrices and cone $\mathcal{R}$ with invariant measures gives a possibility to reduce the problem to the investigation of ergodic action of infinite groups. For example the uniformicity of the distribution of the sequence has no meaning for finite but very useful for infinite sequences. In [16, 17] we use a more general technique which is also based on the ergodic methods in order to prove the analog of uniqueness theorem for the classification of arbitrary measurable functions of two variables (in the case above this was a metric as a function of two arguments).

4.2. PROPERTIES OF MATRIX DISTRIBUTIONS AND EXISTENCE THEOREM

The matrix distribution of a nondegenerated metric triple $T = (X, \rho, \mu)$ is by definition the measure D_T on the cone $\mathcal{R}$. Clearly it can be cosidered as a random (semi)metric on the set of natural numbers. In this section we will characterize those random metrics (or those measures on $\mathcal{R}$) which could be a matrix distribution, in other words those distributions on the cone $\mathcal{R}$ which can appear as a random distance matrices for independent sequences of points $\{x_i\}$ of the metric space (X, ρ) which are distributed with measure μ on that space. To characterize this set is necenssary in orfer to cliam that the classification problem is indeed smooth and the set of the invariants has an explicit description. We will show that the set of matrix distributions is the borel set in the space of probability measures on the cone $\mathcal{R}$.

As we mentioned (Lemma 7) any measure D_T must be invariant and ergodic with respect to action of infinite symmetric group and to NW-shift. But this is not sufficient and below one can find the necessary and sufficient conditions for that (see also [17]). But will start from the counterexamples.

Examples.
   1. A trivial example of an invariant ergodic measure which is not matrix distribution is the following. Denote r^0 a distance matrix: $r^0_{i,j} = \delta(i - j)$

(where $\delta(n) = 1$ if $n = 0$ and $= 0$ otherwise); this is nothing than distance matrix of the countable set such that the distance between two different points is equal to 1. Let a measure μ^0 be a delta-measure at the matrix r^0. Clearly μ^0 is invariant, ergodic and does not correspond to any metric triple.

2. An example of the general type is the following. Firstly note that each symmetric matrix with zeros on the principal diagonal and with entries $r_{i,j}$ from the interval $[1/2, 1]$ when $i \neq j$ is a proper distance matrix; indeed in this case the triangle inequality is valid for each three numbers. For each probability measure m with the support on $[1/2, 1]$ which is not just a single atomic measure consider the product measure m^∞ with the factor m on $\mathbf{M}_N^s(\mathbf{R})$ (it means that all entries upper diagonal are independent and identically distributed). Consequently this measure is concentrated on the cone $\mathcal{R}$. This measure is evidently invariant under permutations and under NW-shift as well as is ergodic measure with respect to those transformation. In the same time this continuous measure is not a matrix distribution for any metric triple. Indeed, with m^∞-probability equal to one all distance matrices define the discrete topology on the set of natural numbers $\mathbf{N}$ because of the absence of nontrivial fundamental sequences in $\mathbf{N}$, and consequently the completion of $\mathbf{N}$ is $\mathbf{N}$, thus a matrix distribution cannot be a continuous measure, but our product measure m^∞ is continious one.

The explanation of those effects will be clear from the proof of the next theorem which gives one of the characterizations o matrix distribution.

Theorem 5 *(Existence of metric triple with given matrix distribution)*
    Let D be a probability measure on the cone $\mathcal{R}$, which is invariant and ergodic with respect to action of infinite symmetric group (=group of all finite permutations of the naturals).

1) *The following condition is necessary and sufficient for D to be a matrix distribution for some metric triple $T = (X, \rho, \mu)$ or $D = D_T$:*
   for each $\epsilon > 0$ there exists integer N such that

$$D\{r = \{r_{i,j}\} \in \mathcal{R} : \lim_{n \to \infty} \frac{|\{j : 1 \leq j \leq n, \min_{1 \leq i \leq N} r_{i,j} < \epsilon\}|}{n} > 1 - \epsilon\} > 1 - \epsilon.$$
(4)

2) *The following stronger condition is necessary and sufficient for D to be a matrix distribution for some metric triple $T = (X, \rho, \mu)$ with compact metric space (X, ρ):*
   for each $\epsilon > 0$ there exists integer N such that

$$D\{r = \{r_{i,j}\} \in \mathcal{R} : \text{for all } j > N, \min_{1 \leq i \leq N} r_{i,j} < \epsilon\} > 1 - \epsilon, \quad (5)$$

Proof. A. Necessity. In the case of compact space the necessity is evident: the condition (5) expresses the fact that sufficiently long sequences of independent (with respect to the nondegenerated μ) points being uniformly distributed with respect to μ contain an ϵ-net of the space. The necessity of conditions (4) follows automatically from well-known property of the borel probability measures on the complete separable metric space: namely a set of full measure is sigma-compact (so called "regularity of the measure"), consequently for each ϵ there exists a compact of measure $> 1 - \epsilon$. Indeed, because of countably additivity of our measure for any $\epsilon > 0$ the exists finite number of the points such that the measure of the union of ϵ-balls with the centers at those points is greater than $1 - \epsilon$ and using a ergodic theorem we can assert that the condition in the brackets in (4) valids for matrix distance from the set of measure more than $1 - \epsilon$.

B. Sufficiency. Suppose now that we have a invariant and ergodic measure D on $\mathcal{R}$ with condition (4). The plan of the proof is the following: we express all the properties of the measure D in terms of "typical" distance matrix r and then we will construct a metric space with measure (metric triple) using only one "typical" distance matrix r. Invariance of D under the group S_∞ (simultaneous permutations of the rows and columns) leads to the invariance of the restrictions of the measure D on the submatrix $\{r_{i,j} : i = 1, 2 \ldots n, j = 1, 2 \ldots\}$ with respect to the shift $j \to j + 1$ for any n. Using ergodic theorem for this shift (which is not ergodic!) we can find the set $F \subset \mathcal{R}$ of full D-measure of such distance matrices $r = \{r_{i,j}\}$ for which the following limits exist for any natural numbers k and positive real numbers $\{h_i\} i = 1, 2 \ldots k$:

$$\lim_{n \to \infty} \frac{1}{n} \sum_{j=1}^{n} \prod_{i=1}^{k} 1_{[0,h_i]}(r_{i,j}) \equiv \mu^{h_1, \ldots h_k} \qquad (6)$$

Now let us use the invariance of the measure D under the action of symmetric group S_∞. By the ergodic theorem (more exactly by the martingale theorem) for the action of S_∞ as locally finite group we can assert that for almost all r and fixed Borel set $B \in \mathcal{R}_n$ the following limits exist

$$\lim_{N \to \infty} 1/N! \sum_{g \in S_N} 1_B(g(r^{(n)})) \equiv \Lambda_r^{(n)}(B) \qquad (7)$$

where $g(r^{(n)}) = \{r_{g(i),g(j)}\}_{i,j=1}^{n}$, g is a permutation i.e. element of S_N, which permutes the first N naturals numbers, 1_B is a characteristic function of the Borel set $B \subset \mathcal{R}_n$; the measure $\Lambda^{(n)}r(.)$ in $\mathcal{R}_n$ is called the empirical distribution of matrix $r \in \mathcal{R}_n$. These empirical distributions as a family of measures on the cones $\mathcal{R}_n$ are concordant with respect to the projections

p_n (see section 2) and consequently define an S_∞-invariant measure on $\mathcal{R}$. Our assumption about the matrix r is that this measure coincides with the initial measure D; it is possible to assume this because of ergodicity of action of S_∞. If we choose a countable basis of the Borel sets $\{B_i^n\}_{i=1}^\infty$ in $\mathcal{R}_n; n = 1, 2\ldots$ then for D-almost all r the existence is valid for all $B_i^n, i, n = 1, 2\ldots$.

Finally let us restate the condition (4) in terms of the distance matrix r. It follows from (4) that the for D-almost all r the following is true: for each k there exist integer N such that

$$\lim_{n \to \infty} \frac{|\{j : 1 \leq j \leq n, \min_{1 \leq i \leq N} r_{i,j} < k^{-1}\}|}{n} > 1 - k^{-1}. \tag{8}$$

Let us fix one such distance matrix $r = \{r_{i,j}\}$ which satisfies to the conditions (6-8) and consider it as a metric on the set of natural numbers. Denote by X_r of the completion of the metric space $(\mathbf{N}, r)$, denote the metric in this completion by $\rho_r \equiv \rho$, and the natural numbers as a dense countable set in this completion by X_r by $x_1, x_2, \ldots$. Denote by $B^h(x)$ the ball of the radius h with the center at the point x in the space X_r and let $\mathcal{A}$ be the *algebra of subsets* of X_r generated by all the balls with the center at the points $x_i, i = 1, 2\ldots$ and arbitrary radius. Let by definition the measure μ_r of the finite intersections of the balls be as follows:

$$\mu_r(\cap_{i=1}^k B^{h_i}(x_i)) = \mu^{h_1, \ldots h_k} \tag{9}$$

It is easy to check that this equality correctly defines nonnegative finitely additive normalized measure μ_r on the *algebra* $\mathcal{A}$ of the sets generated by the mentioned balls, but in general it is NOT sigma-additive and consequently can not be extended to sigma-algebra of all Borel set in X_r as true probability measure. This is just the case in our counterexamples above: we have had a countable space and the definition above gave a measure which takes value zero on each finite set but equal to 1 on the whole space. [1]

Now we will use the condition (4) in the form (8) for r. Choose $\epsilon > 0$, ondition (8) allows to find for each k a finite union of the balls, say, C in X_r of measure more than $1 - \epsilon$. Normalize our measure on C to 1, denote it as $\bar{\mu}_r$. Using induction on k we can construct a set of balls with radius 2^{-k} such that the intersection of union of C_k and C has the $\bar{\mu}_r$-measure more

---

[1] In a sense we are in the situation of the classical Kolmogorov's theorem about extension of the measures and its generalizations: the measure is defined on the algebra of the cylindric sets and after the test on countable additivity we can extend a measure on sigma-algebras. This is possible for each measures in the linear space R^∞ (Kolmogoroff's theorem), but not possible in general in other spaces. In our case the measures are defined on the algebra generated with balls and condition (4) guarantees the countable additivity; in our cases also for some spaces countable additivity takes place automatically.

than $1 - 2^{-k}\epsilon$; $k = 1, 2 \ldots$. This means that the intersection of all these sets $C \cap (\cap_k C_k)$ has $\bar{\mu}_r$-measure more than $1 - 2\epsilon$ and is a totally bounded set (i.e. has an ϵ-net for all ϵ); because of completeness of X_r this intersection is a compact. But any finite additive measure which is defined on an algebra of the sets dense in the sigma algebra of the Borel sets in the compact is countable additive. So we have found a compact C in X_r whose μ-measure is not less than $1 - 3\epsilon$. Because ϵ is arbitrary we have constructed a true probability measure μ in X_r with sigma-compact support. If we use instead of conditions (4) and (8) the condition (5) and its individualization for r we obtain along the same construction a compact of full measure in X_r.

We have constructed a metric triple $T_r = (X_r, \rho_r, \mu_r)$ where the measure μ is probability measure on the Polish space (X_r, ρ_r) with full support and with distinguished dense countable subset $\{x_i\}$ which is *uniformly distributed (with respect to measure μ_r)*, and also satisfes the condition (7). The final part of the proof consists in the verification of the fact that matrix distribution D_{T_r} of the metric triple $T_r = (X_r, \rho_r, \mu_r)$ and initial measure D are equal as the measures on the cone $\mathcal{R}$. We formulate this as a Lemma which is useful in more general situations. This completes the proof of the theorem. □

Lemma 8 *Suppose $r \in \mathcal{R}$ is a matrix for which all the limits (6) exist and equation (7),(8) is also valid. Construct the metric triple $T_r = (X_r, \rho_r, \mu_r)$ using the equations (6) and (8). Then the matrix distribution D_{T_r} of this triple is equal to the S_∞-invariant measure which is generated from the matrix r by formula (7).*

Proof. For the proof we must check the coincideness of the finite dimensional distributions of both measures. Let us illustrate this for the case of the distribution of the element $r_{1,2}, (n = 2)$; for general n the verification is similar.

$$\int_{X_r} \int_{X_r} \mathbf{1}_B(\rho_r(u, z)) d\mu_r(u) d\mu_r(z) = \lim_{n \to \infty} n^{-1} \sum_{i=1}^{n} \lim_{m \to \infty} m^{-1} \sum_{j=1}^{m} \mathbf{1}_B(r_{i,j})$$

$$= \lim_{n \to \infty} n^{-2} \sum_{i,j=1}^{n} \mathbf{1}_B(r_{i,j}) = \lim_{N \to \infty} (N!)^{-1} \sum_{g \in S_N} \mathbf{1}_B(r_{g(1),g(2)}) = \Lambda_r^{(2)}(B).$$

Here $B \subset \mathbf{R}_+$; the last equality follows from (7); the equalities above used the uniformity of the distribution of the sequence $\{x_i\}$ in the space (X_r, μ_r). This concludes the proof of the theorem, because by the condition (7) the S_∞-invariant measure on $\mathcal{R}$ which is generated by matrix r is just the measure D. □

Remark 2

1. The structure of the conditions on the measure in the thoerem shows that the set of matric distributions is indeed a borel set in the space of all borel probability measures on the cone $\mathcal{R}$.

2. The condition (4) could be replaced by another condition from the paper [17] (simplicity of S_∞-invariant measure). That condition guarantees that measure D appeared from *some* measurable function of two variables as matrix distribution which is sufficient for our goals.

4.3. THE SPACE OF MEASURE-THEORETICAL METRIC TRIPLES

We can extend the notion of the space of metric spaces (see section 1) and introduce a similar space for the metric triples. Instead of the ordinary point of view where one considers the set of all Borel measures on the given topological space, we in opposite, consider the set of all *measurable (semi)metrics* on a fixed Lebesgue space with continuous measure. (see [18], par.6).

Let (X, μ) be a Lebesgue space with measure μ finite or sigma-finite (say, interval [0,1] with Lebesgue measure or natural numbers with the uniform mreasure), and $S_\mu(X)$- the space of all classes $mod0$ of measurable functions; define $\mathcal{R}^c \subset S_\mu(X)$ as a cone of measurable metrics e.g. the cone of the classes $mod0$ of symmetric measurable functions $\rho : (X \times X, \mu \times \mu)) \to \mathbf{R}_+$ with the triangle inequality:

$$\rho(x,y)+\rho(y,z) \geq \rho(x,z) \text{ for } (\mu \times \mu \times \mu)- \text{ almost all } (x,y,z) \in (X \times X \times X).$$

It is natural to assume that $\mu \times \mu\{(x,y) : \rho(x,y) = 0\} = 0$.

Remark that ρ is not individual function but the class of $mod0$ equivalent functions, so it is not evident a priori that such ρ defines the structure of metric space on X in the usual sense.

If measure μ) is discrete one then the cone $\mathcal{R}^c$ is the cone of ordinary (semi)metric on the finite or countable set e.g. they coinside with $\mathcal{R}$ or $\mathcal{R}_n$ (see section 1), Thus the cone $\mathcal{R}^c$ is a continuous generalization of the cone $\mathcal{R}$ where instead of the set of natural numbers $\mathbf{N}$ with counting measure we consider the space (X, μ) with continuous measure.

Suppose now that measure μ) is finite and continuous and $\rho \in \mathcal{R}^c$ is a *pure* function (see [17]) [2], and the measure D_ρ on the space $\mathbf{M}_\infty(\mathbf{R})$ is a matrix distribution of the measurable function ρ. (see definition in the

---

[2]A measurable function $f(x,y)$ of two variables on the unit square with Lebesgue measure calls pure if for almost all pairs (x_1, x_2) (corresp. (y_1, y_2)) the functions of one variable $f(x_1,.)$ and $f(x_2,.)$ (corresp. $f(., y_1)$ and $f(., y_2)$) are not coniside everywhere. Evidently, a proper metric on the measure space is pure function.

previous section). Using the ergodic theorem we can prove that $D_\rho(r \in \mathcal{R} : r_{i,k} + r_{j,k} \geq r_{i,k}) = 1$ for each $i, j, k \in \mathbf{N}$ and consequently $D_\rho(\mathcal{R}) = 1$. From this using characterization of matrix distributions from [18] we conclude that the following assertion is true:

Lemma 9 *The measure D_ρ concentrates on the cone $\mathcal{R}$ (e.g. $D_\rho(\mathcal{R}) = 1$) and is an ergodic S_∞-invariant measure. Consequently, each pure function $\rho \in \mathcal{R}^c$ on $(X \times X, \mu \times \mu)$ defines a true metrics mod0 on the space (X, μ).*

Corollary 7 *The class of measurable (semi)metrics on the Lebesgue space with continuous measures coinsides with the class of the metric triples with finite continuous measures.*

This corollary shows that the language of the matrix distributions which are concentrated on the cone $\mathcal{R}$, is an invariant manner to study (semi)metric triples. It seems that sometime it is convinient to fix the measure space and to vary in measurable manner the metrics instead of consideration of the fixed metric spaces with various measures - the generality of the objects is the same. We have used this way in [18]).

5. General Classification of the Measures on the Cone of Distance Matrices, Examples

5.1. DEFINITIONS

Let us consider arbitrary measures on the cone $\mathcal{R}$, or - arbitrary random metrics on the naturals and choose some notations. Remark that the cone $\mathcal{R}$ with weak topology is metrizable separable space (e.g. is the Polish space if we fix a metric which is compatible with weak topology).

Notation. Denote by $\mathcal{V}$ the set of all probability Borel measures [3] on the cone $\mathcal{R}$ and endow it with weak topology,- this is the topology of inverse limit of the sets of probability measures on the finite dimensional cones $\mathcal{R}_n$ with its usual weak topology. The convergence in this topology is convergence on the cylindric sets with open bases. All classes of measures which we define below are the subsets of $\mathcal{V}$ with induced topology. Remark that the set of non-degenerated (=positive on the nonvoid open sets) measures is of course everywhere dense G_δ set in $\mathcal{V}$.

Let $\mathcal{D}$ be the subset in $\mathcal{V}$ of the matrix distributions; as we proved (Theorem 4,orollary 7) this set is in the bijective correspondense with the set of all classes of isomorphic all metric triples. The constructive description of $\mathcal{D}$ follows from the existence theorem (section 4).

The subset $\mathcal{P}$ of $\mathcal{V}$ is the set of measures which are concentrated on the set of universal distance matrices: $\nu \in \mathcal{P}$ iff $\nu(\mathcal{M}) = 1$ - see section 3.

---

[3] We use later the term "measure" in the meaning "Borel probability measure"

The subset W of V is the set of measures which are concentrated on the set of weakly universal distance matrices Both sets are convex (not closed) subspaces of the simplex of all measures on the cone R. It is possible to give direct characterization of those measures which analogous to the criteria of universality from Statement 1.

Statement 2. *Let us define for each natural n a partition of the cone R on the subsets $R^n(q), q \in R_n$ (see denotation before Statement 1 in the section 3). Measure μ belongs to the cone P, (e.g. measure is concentrated on the set of universal matrices iff for each n and for almost all elements of those partitions (or for almost all finite distance matrix q) the support of conditional measure is whole set $R^n(q)$ (conditional mesusre are not degenerated). Belonging of the measure μ to the set $W \subset V$ is equivalent to the following: support of measure μ is whole cone R.*

The set $Q \subset D$ consists of the measures which corresponds to the metric triples $T = (U, \rho, \mu)$ in which U is Urysohn space.

Finally denote by H the set of measures μ in V which have the following property: μ-almost all distance matrices generate the isometric metric spaces. A measure $\mu \in H$ generates a *random everywhere dense sequence of points* on the given space. From this point of view the elemetns of D induced *a random everywhere dense subset of the special type*, namely infinite independent sampling of the points of the given metric triple; and the elements of Q induced a random independent sampling of the points in the Urysohn space with nondegenerated measure.

We have embeddings:

$$V \supset H \supset D \supset Q \subset P \subset H \subset V; \qquad Q = P \cap D$$

The Theorem 4 shows that each measure from the set D defines a class of isomorphic metric triples and in particular the set Q is the set of classes isometric nondegenerated measures on the Urysohn space (e.g. the orbit of the group all isometries of the Urysohn space on the set of nondegenrated measures on that spaces.

Theorem 6

1. *The subset $P \subset V$ is an everywhere dense G_δ subset in V. This means that for generic measures ν on R ν-almost all distance matrices are universal, and consequently tance matrix r defines a metric on the naturals such that completion of naturals with respect to this metric is the Urysohn space.*

2. *The subset $Q \subset D$ is an everywhere dense G_δ in D. This means that the generic metric triple $T = (X, \rho, \mu)$ has Urysohn space as the space (X, ρ).*

Proof. The first claim follows from Theorem 1 which states in particular that the set of universal distance matrices is a G_δ in $\mathcal{R}$, and from a general fact we can deduce that the set of all measures on separable metrizable space such that some fixed everywhere dense G_δ (in our case - $\mathcal{M}$) has measure 1, is in its turn itself an everywhere dense G_δ in the space of all measures in the weak topology. The second claim follows from the fact that the intersection of G_δ-set with any subspace in a Polish space is G_δ in induced topology.   □

5.2. EXAMPLES OF THE MEASURES WHICH ARE CONCENTRATED ON THE UNIVERSAL MATRICES

Now we can give a probabilistic (markov) construction of the measures on the cone $\mathcal{R}$ and in particular to represent the examples of the measures from the set $\mathcal{P} \subset \mathcal{H} \subset \mathcal{V}$. This gives a new proof of the existence of the Urysohn space. In fact we use the arguments from the section 3 but in the probabilistic interpretation. Also the method gives a concrete illustration how to construct a random metric space.

Let γ be an arbitrary continuous measure on half-line $\mathbf{R}^1_+$ with full support - for example Gaussian measure on the half-line. We will define inductively the measure ν on the cone of distance matrices $\mathcal{R}$ by construction of its finite dimensional projection on the cones $\mathcal{R}^0_n$ or in other words - joint distributions of the elemetns of random distance matrices.

The distribution of the element $r_{1,2}$ of the random matrix is distribution γ. So we have defined the measure on $\mathcal{R}_2$, denote it as ν_2. Suppose we have already defined the joint distribution of the entries $\{r_{i,j}\}^n_{i,j=1}$, which means that we have defined a measure ν_n on $\mathcal{R}^0_n$. By Lemma 4 the cone $\mathcal{R}_{n+1}$ is a fibration over cone $\mathcal{R}_n$ with the fibers $A(r)$ over matrix $r \in \mathcal{R}_n$ We use only the structure of this fibration: projection $\mathcal{R}_n \xleftarrow{p_{n+1}} \mathcal{R}_{n+1}$ - in order to define a measure on $\mathcal{R}_{n+1}$ with given projection.

So we need to define a conditional measure on $A(r)$ for all $r \in \mathcal{R}$ which are measurably depend on r. From probabilistic point of view this means that we want to define transition probabilities from given a distance matrix r of order n to the distance matrix r^a (see section 2) of order $n + 1$.

Let us recall the geometrical structure of the set of admissible vectors $A(r)$. It is Minkowski sum:

$$A(r) = M_r + \Delta_n,$$

(see 2.2) or as projection of the direct product $\pi : M_r \times \Delta_n \to M_r + \Delta_n = A(r)$. Consider product measure on $M_r \times \Delta_n : \gamma_r = m_r \times \gamma$ where m_r is for example normalized Lebesgue measure on the compact polytope M_r or another measure with full support on M_r, with the conditions which we formulate below.

Let $\pi\gamma_r$ be its projection on $A(r)$. We define the *conditional measure* on $A(r)$ as $\pi\gamma_r$. So, we have

$$\text{Prob}(r^{da}|r) = \pi(m_r \times \gamma)(da).$$

The conditions on the measures m_r are the following: at each step of the construction for each N and $n > N$ the projection of the measure $m_r, r \in \mathcal{R}_n$ to the set of admissible vectors $A(p_N(r))$ are uniformly positive on the open sets; this means that for any open set $B \subset A(p_N(r))$ there exist $\epsilon > 0$ such that for any $n > N$ the value of projection of the measure $m_r, r \in \mathcal{R}_n$ on the set B more than ϵ. Thus we define a measure L_n on $\mathcal{R}_{n+1}$. By construction all these measures are concordant and define the a measure L on $\mathcal{R}$. Denote this measure as $L = L(\gamma, \{m_r : r \in \mathcal{R}_n, n = 1, 2 \ldots\})$.

A more intuitive and combinatorial variant of this description is the following: to the given n-point metric space we randomly add a $n + 1$-th point choosing the vector of the distances between the new and the previous points (admissable vector), with the natural probability which is positive on all open sets of admissible vectors.

Theorem 7 *The constructed measure L belongs to the cone $\mathcal{P}$ which means that L is concentrated on the set of universal matrices. and therefore the completion of $(\mathbf{N}, r)$ is Urysohn space L-almost sure.*

Proof. It is sufficient to check the conditions of the Statement 2. By the theorem about convergence of the martingales the conditional measure on almost all elements of the partitions - $\mathcal{R}^n(q)$ - is the limit of the conditional measures on the elements of the partitions $\mathcal{R}^n(q) \cap \mathcal{R}_N$ when $N \to \infty$. Thus the claim that almost all conditional measures are nondegenerated is the consequence of the condition about uniform positivity of the probability on the set of admissable vectors.   □

We can say that *a random countable metric space is an everywhere dense subset of the Urysohn space* or equivalently completion of the random countable metric space with probability one is the the Urysohn space. Here "random" means randomness respectively to that natural procedure which was defined above and which is in a sense very close to independence and has a very wide variations which allow to define the measures on $\mathcal{R}$.

A much more complicated problem is to construct a measure on $\mathcal{R}$ from the set $\mathcal{Q}$ i.e. which is a matrix distribution for some measure on $\mathcal{U}$. The properties of the measures on the Urysohn space are very intriguing. But becuase we have no useful model for thisspace, it is natural to use indirect way for the definition and studying of such measures: to define matrix distribution as a measure on the cone $\mathcal{R}$ which belongs to the set $\mathcal{Q}$ which defined an isometric class of the measures on $\mathcal{U}$. In its turn for this we can take any measure from the set $\mathcal{P}$ and then to construct the S_∞-hull of its -S_∞-invariant ergodic measure The simplification is that we can omit the condition (4) from the theorem which guarantees the fact that the measure on $\mathcal{R}$ is matrix distribution:

Statement 3. *Each S_∞-invariant ergodic measure from the set $\mathcal{P}$ is matrix distribution (belongs to $\mathcal{Q}$). This means that S_∞-hull of measures which we had constructed above defines the isometry class of the measures on Urysohn space.*

The proof is based on the criteria of simplicity of the measures on the set of infinite matrices from [17], and we will discuss it elsewhere.

The probabilistic analysis on the distance matrices is useful for integration over set of metric spaces in the spirit of statistical physics. The measures which we had considered here are interesting from the point of view of the modern theory of random matrices. It is natural to study the spectra of the random distance (symmetric) matrices. The simpliestcase is to calculation of the joint distributions of distances between independent random points of the homogeneous manifolds spaces (spheres, for example), this is interesting and complicate new problem.

Returning back to the Theorem 7, I must recall a very interesting analogy with the old and simple theorem by Erdös-Rényi [6] about the random graphs. It asserts that with probability one a random graph is the universal graph, see [12, 3]. This is the simplest case of the the theory which we developed here because each infinite graph defines the distance onthe set of vertices and in our case ditance takes only two nonzero values - 1 and 2. The random graph in the sense of [6] defines the measure on the distance matrices such that all the entries of matrix are indepedent and have unifirm distribution on $\{1, 2\}$ (of more general). All the matrices belong to the cone $\mathcal{R}$. In the same time almost all matrices are universal in our sense (see paragraph 3) if we consider only two values $\{1, 2\}$ of distances (instead of values from $\mathbf{R}_+$).

Acknowledgement. Partially supported by RFBR, grant 02-01-00093, and by CDRF, grant RMI-2244.

References

1. Avis, D. (1980) On the Extreme Rays of the Metric Cone, *Canad. J. Math.*, **Vol.8 no. 1**, pp. 126–144.
2. Bogatyi, S.A. (2000) Compact Homogeneus of the Uryson's Universal Metric Space, *Russian Math. Surveys.*, **Vol.55 no. 2**, pp. 131–132.
3. Cameron, P. (1996) The Random Graphs, The Mathematics of Paul Erdos, (Nesetril, J. and Graham, R.L. Eds.), Springer-Verlag, pp. 331–351.

4. Clemens, J., Gao, S. and Kechris, A. (2001) Polish Metric Spaces: their Classification and Isometry Groups, *Bull. Symb. Logic*, **Vol.7 no. 30**, pp. 361–375.

5. Deza, M. and Laurent, M. (1997) Geometry of Cuts and Metrics, Springer-Verlag, Berlin.

6. Erdös, P. and Rényi, A. (1963) Asymetric Graphs, *Acta Math. Acad. Sci. Hungar*, **Vol.14**, pp. 295–315.

7. Gromov, Misha (1998) *Metric Structures for Riemannian and Non-Riemannian Spaces*, Birkhuaser.

8. Katetov, M. (1988) On Universal Metric Spaces, *Proc. Sixth Prague Toplogical Symposium 1986*, (Frolik Ed.), Helderman Verlag Berlin, pp. 323–330.

9. Lindensrauss, J., Olsen, G. and Sterfeld, Y. (1978) The Poulsen Simplex, *Ann. Inst. Fourier, Grenoble*, **Vol.28 no. 1**, pp. 91–114.

10. Lusku, W. (1976) The Gurarij Spaces are Unique, *Arch der Mathematik*, **Vol.27 no. 6**, pp. 627–635.

11. Pestov, V. Ramsey-Molman Phenomen, Urysohn Metric Spaces, and Extremely Amenable Groups, *Israel J. Math.*, to appear, e-print http://xxx.lanl.gov/abs/math.FA/9810168.

12. Rado, R. (1964) Universal Graphs and Universal Functions, *Acta Arith.*, **Vol.9**, pp. 331–340, 1964.

13. Urysohn, P.S. (1927) Sur un Espace Metrique Universel, *Bull. Sci. Math.*, **Vol.51**, pp. 1–38.

14. Uspenskij, V.V. (October 2001) Compactification of Topological Groups, Preprint.

15. Vershik, A. (1998) The Universal Urysohn Space, Gromov's Triples, and Random Metrics on the Series of Natural Numbers, *Russian Math. Surveys*, **Vol.53 no.5**, pp. 57–64.

16. Vershik, A. (2001) Classification of the Measurable Functions of Several Arguments and Invariant Measure on the Space of the Matrices and Tensors, 1. Classification, Preprint of Erwin Schrodinger Institute, No. 1107.

17. Vershik, A. (2002) Classification of the Measurable Functions of Several Arguments and Invariant Distributions of Random Matrices, *Functional Anal. and its Appl.*, **Vol.2 no.36**, pp. 1–16.

18. Vershik, A. (2000) Dynamical Theory of the Growth in the Groups: Entropy, Boundary, Examples, *Russian Mathematical Surveys*, **Vol.55,4 no.434**, pp. 60–128.

19. Vershik, A. (2002) Random Metric Space is Urysohn Space, Doklady of Russian Academy of Science, to appear.

Nonlinear Phenomena and Complex Systems

KLUWER ACADEMIC PUBLISHERS – DORDRECHT / BOSTON / LONDON